Preface

MOLECULAR ENZYMOLOGY, BECAUSE OF ITS CHEMICAL AND MATHEMATICAL content, is often regarded as a formidable and forbidding topic by undergraduates on a biology or biochemistry course. As a result of teaching enzymology to undergraduates for a number of years, we recognize the areas which appear to cause the most common difficulties in conceptual understanding. We feel that a book treating those areas by means of a logical approach carefully developed from basic principles fills a gap in the multiplicity of enzymology texts currently available. In writing this book we have had in mind the needs of Honours Biochemistry students, in particular those who may take a special interest in enzymology. The text covers the main bulk of the material required in the second and third years of such courses. In addition, those taking courses in Biological Chemistry may well find the book to be of central interest.

The book begins with a description of the fundamentals of catalysis, illustrating these with simple chemical reactions which may be supposed to serve as models of catalytic processes. Protein structure is discussed in terms of the fundamental forces which determine the shape and dynamic behaviour of protein molecules. The approach emphasizes those features thought to be most intimately involved in the catalytic function of enzyme molecules, and is illustrated with specific examples.

Although a number of excellent texts dealing specifically with enzyme kinetics have appeared over the past four years, the subject has not been adequately treated in textbooks which purport also to cover the mechanism of enzyme catalysis. We have therefore emphasized in our chapters on enzyme kinetics the aspects required for a sound understanding of enzymic catalysis. The fundamental kinetic equations which describe both the transient and the steady state are fully derived, and the equations are extended to the description of inhibition, pH-dependence, multisubstrate kinetics and cooperativity. Stress is laid upon the practical aspects of methods of assay and analysis and interpretation of kinetic data.

In discussing the classical mechanisms of reactions involving coenzymes, the prime focus is upon the coenzyme chemistry but attention is paid to the role of the enzyme molecule in the catalytic process. The penultimate chapter treats in depth several enzymes whose mechanisms have been extensively studied—here detailed consideration is given to the role played by the amino-acid side-chain assemblies at the active centre. Our choice of examples is governed by the need to expose the underlying principles in the clearest possible fashion.

Recent advances in our understanding of the origin and evolution of enzyme catalytic power are discussed. Model systems based on polymers and macrocycles are described in an attempt to fuse the structural and mechanistic aspects of enzymic catalysis. The bibliographic references include both leading review articles and seminal papers from the primary literature and should provide a comprehensive basis for further reading.

We are grateful to Keith Brocklehurst, Athel Cornish-Bowden, Mike Danson, Roger Harrison, Bruno Orsi and Keith Tipton for their helpful comments and to Pat Wharton for preparing many of the diagrams.

C.W.W.
R.E.

Molecular Enzymology

TERTIARY LEVEL BIOLOGY

A series covering selected areas of biology at advanced undergraduate level. While designed specifically for course options at this level within Universities and Polytechnics, the series will be of great value to specialists and research workers in other fields who require a knowledge of the essentials of a subject.

Titles in the series:

Experimentation in Biology	Ridgman
Methods in Experimental Biology	Ralph
Visceral Muscle	Huddart and Hunt
Biological Membranes	Harrison and Lunt
Comparative Immunobiology	Manning and Turner
Water and Plants	Meidner and Sheriff
Biology of Nematodes	Croll and Matthews
An Introduction to Biological Rhythms	Saunders
Biology of Ageing	Lamb
Biology of Reproduction	Hogarth
An Introduction to Marine Science	Meadows and Campbell
Biology of Fresh Waters	Maitland
An Introduction to Developmental Biology	Ede
Physiology of Parasites	Chappell
Neurosecretion	Maddrell and Nordmann
Biology of Communication	Lewis and Gower
Population Genetics	Gale
Structure and Biochemistry of Cell Organelles	Reid
Developmental Microbiology	Peberdy
Genetics of Microbes	Bainbridge
Biological Functions of Carbohydrates	Candy
Endocrinology	Goldsworthy, Robinson and Mordue
The Estuarine Ecosystem	McLusky
Animal Osmoregulation	Rankin and Davenport

TERTIARY LEVEL BIOLOGY

Molecular Enzymology

CHRISTOPHER W. WHARTON, B.Sc., Ph.D.
Lecturer in Biochemistry
University of Birmingham

and

ROBERT EISENTHAL, A.B., Ph.D.
Senior Lecturer in Biochemistry
University of Bath

Blackie
Glasgow and London

Blackie & Son Limited
Bishopbriggs
Glasgow G64 2NZ

Furnival House
14–18 High Holborn
London WC1V 6BX

First published 1981

British Library Cataloguing in Publication Data
Wharton, Christopher W.
Molecular enzymology.—(Tertiary level biology)
1. Enzymes
I. Title II. Eisenthal, Robert. III. Series
574.19'25 QP601

ISBN 0-216-91012-9.

Filmset by Advanced Filmsetters (Glasgow) Ltd.

Printed in Great Britain by
Thomson Litho Ltd, East Kilbride, Scotland

Contents

CHAPTER ONE

THE NATURE OF CATALYSIS

Chemical reaction rates and their description

We face a daunting situation whenever we attempt to describe a system in which chemical reaction takes place. At present, we can describe the molecular dynamics of only simple gas phase reactions in a quantitative fashion. In such reactions, the forces which influence atomic and molecular interactions are short-range; in solution, where solvation effects are important, long-range electrostatic forces which can operate at larger interatomic distances are dominant.

To attempt the deduction of a reaction mechanism, we must first have a conceptual model of the nature of chemical reactions. It is convenient to consider the details of reaction in solution in qualitative terms, and to apply phenomenological (or empirical) kinetic treatments. The interactions which must be considered include differential solvation, ion–ion dipolar interaction, hydrophobic interactions and hydrogen bonding. Each of these is likely to be complex in physiological systems, but it is possible to devise appropriate systems *in vitro* which may provide a realistic analogy.

One such conceptual model which has found wide application is known as transition state theory, which we will use as the basis for our discussion. Description of a reaction in terms of transition state theory ideally requires the computation of a potential energy surface, which describes the set of potential energy paths that must be followed by the reaction constituents. The lowest energy transition state is here represented by a "saddle point" on the lowest energy pathway.

As the construction of a potential energy surface is a very laborious process even for a simple reaction such as gas phase proton transfer, these surfaces have been computed for only a limited number of reactions and it is unlikely that they will be available in the near future for reactions in

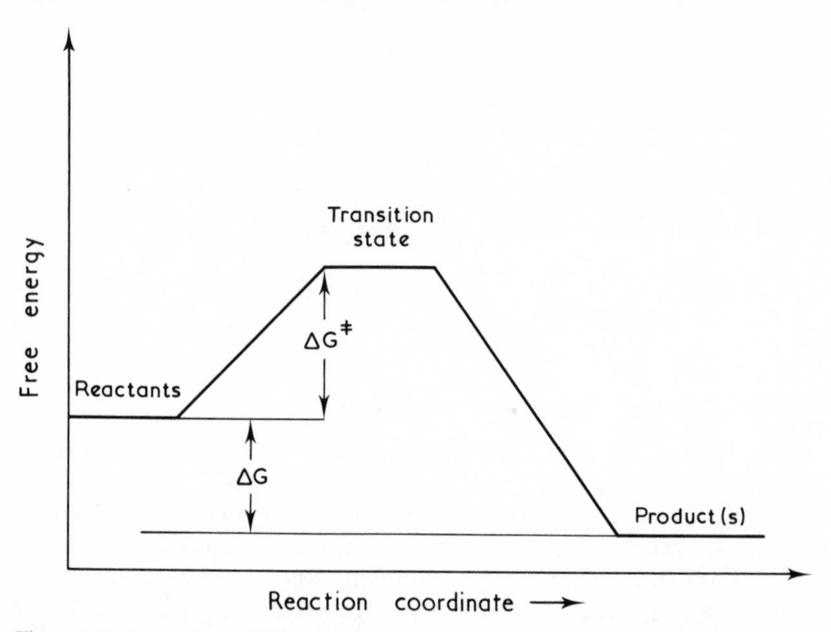

Figure 1.1 A much simplified reaction diagram of the type commonly used for the interpretation and representation of more complex reactions. Note that the ordinate is given in terms of Gibbs free energy rather than potential energy and no indication is given concerning barrier shape.

solution. Thus an approach which retains as many features of the potential energy surface method as possible is desirable. A free energy diagram (figure 1.1) albeit a much simplified version of a three-dimensional potential energy surface, is very useful in practice.

The free energy of activation of the reaction, $\Delta G^{\ddagger}$ in figure 1.1, is determined from the temperature dependence of the reaction rate, and the overall free energy change ΔG is calculated from the reaction equilibrium constant. In the case of second order association reactions between enzyme (E) and substrate (S) (i.e. $E+S \rightleftharpoons ES$), the values of $\Delta G^{\ddagger}$ and ΔG are dependent upon the choice of standard state. Chemists traditionally work with standard states of 1 molar, but this is regarded as unrealistic in the case of enzyme reactions. Substrate concentrations *in vivo* and *in vitro* are frequently of the order of 10^{-3} M, this value being a suitable "enzymic" standard state for the substrate. This conversion to a 1 mM standard state is achieved simply by expressing the equilibrium constant in millimolar rather than in molar units. The free energy difference between initial and

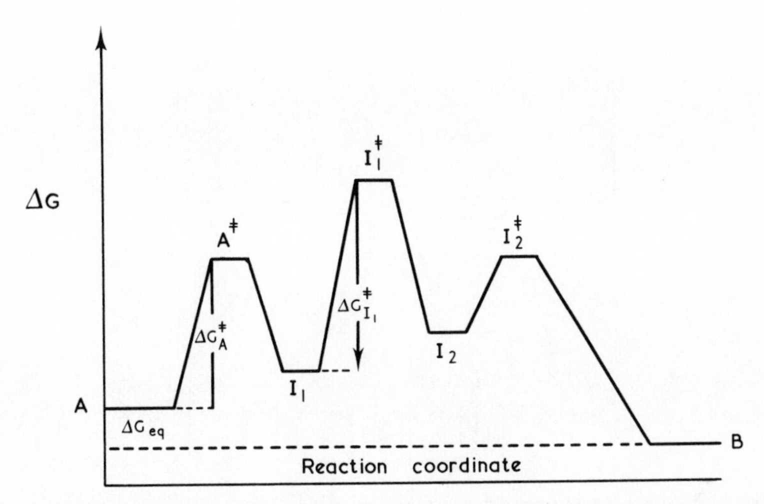

Figure 1.2 An example of a free energy diagram of a unimolecular reaction having two intermediates upon the reaction pathway. Step $I_1 \rightarrow I_2$ will have the lowest rate constant in the forward direction since $\Delta G^{\ddagger}_{I_1}$ is the largest activation energy. I_1 will accumulate more than I_2 due to lower relative ΔG value. At equilibrium (B) > (A) > (I_1) > (I_2) *all* barriers being crossed in *both* directions at *equal* rates. In the reverse direction $B \rightarrow I_2$ will have the lowest rate constant. Note that *rate constants* are found directly from free energies of activation but that *rates* are the product of rate constant and reactant concentration (for that particular step).

final states (i.e. enzyme + substrate, E + S and enzyme + product, E + P) is unaffected by standard state definition when the stoichiometry is conserved in forward and reverse directions. A free energy diagram of the type shown in figure 1.1 can be expanded to take account of more complex reactions as shown in figure 1.2. If any one transition state has an energy significantly higher than the others, the overall reaction rate will be limited by the rate of passage over this barrier, the other steps having the potential of greater rates but being restricted to that of the slowest step. The one or more transition states that occur in the rate-limiting reaction step(s) can be described as *kinetically significant.*

The valleys in the reaction coordinate represent intermediates whose stability depends upon their free energy relative to reactants and products. Intermediates having free energies similar to reactants or products will accumulate during the course of the reaction, and can thus be termed kinetically significant intermediates. Clearly the existence of very stable intermediates will lead to significant trapping of reactant (in the form of such an intermediate), and will thus lead to a less rapid reaction than if the

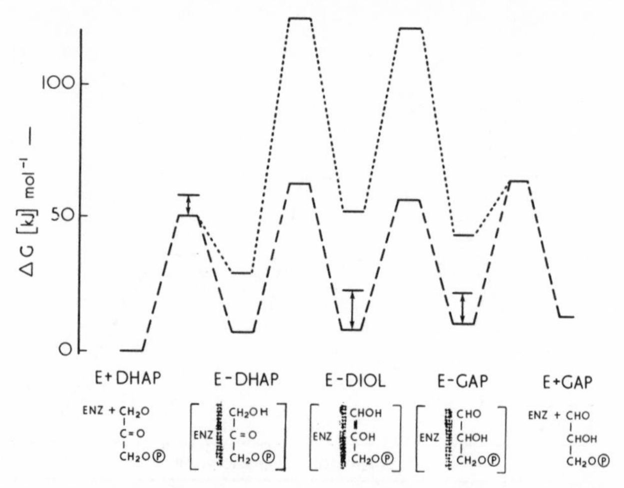

Figure 1.3 A complete free energy diagram of an enzyme-catalysed reaction. The reaction is the triosephosphate isomerase-catalysed interconversion of glyceraldehyde-3-phosphate (GAP) and dihydroxyacetone phosphate (DHAP) which is represented by the dashed line. The "enediol" intermediate (DIOL) is also included. The dotted line represents the acetate-catalysed reaction. The arrows represent the limits of uncertainty in the assignment of some of the free energy levels. Redrawn from Albery and Knowles (1976).

intermediate were less stable. A complete free energy diagram has been constructed in the case of triose phosphate isomerase and is shown in figure 1.3. Typically a large number of experiments is required in order to achieve definition of a diagram of this type and in the case of most enzyme reactions this has not yet been accomplished.

The relationship between free energy diagrams and rate constants

If we make the postulate that each transition state is in (pseudo)-equilibrium with its relevant ground state(s), it must be possible to express the free energy of activation of each step in either direction in terms of an equilibrium constant. This is achieved by means of equation 1.1.

$$\Delta G^{\ddagger} = -RT \ln K^{\ddagger} \tag{1.1}$$

where R = the gas constant
T = temperature (°K)
$K^{\ddagger} = C^{\ddagger}/(a)(b)$, where for the second order reaction (1.2),

$$a + b \xrightarrow{k_{(2)}} e + f \tag{1.2}$$

$C^{\ddagger}$ = concentration (more strictly, activity) of the transition state, and (a) and (b) are the concentrations of species a and b.

For the reaction (1.2), the second order rate constant $k_{(2)}$ is given by

$$k_{(2)} = \left(\frac{\mathrm{k}T}{\mathrm{h}}\right) \exp\left(-\Delta G^{\ddagger}/RT\right) \quad (1.3)$$

where k = Boltzmann's constant
T = temperature (°K)
h = Planck's constant

(The factor $\mathrm{k}T/\mathrm{h}$ is independent of the nature of the reactants.)

Thus equation (1.3) demonstrates the simple natural logarithmic relationship between the rate constant and the free energy of activation $\Delta G^{\ddagger}$. This logarithmic relationship means that fairly small changes in $\Delta G^{\ddagger}$ represent quite large changes in the values of the rate constants. Unless this point is remembered, it is easy to misinterpret free energy diagrams.

The free energy of activation $\Delta G^{\ddagger}$ can be expanded to yield

$$\Delta G^{\ddagger} = \Delta H^{\ddagger} - T\Delta S^{\ddagger} \quad (1.4)$$

where $\Delta H^{\ddagger}$ is the enthalpy of activation due to heat content changes, whilst $\Delta S^{\ddagger}$ is the entropy of activation which reflects changes in the ordering of the transition state relative to the ground state.

The entropy of activation is related to the probability that a molecule which possesses the required activation energy to reach the transition state has the energy in the appropriate molecular coordinate (mode) for reactive chemical bond rearrangement to take place. $\Delta S^{\ddagger}$ thus involves solvation, steric and orientation factors which can be interpreted relatively simply in qualitative terms. An important consequence of these considerations is that unimolecular reactions will have small (often near zero) entropies of activation, while bimolecular reactions will have considerable negative entropies of activation reflecting the need to bring two molecules together from standard state concentrations. When two molecules contrive to form an activated complex, three degrees of translational freedom are lost (each of the reactants has three, the complex has three), and this represents a considerable negative entropy change. The loss of entropy in activated complex formation can be compensated to some extent by an increased entropy due to internal vibrational and rotational modes, made possible by the increased molecular weight of the complex relative to the reactants.

Thus any process capable of providing the free energy necessary for the approximation of reactants in a bimolecular reaction is capable of greatly enhancing the reaction rate (up to 10^8-fold) since the entropic barrier to complex formation can be surmounted by the enthalpic binding forces provided by the approximation process. This is an important factor in catalysis and will be discussed in more detail in chapter 2.

The principle of microscopic reversibility (detailed balance)

A priori it might seem plausible that a chemical reaction, catalysed or otherwise, could proceed in the forward direction by one mechanism and return to reactants by another mechanism, provided that the overall stoichiometric and equilibrium constraints were satisfied. The principle of microscopic reversibility states that this is not so—a given reaction must proceed in forward and reverse directions by the same pathway. This conclusion applies rigorously only at equilibrium but it is generally accepted that it may be applied to kinetic (at least steady-state) situations. It is therefore important to ensure that any mechanism which is proposed to describe a particular reaction is consistent with this principle. The implications of the principle are useful in the deduction of mechanisms and in the planning of experiments which solve mechanistic problems; for instance, in a catalytic system a forward step which is base-catalysed will be acid-catalysed in the reverse of this step.

Limiting factors in reaction rates

If the rate of a reaction is limited by the rate at which the molecules encounter each other to form activated complexes, the reaction is said to be *diffusion-limited*. Any other reaction steps (which will of course be intramolecular) must be fast or commensurate with the encounter rate. Consequently the energy barrier for the interaction that results on diffusive encounter must be small, otherwise only a small proportion of encounters will lead to reaction. It is apparent that the collision theory implies that all reaction rates are limited by encounter rates, or, more strictly, by the number of encounters which are sufficiently energetic to lead to reaction. In the case of diffusion limitation all encounters are considered, since all (or very nearly all) lead to reaction. The equation for the rate constant of bimolecular diffusive encounter is

$$k_{(2)} = a(D_A + D_B)F_p 4\pi \quad (N/1000) \tag{1.5}$$

where a is the distance of closest approach (cm), N is Avogadro's number, F_p is a potential energy function for the energy of interaction and D_A and D_B are the mutual self-diffusion constants of the reactants A and B. If the various parameters are substituted with values typical for aqueous solution, the value of $k_{(2)}$ is approximately 4×10^9 M^{-1} sec^{-1}. This value is calculated for neutral molecules where $F_p = 1$ but would be increased in the case of reaction between oppositely charged ions since the interaction forces in solution are predominantly electrostatic. The collision rate calculated from collision theory for both gases and liquids is ca. 10^{11} M^{-1} sec^{-1}.*

The difference between this latter value and that calculated using the diffusion equation is due to the inclusion of secondary collisions in the collision theory model; these are not counted in the diffusion theory model. The existence of "solvent cages" in liquid ensures that a significant number of non-reactive secondary collisions occur but in homogeneous gases the collision number is virtually the same as the rate constant calculated from diffusion theory. The maximum rate of an intramolecular or unimolecular process is limited by the coupled vibrational frequency of the nuclei on the reaction coordinate and is thus of the order of 10^{14} sec^{-1}.

A reaction may be limited by the rate at which the products diffuse apart, particularly if ion pair formation occurs on the reaction pathway and the bond rearranging step is very fast. Proton transfer reactions in aqueous solvent can have rate constants up to 10^{11} M^{-1} sec^{-1} ($H_3O^+ + OH^- \rightarrow 2H_2O$) due to the favourable electrostatic interactions and the elaborate hydrogen-bonded "flickering cluster" structure of water. It is interesting to note that the maximal possible value of a bimolecular rate constant is largely unaffected by the size of the molecules involved, since the lower diffusion constant of a large molecule is compensated by the larger collision cross-section. Desolvation is likely to be an important factor in collisions between large molecules in water, and enzyme–substrate collisions are likely to involve desolvation followed by reorientation until the substrates "find" the most energetically favourable interactions with the enzyme. This type of reorientation effect is most likely to predominate when the substrate is polymeric, for example, in the binding of a peptide to a proteolytic enzyme.

Some values of rate constants for enzyme–substrate complex formation and decomposition are given in chapter 8. The rate constants for complex

* $\left(\frac{kT}{h}\right) \simeq 6 \times 10^{12}$ sec^{-1} at normal temperatures (300°K).

dissociation are numerically much smaller than those for formation since the favourable interactions that allow the complex to accumulate in significant quantities must be overcome if the complex is to dissociate.

The nature of catalysis

We start by defining in broad terms what we mean by catalysis. A catalyst is a material capable of accelerating a chemical reaction but which does not disturb the equilibrium position of the reaction when present in a quantity which is small compared with that of the reactants. The catalyst must remain unchanged by the reaction process and thus be capable of recycling. The rate is accelerated as a result of a reduction in the free energy change required to pass from the ground state to the transition state (of highest free energy).

The catalytic process must be associative in nature, as will be realized by

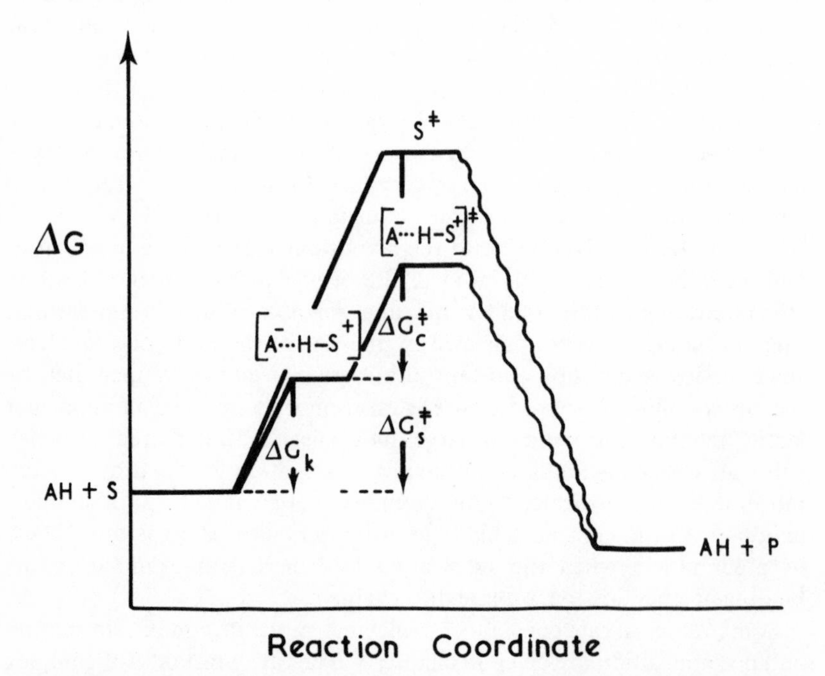

Figure 1.4 A free energy diagram which demonstrates the essential difference(s) between a catalysed $[A^- \ldots H\text{–}S]^{\ddagger}$ and an uncatalysed $[S]^{\ddagger}$ reaction. This example shows acid catalysis in which the protonation of S is energetically unfavourable but which results in a massive decrease in activation energy. Thus $\Delta G_c^{\ddagger} \ll \Delta G_s^{\ddagger}$ and $(\Delta G_k^{\ddagger} + \Delta G_c^{\ddagger}) < \Delta G_s^{\ddagger}$.

considering the forces involved in bond exchange processes. The magnitude of these strong short-range forces is dependent, as mentioned earlier, on a high ($\geqslant 6$) inverse power of the internuclear separation. These forces are significant only at internuclear separations of the order of the sum of the atomic radii of the reacting atoms, which in turn, are of the order of bond lengths.

The simplest model for a catalytic process is represented by a simple bimolecular process which involves a single reagent and catalyst. The catalysed reaction is effected by direct collisional activation of the reagent upon contact with the catalyst. Only those complexes that directly form activated complex pass through the transition state and so react. No provision for pre-transition state complex formation prior to activation, such as is postulated in enzymic catalysis, is provided. This type of catalysis could be provided by the presence of an inert gas in a gas phase reaction, the inert gas providing collisional activation energy. A rather more complex catalytic process might involve direct chemical participation of the catalyst, e.g. proton donation from catalyst to substrate. The free energy diagram for such a process relative to the uncatalysed process is shown in figure 1.4. Plainly the act of proton donation could occur prior to, or concerted with, activation. This type of process can be represented by the scheme

$$\text{A—H} + \text{S} \underset{k_{-1}}{\overset{k_1}{\rightleftharpoons}} \text{A}^- + \text{HS}^+ \underset{k_{-2}}{\overset{k_2}{\rightleftharpoons}} \text{P} + \text{A—H}$$

where A—H is the catalyst, S the substrate and P the product. Note that in order for AH to be a catalyst, A^- must recover a proton as a result of the transformation of HS^+ to P.

The catalytic process can occur by means of equilibrium protonation prior to the chemical step, in which case k_1 and $k_{-1} \gg k_2$, or by means of concerted proton transfer where the proton is transferred as reaction takes place with $k_2 \gg k_1$ and k_{-1}, or by any pathway intermediate between these extremes. In circumstances where equilibrium pre-transition state protonation occurs, HS^+ will accumulate to an extent dependent upon the relative values of the rate constants. In the particular case of acid catalysis given here, the reaction is most likely to proceed via pre-transition state protonation, since proton transfer reactions are frequently very fast. The proton transfer step can be rate-limiting, particularly if the equilibrium for protonation of S is very unfavourable. The putative intermediate (HS^+) could exist in the form of a hydrogen bonded complex

$$[\text{A}^- \ldots \text{H—S}^+]$$

Alternatively, the degree of proton transfer could be less and the complex have the form [A—H ... S], charge development having occurred either partly or completely in the transition state. It can be shown that the overall second order rate constant for the forward direction of the catalytic reaction in the absence of P is given by

$$k_0 = k_1 k_2/(k_{-1} + k_2)$$

(see chapter 4, and compare with the expression for a simple enzyme reaction) provided that the reaction is in a steady state (intermediate concentration constant) and the substrate is in excess of the catalyst. Clearly if $k_{-1} \ll k_2$ then $k_0 = k_1$ while if $k_{-1} \gg k_2$,

$$k_0 = \frac{k_2 k_1}{k_{-1}} = \frac{k_2}{K_S}$$

where K_S is the dissociation constant of the intermediate. The catalytic reaction will be subject to saturation, provided (i) that it does not proceed via a concerted mechanism and (ii) that k_{-1} is not so large as to render HS^+ accumulation insignificant at realistic reactant concentrations. The proton may be returned to A in the reaction complex $[A^- \ldots H{-}P^+]$ or by an equilibrium proton transfer from non-complexed $H{-}P^+$. Two rather extreme variations of the pathway are given below.

$$A{-}H + S \rightleftharpoons [A{-}H\cdots S] \underset{FAST}{\rightleftharpoons} [A^{\ominus}\cdots H{-}P^{\oplus}]$$

$$[A{-}H\cdots S] \overset{FAST}{\rightleftharpoons} [A^{\ominus}\cdots H{-}S^{\oplus}] \rightleftharpoons A^{\ominus} + HS^{\oplus} \underset{SLOW}{\rightleftharpoons} A^{\ominus} + HP^{\oplus} \rightleftharpoons [A^{\ominus}\cdots H{-}P^{\oplus}]$$

$$[A^{\ominus}\cdots H{-}P^{\oplus}] \rightleftharpoons [A{-}H\cdots P] \rightleftharpoons A{-}H + P$$

The general nature of enzyme catalysis

As the structure of enzyme proteins will be discussed in some detail in chapter 3, a short description will suffice at this point. Enzymes may be described as globular structures bearing considerable similarity to micelles. The surface, which interacts with water, consists predominantly of polar groups while the interior is predominantly apolar. This presents the possibility of creating binding sites in the surface region which may have steeply graded polarity. In addition, proteins rely upon hydrogen bonding together with hydrophobic and charge pairing interactions for the maintenance of a discrete three-dimensional structure.

All of these types of interaction can be used to bind a substrate molecule in a highly specific fashion by means of a cleft in the surface region of the enzyme. The base of the cleft can be hydrophobic (since it may protrude into the interior of the protein) and it may have polar groups disposed in a suitable fashion for hydrogen bonding and charge pairing interactions with the substrate. It is plain that in order to achieve specificity, the detailed disposition of the interacting groups provided by the protein in the cleft must match a set of groups or atoms on the substrate (in particular the transition state) with which they may interact favourably.

In the absence of substrate the cleft or "active site" will be occupied by water molecules which must be expelled (except one water molecule in hydrolytic reactions) consequent upon enzyme–substrate complex formation. The favourable energy of interaction must be sufficient to ensure water expulsion and to overcome the considerable negative entropy change that results from bimolecular combination, particularly if a tight complex is to be achieved.

The high intrinsic reactivity of the productive enzyme–substrate complex (i.e. a complex in which the substrate is bound to the enzyme in a mode that leads to optimum reactivity) is made possible by the presence in the cleft of catalytic groups provided by the enzyme protein. The nature of the catalytic groups and the chemistry of the catalytic processes will be more fully covered in chapters 2 and 9, but it is apparent at this stage that the alignment of the substrate with respect to the catalytic groups must be accurately defined. The catalytic groups must be placed immediately adjacent to the sites upon the substrate at which chemical interaction is to take place; angular relationships being defined to the order of $\pm 10°$ but distance relationships must be organized very accurately owing to the high inverse power of the atomic dispersion forces that are involved in bonding exchange interactions. A very much simplified scheme representing a generalized enzyme–substrate interaction is shown in figure 1.5. In figure 1.5 the substrate is shown as having dimensions of the same order of magnitude as the enzyme. This is not usually the case, enzymes frequently having a mass of up to three orders of magnitude greater than the substrate. One of the important questions we shall tackle in the course of this book concerns the reason(s) for this enormous difference in mass, since

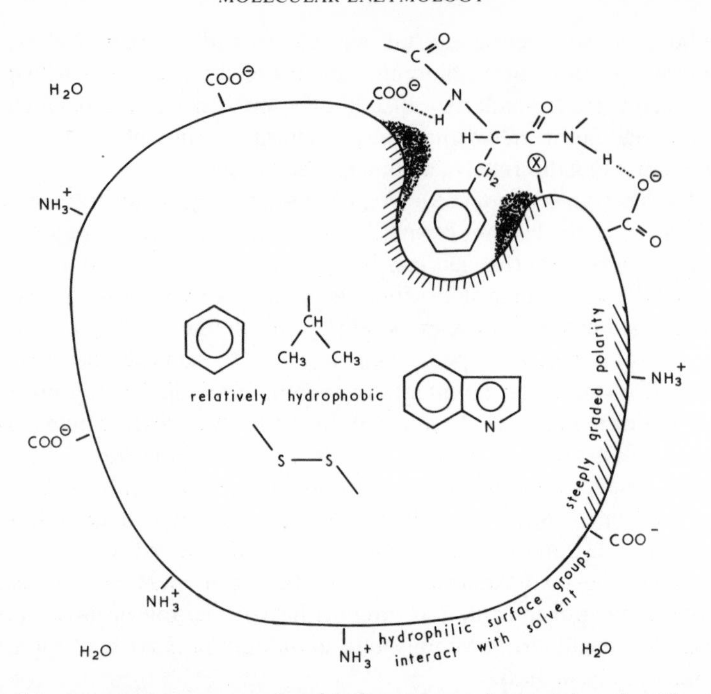

Figure 1.5 A generalized conceptual model of an enzyme-substrate interaction. The phenyl group of the substrate is buried in the hydrophobic cleft. The shaded areas represent steric restriction in the benzyl methylene and α-hydrogen regions. The enzyme must be able to interact with the substrate at a minimum of three points in order that it may be optically stereospecific. The chemical catalytic apparatus (×) is juxtaposed close to the carbonyl carbon atom of the C–N bond that will be cleaved as a result of the hydrolytic catalysis. Note that the size of the substrate is grossly exaggerated for the purposes of illustration; the hydrogen-bonding interactions will be much closer to, or in the cleft.

superficially at least it would seem to be highly inefficient for a biological system to have to synthesize such large molecules. There are several groups of enzymes which act upon polymeric substrates which may have molecular weights greater than the enzyme but the site of interaction between enzyme and substrate is restricted to a small region of the polymer substrate (up to a mol. wt. of about 800); the enzyme molecular weight being not less than about 11 000. Although the enzyme may interact with substrates over a region of molecular weight of up to 800 (six amino acid residues of a protein) the site at which the catalytic action will take place is uniquely defined when the substrate is productively (correctly) bound.

Enzymes are capable of achieving enormous rate enhancements (of the order of 10^6–10^{12} fold) relative to the equivalent uncatalysed reaction. These remarkable rate enhancement factors imply that enzymes are able to greatly stabilize the transition state of the catalysed reaction relative to that of the uncatalysed reaction. Stabilization of a transition state is one way of expressing the concept of loss of free energy of activation; it can also be expressed in terms of ground state destabilization. We shall discuss these matters at some length later in the book, but as a result of the considerations covered in this chapter we suggest that the catalytic power of enzymes may be explained in terms of highly favourable specific enzyme–substrate interaction which can be used to

(1) overcome the unfavourable entropy change that results upon enzyme–substrate complex formation which in turn allows unimolecular chemical transformations;
(2) introduce strain into the substrate such that it is distorted towards the transition state;
(3) ensure optimum binding of the transition state rather than the substrate ground state;
(4) provide suitable catalytic groups optimally oriented with respect to the site of chemical action upon the substrate.

We can infer from these considerations that the system will tend to "slide" to the transition state (which is the criterion required for rapid reaction) and will show a high degree of specificity as a result of intimate well-defined interactions.

CHAPTER TWO
CHEMICAL CATALYSIS

IN THIS CHAPTER WE ASSESS THE DEGREE TO WHICH OUR KNOWLEDGE OF SIMPLE chemical catalysis can contribute to an understanding of the efficiency of enzyme-catalysed reactions. Enzyme-catalytic mechanisms must make use of the groupings found in proteins; the groups that must be considered as having catalytic potential are the imidazoyl, hydroxyl, thiol, carboxyl and amino groups (see table 6.1). (Note that all of these, with the exception of the hydroxyl group, ionize within the pH range that exists within living organisms. The carboxyl group for instance will be significantly protonated in the acidic pH regime of the stomach, and at the other extreme some deprotonation of amino groups will occur in the mildly alkaline environment of the small intestine. This means that each of these groups may potentially act as either an acid or a base.) In this chapter we shall describe some well-known examples of simple catalysis, frequently involving the species mentioned above, and this will allow us to delineate a range of catalytic mechanisms that *may* be employed in enzymic catalysis. We shall, however, have some surprises and will find that the knowledge gained from the study of simple chemical catalysis is not completely adequate for the description of enzymic catalysis. We shall then, in subsequent chapters, turn to the study of enzymes themselves in order to attempt to complete the picture.

Nucleophilic catalysis

The amino acid histidine is frequently found at the active (catalytic) site of enzymes and has been implicated as having a chemical catalytic role in a number of cases (see chapter 9). The reactive moiety is the resonance stabilized imidazole ring (2.1) which has a pK of ca. 7.0.

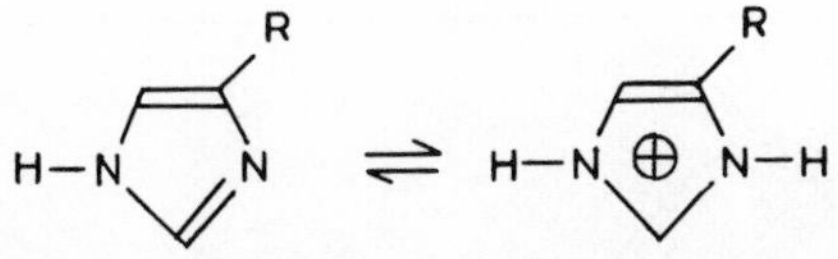

(2.1)

Imidazole is thus ideally suited to act both as an acid and as a base or nucleophile at neutral (physiological) pH. The ester *p*-nitrophenyl acetate (PNPA) (2.2) has been much studied because it is reactive as an electrophile towards nucleophilic species such as imidazole.

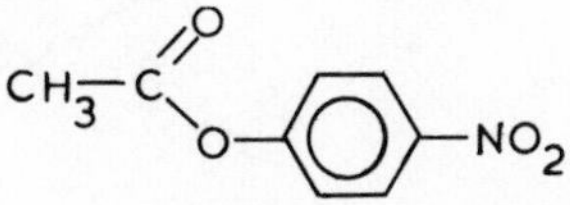

(2.2)

This is due to the electrophilicity of the carbonyl carbon induced by the electron-withdrawing *p*-nitrophenyl group. The *p*-nitrophenolate leaving group has a pK of 7, due to the favourable electronic transition of the nitrophenolate ion (2.3)

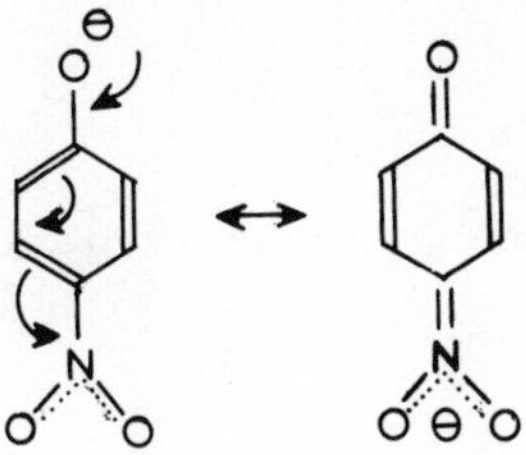

(2.3)

which renders the ion a bright yellow colour with an absorbance maximum at 400 nm. Thus the loss of *p*-nitrophenol anion from PNPA can be followed in a simple fashion by measurement of the increase in the absorbance (A) at 400 nm. Both *p*-nitrophenol and PNPA have λ_{max} at ca. 320 nm, well clear of the 400 nm band.

When added to an aqueous solution of PNPA at neutral pH, imidazole greatly increases the rate of *p*-nitrophenolate loss from PNPA. At neutral pH the hydroxide ion hydrolysis rate is quite slow due to the low (10^{-7} M) concentration of hydroxide ion, and the neutral water rate is also very small. The reaction of imidazole (in excess) with PNPA can be followed by measurement of the absorbance at 245 nm, (figure 2.1), and the biphasic behaviour observed indicates the formation and decay of an intermediate on the reaction pathway. The first step of the imidazole–PNPA reaction involves imidazole attack at the carbonyl carbon atom (2.4).

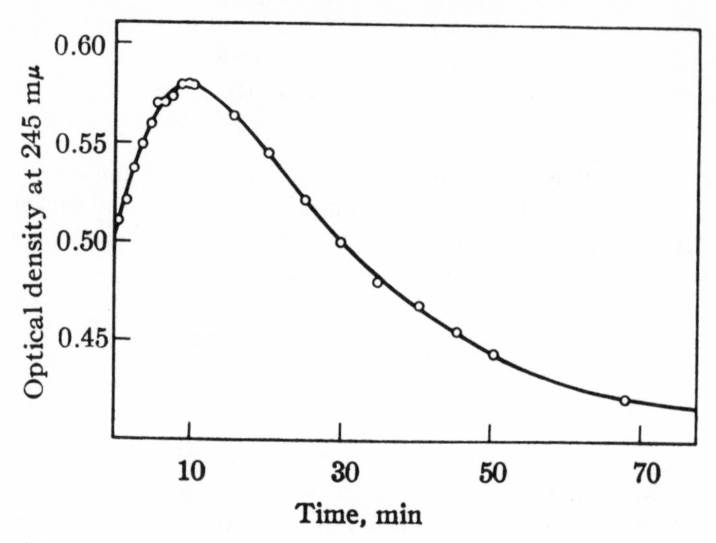

Figure 2.1 The formation and decay of *N*-acetyl imidazole in the imidazole-catalysed hydrolysis of *p*-nitrophenyl acetate. From Bender and Turnquest (1957).

$$\text{H—N}\langle\text{imidazole}\rangle\text{N:} \curvearrowright \underset{\text{O}\phi\text{NO}_2}{\overset{\text{CH}_3}{\text{C}}}\text{=O} \rightleftharpoons \text{H—N}\langle\text{imidazolium}^{\oplus}\rangle\text{N—}\underset{\text{O}\phi\text{NO}_2}{\overset{\text{CH}_3}{\text{C}}}\text{—O}^{\ominus} \qquad (2.4)$$

The addition compound is known as a *tetrahedral intermediate* since the carbonyl carbon atom hybridization changes from sp^2 to sp^3. The intermediacy of tetrahedral adducts in ester hydrolysis has been established by the observation that ^{18}O is *back* incorporated into the ester reactant during hydrolysis in water containing $H_2{}^{18}O$. The kinetic significance of the tetrahedral intermediate varies from one reaction to another (see chapter 1) but the carbonyl carbon atom must be at least transiently sp^3 hybrid for expulsion of the leaving group to occur, since the orbital geometry is not suitable for direct displacement of the leaving group from the sp^2 hybrid form. The tetrahedral intermediate may revert to reactant by loss of imidazole or proceed along the reaction pathway by expulsion of *p*-nitrophenolate. Expulsion of *p*-nitrophenolate leads to acetyl imidazole formation (2.5), the driving force for the expulsion being the oxyanion electron pair. The rate at which the reaction proceeds to acetyl imidazole is determined by the *relative* leaving ability of *p*-

nitrophenol(ate) as compared with that of imidazole. The imidazolium of the tetrahedral intermediate will have a p*K* between 5 and 7 and will be partly deprotonated at neutral pH; indeed the imidazole may lose a proton coincident with its nucleophilic insertion at the carbonyl carbon as shown in (2.6).

$$\text{H—N}\oplus\text{N—C(CH}_3\text{)(O}^\ominus\text{)(O}\phi\text{NO}_2\text{)} \rightleftharpoons \text{H—N}\oplus\text{N—C(=O)CH}_3 + \text{O}_2\text{N}\phi\text{O}^\ominus \quad (2.5)$$

$$\text{H—N}\diagdown\text{N} + \text{C(CH}_3\text{)(=O)(O}\phi\text{NO}_2\text{)} \rightleftharpoons \text{H}^\oplus + \text{N}\diagdown\text{N—C(CH}_3\text{)(O}^\ominus\text{)(O}\phi\text{NO}_2\text{)} \quad (2.6)$$

However the imidazole group will not leave as such, since the electron affinity of the imidazole anion is far too high for this to happen (the second p*K* of imidazole is 14.5). Since the reaction is conducted at neutral pH the *p*-nitrophenol will happily leave in the anionic form since it has a p*K* of 7. There will be virtually no advantage to be gained from acid catalysis (2.7) of this step.

$$\text{H—N}\oplus\text{N—C(CH}_3\text{)(O(H))(HO}^\oplus\phi\text{NO}_2\text{)} \quad (2.7)$$

Since the pK_a values are similar, the expulsion rates from the tetrahedral intermediate are expected to be similar, so the intermediate will accumulate only if the attack step is faster than the sum of the depletion steps. Present evidence suggests that the attack step (2.4) is rate-limiting in acetyl imidazole formation so the intermediate will not accumulate. The biphasic curve of figure 2.1 measured at 245 nm cannot represent the formation and decay of tetrahedral intermediate for the simple reason that the rate of appearance of *p*-nitrophenolate measured at 400 nm is monotonic and shows simple second order dependence upon imidazole and PNPA concentrations. A lag phase in the $A_{400\,nm}$ *vs.* time curve would be seen if the curve of figure 2.1 represented tetrahedral intermediate formation and decay, since the change in spectrum on going from ester to tetrahedral intermediate will be small. As the overall observed reaction is hydrolysis, the only plausible candidate for the intermediate indicated by the curve of figure 2.1 must be acetyl imidazole. It is clear that the rate of acetyl imidazole formation is faster than the rate of breakdown at neutral pH and that the intermediate acetyl imidazole accumulates at the high imidazole concentration employed in this experiment. The hydrolysis of acetyl imidazole proceeds via the cationic species which is hydrolysed by hydroxide ion and neutral water (2.8). These terms in the acetyl imidazole hydrolysis rate equation (2.9) correspond to neutral water- and acid-catalysed specific base hydrolysis of acetyl imidazole

$$CH_3C(=O)\text{–}\overset{\oplus}{\text{Im}}\text{–}H \xrightarrow[OH^{\ominus}]{H_2O} CH_3COO(H) + Im(H) \qquad (2.8)$$

$$\text{Rate} = k(\text{AcImH}^+)(\text{OH}^-) + k'(\text{AcImH}^+)(\text{H}_2\text{O})$$
$$k''(\text{AcIm})(\text{H}_2\text{O}) = k(\text{AcImH}^+)(\text{OH}^-) \qquad (2.9)$$

respectively. Thus $k(\text{OH}^-)(\text{AcImH}^+)$ is kinetically equivalent to $k''(\text{AcIm})(\text{H}_2\text{O})$ since pH-dependence studies will not distinguish between the two. However it can safely be assumed that neutral water will not be able to expel imidazole anion and thus the former pathway of these two will occur. PNPA is, of course, susceptible to simple hydroxide ion hydrolysis and for overall catalysis to be observed, the reaction in the presence of imidazole must be faster than the reaction in a non-nucleophilic buffer (extrapolated to zero buffer concentration) at the same pH.

Thus, criteria for nucleophilic catalysis are that the attacking catalytic nucleophile must be able to expel the leaving group from the tetrahedral intermediate to a significant extent, and that the intermediate so created must be *less* stable (i.e. more reactive) towards hydrolysis in this instance, than is the starting material. In the above case the reaction pathway has been substantiated by the independent synthesis of acetyl imidazole and study of its hydrolysis. The fact that *N*-methylimidazole is essentially as effective a catalyst as imidazole indicates that the attack step proceeds with proton retention and that hydrolysis of acetyl imidazole proceeds via the acetyl imidazolium cation. As mentioned earlier, PNPA is a reactive ester due to the electron affinity (relatively low pK) of the leaving group. Only compounds of a similarly reactive nature will be subject to nucleophilic catalysis by imidazole since the nucleophile must be able to expel the leaving group from the tetrahedral intermediate. Imidazole is quite incapable of catalysing the hydrolysis of an unreactive ester such as ethyl acetate. In order to achieve loss of a leaving group of high pK (relatively low electron affinity) either the nucleophile must have a similar pK, e.g. $H_2O(OH^-)$, or acid catalysis must be used to facilitate departure of the leaving group. Clearly, significant protonation of the ester oxygen atom will occur only in strong acid under circumstances in which imidazole will be totally protonated and hence totally unreactive as a nucleophile. Enzymes such as chymotrypsin do achieve the catalytic hydrolysis of unreactive esters at neutral pH via nucleophilic attack and therefore must employ features that are seemingly irreconcilable when considered in terms of the features of this simple catalytic reaction.

The thiol anion is a very effective nucleophile and in suitable circumstances will fill a catalytic role although the intermediate thiol esters tend to be more stable than are acyl-imidazoles. Neutral amines are also effective nucleophiles but, at least in hydrolytic reactions, cannot play a catalytic role at neutral pH, since the intermediate amides are quite stable.

The discussion of the imidazole-catalysed hydrolysis of PNPA has been rather extended. However, it has been used to introduce a number of important concepts that can now be applied to catalytic reactions in general.

General catalysis

General catalysis is distinguishable from nucleophilic catalysis in that the catalytic species does not become directly attached to the substrate molecule. Specific catalysis involves catalysis by H_3O^+ (or other hydrated forms of the proton) and OH^- and is thus usually expected to occur at either low or high pH values. General catalysis occurs as a result of contributions of other buffer species that may be present in the reaction mixture. Thus general catalysis is observed as a dependence of the rate of reaction upon the buffer concentration *at constant pH*. The requirement that the rate measurements be made at constant pH allows the elimination of any variable contribution from specific catalysis, this being measured by extrapolation of the buffer concentration to zero. If the rate constant k_{obs} observed at constant pH in a reaction which is subject to general catalysis is plotted against the buffer concentration, the results are as shown in

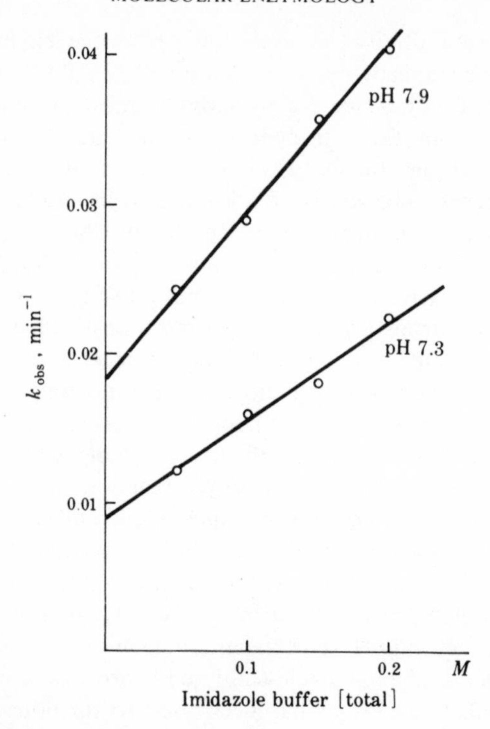

Figure 2.2 General base-catalysed hydrolysis of *N*-acetyl imidazole by imidazole. From Jencks, W. P. (1969).

figure 2.2. If the slope increases with increasing pH then the catalysis has a general base component; a general acid contribution is indicated by an increase in slope when the pH is decreased, the pH variation being within the buffering range of the species involved. Clearly if pH values outside the buffering range are chosen (beware pH changes during reaction) then the general acid and general base contributions can be measured independently of each other. The rate equation for general acid, general base and general acid–base catalysis by the components of a single buffer system takes the form of equation (2.10).

$$\frac{dP}{dt} = k_{obs}(\mathrm{S})(\mathrm{B_1}) = k_{\mathrm{H_2O}}(\mathrm{S}) + k_{\mathrm{H_3O^+}}(\mathrm{S})(\mathrm{H_3O^+}) + k_{\mathrm{OH^-}}(\mathrm{S})(\mathrm{OH^-}) \\ + k_{\mathrm{GA}}(\mathrm{S})(\mathrm{HB^+}) + k_{\mathrm{GB}}(\mathrm{S})(\mathrm{B}) + k_{\mathrm{GAB}}(\mathrm{S})(\mathrm{HB^+})(\mathrm{B}) \tag{2.10}$$

In this equation (S) is the substrate concentration, $(\mathrm{B_t})$ is the total buffer concentration, $(\mathrm{HB^+})$ is the concentration of protonated buffer species, (B) is deprotonated buffer species; the meaning of the other expressions in the rate equation being self-evident. The general approach towards the determination of each of the rate constants has been indicated earlier and we leave it to the reader as a useful exercise to work out a scheme for the determination of each term of equation (2.10).

Occasionally it may be found that an equation of this type does not completely describe the overall reaction rate and it may prove necessary to introduce terms having a higher power in one of the species, e.g. $k_{GB_2}(S)(B)^2$—this is general base catalysis *of* general base catalysis. Note that the term $k(S)(B)^2$ *could* represent general base catalysis of nucleophilic attack. It can be rather difficult to rule out a nucleophilic mechanism particularly if a trapping reaction for the putative nucleophile-containing intermediate is not available.

A good example of general base catalysis is provided by the imidazole buffer-catalysed hydrolysis of acetyl imidazole. This example unambiguously represents general base catalysis since nucleophilic attack by imidazole will lead to regeneration of substrate. Figure 2.2 shows the linear relationship between imidazole buffer concentration and the observed rate constant k_{obs}. A plausible mechanism for this reaction is given by (2.11).

$$\text{H—N} \diagup\!\!\!\diagdown \text{N:} \curvearrowright \text{H—}\underset{\text{H}}{\text{O}} \curvearrowright \underset{\text{Im}\overset{\oplus}{\text{H}}}{\overset{\text{CH}_3}{\text{C}}}\text{=O} \rightleftharpoons \text{H—O—}\underset{\text{Im}\overset{\oplus}{\text{H}}}{\overset{\text{CH}_3}{\text{C}}}\text{—O}^{\ominus} \longrightarrow \begin{matrix}\text{CH}_3\text{COO(H)} \\ + \\ \text{Im (H)}\end{matrix} \qquad (2.11)$$

The ionic species will rapidly equilibrate with solvent in proportions relevant to the particular pH value but these equilibria are not included in the reaction pathway shown above.

General catalysis is frequently found in reactions of moderately reactive substrates. Thus PNPA, which is highly reactive, tends to undergo nucleophilic or specific catalysis in aqueous solution whilst ethyl acetate, which is unreactive, is subject to specific catalysis. Ethyl dichloroacetate is a good example of a substrate that readily undergoes general base catalysis and is moderately reactive. It is acyl-activated which enhances the initial water attacking step, a step which will be very unfavourable in the case of ethyl acetate. Once the tetrahedral intermediate has formed, hydroxide and ethoxide (relevant pK_a values are similar) will leave at similar rates and so the reaction can proceed in the forward direction. Note that general base catalysis *will* occur in the case of PNPA but that the rate of nucleophilic catalysis is so much greater that it swamps general base catalysis.

A good example of general acid catalysis is provided by the dehydration of acetaldehyde hydrate shown in (2.12).

$$CH_3-\underset{\underset{H}{|}}{\overset{O-H}{C}}\!-H \;\; (\text{with } O-H \text{ and } H-A) \rightleftharpoons CH_3-C\overset{\displaystyle O}{\underset{\displaystyle H}{\diagdown}} + H_2O + HA \qquad (2.12)$$

There is a potential for general base reactivity in this reaction which could provide assistance with proton removal from the aldehydic oxygen. A wide range of oxyacids and phenolic compounds act as general acids in this reaction (see figure 2.3).

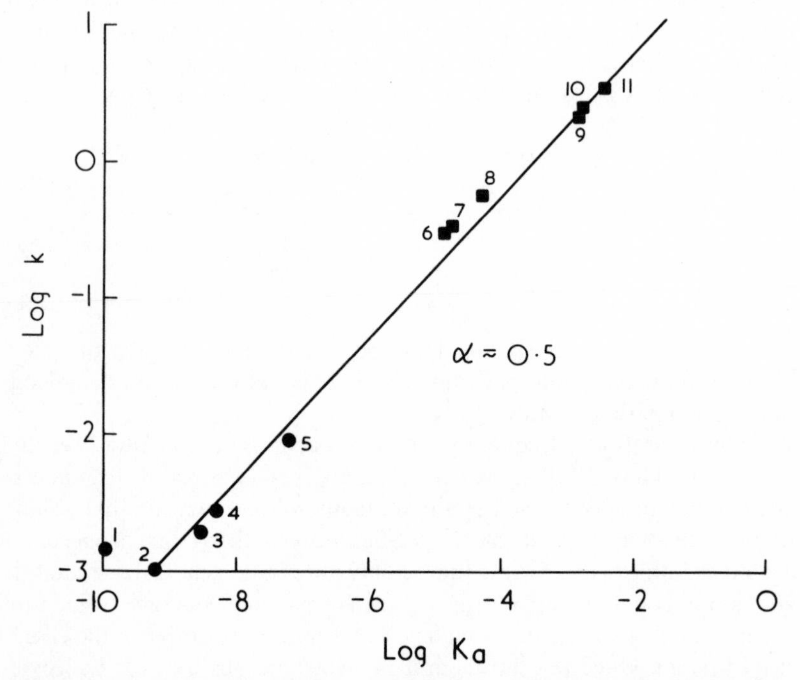

Figure 2.3 Brønsted plot of the general acid catalysis of acetaldehyde hydrate dehydration. 1. Phenol; 2. *p*-chlorophenol; 3. *o*-chlorophenol; 4. *m*-nitrophenol; 5. *p*-nitrophenol; 6. propionic acid; 7. acetic acid; 8. phenylacetic acid; 9. bromoacetic acid; 10. chloroacetic acid; 11. cyanoacetic acid. Adapted from Bell and Higginson (1949).

Reactivity correlations in catalysis

The leaving ability of a particular group is directly related to the intrinsic electron affinity of the group, or inversely to the potential for sharing its electrons with a proton. A group of relatively low pK has a relatively low potential for sharing its electrons with a proton or similarly with an acyl group and is thus a good leaving group.

There is no obvious reason why we should not reverse this argument and apply it to a consideration of the attacking ability of basic species and protonation ability of acidic species. Thus a correlation between pK_a values and general basic or acidic catalytic reactivity might be expected, particularly among a series of structurally related compounds. Since the pK is a logarithmic function the expected correlation will be of the form

$$\log k = \beta \,.\, \mathrm{p}K + \text{const.}$$

for general basic catalysis and

$$\log k = -\alpha \,.\, \mathrm{p}K + \text{const.}$$

for general acid catalysis.

These equations are known as the *Brønsted catalysis laws* and may be described as linear free energy relations. Log k is related to the free energy of activation of the reaction and pK = log $1/K_a$ is related to the free energy of deprotonation of the catalytic species. Since general catalysis involves proton removal or donation it is apparent that the two free energy terms are related. In many situations, linear Brønsted plots of log k *vs.* pK result which frequently cover a wide range of general catalytic species (as in figure 2.3). If α or β is unity, this represents specific catalysis. If either is zero, the reaction is quite insensitive to general catalysis—this is usually due to high solvent reactivity towards the particular substrate or a general lack of reactivity of the substrate. Most easily observed general catalysed reactions will have α or β values in the range 0.3–0.7 as competitive effects will be minimized in this range. The above arguments lead to the conclusion that the value of α or β may be interpreted in terms of the degree of proton transfer that has occurred in the transition state of the reaction (2.13).

$$\mathrm{B}\overset{\beta}{\cdots\cdots}\mathrm{H}\overset{(1-\beta)}{\cdots\cdots}\underset{\mathrm{H}}{\mathrm{O}}\overset{\beta}{\cdots\cdots}\overset{\mathrm{R}}{\underset{\mathrm{O}\ (1-\beta)}{\mathrm{C}}}\text{——}\mathrm{OR'} \qquad (2.13)$$

The bond order interpretation shown above (2.13) assumes that charge is smoothly transferred through the reaction coordinate during transition state formation and does not accumulate in a fashion which would interfere with smooth bond order changes.

Values of α and β are most reliably interpreted in terms of the charge change experienced by the acid or base catalyst between ground and transition states since compensating bond changes (particularly in acid-base catalysis) can interfere with the bond order interpretation. Thus the α value for acetaldehyde hydrate dehydration (figure 2.3) may tentatively be interpreted in terms of a transition state structure in which the proton is $\sim 50\%$ transferred from the acid to the hydrate, but most reliably in terms of a structure which implies accumulation of $\sim 50\%$ of a formal negative charge upon the general acid (assuming no base catalysis).

The general base-catalysed hydrolysis of ethyl trifluorothiolacetate obeys the Brønsted Law quite well (figure 2.4) although O-methyl hydroxylamine is well off the line. Such deviations may often be related to solvation effects in the absence of super-reactive α-effects and catalyses that require statistical correction. These factors are considered in detail in the reference text mentioned in the bibliography. The value of β calculated from the slope of the plot in figure 2.4 is 0.33, and so the structure of the

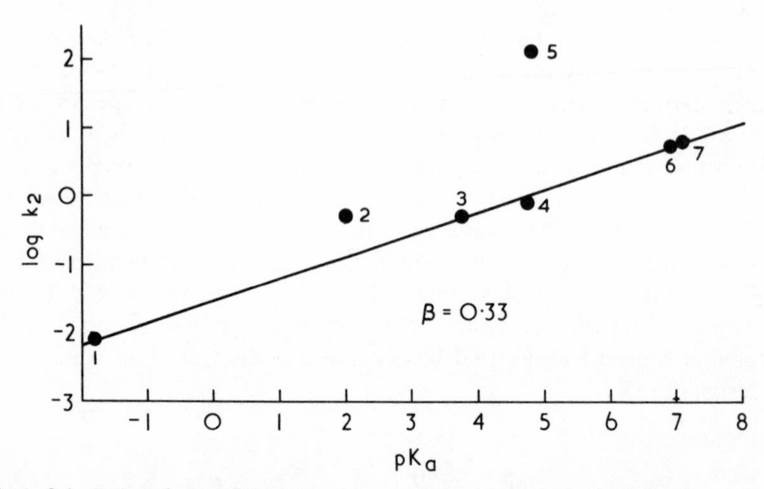

Figure 2.4 Brønsted plot of the general base-catalysed hydrolysis of ethyl trifluorothiolacetate. (1) H_2O; (2) $H_2PO_4^-$; (3) HCO_2^-; (4) CH_3COO^-; (5) CH_3ONH_2; (6) $HPO_4^=$; (7) imidazole. That the catalysis is general base in all cases was confirmed by product analysis. Redrawn from Fedor and Bruice (1965).

transition state for tetrahedral intermediate formation may tentatively be interpreted as:

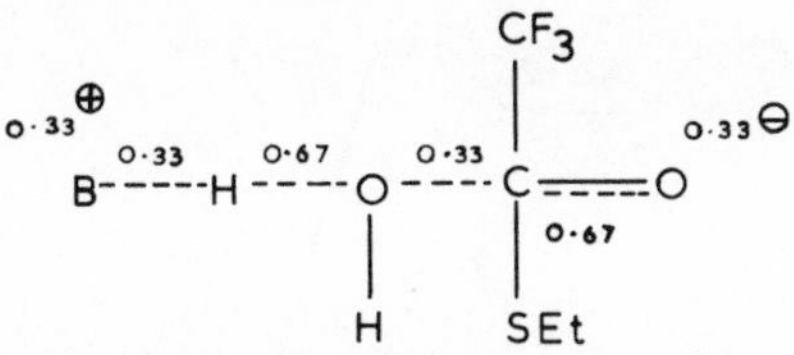

$$\begin{array}{ccccccccc} & & & & & & CF_3 & & \\ & & & & & & | & & \\ \overset{0.33\oplus}{B} & \overset{0.33}{----} & H & \overset{0.67}{----} & O & \overset{0.33}{----} & C & \underset{0.67}{=\!=\!=} & \overset{0.33\ominus}{O} \\ & & & & | & & | & & \\ & & & & H & & SEt & & \end{array} \qquad (2.14)$$

The same considerations apply in the interpretation of Brønsted plots of nucleophilic reactions as in true general base catalysis except that the values of β can be outside the range 0–1.0.

The Hammond postulate

The Hammond postulate states that an intermediate on a reaction pathway will resemble the transition state more closely than either reactant or product ground states. Clearly the more unstable the intermediate the more closely it will resemble the transition state, which must, of course, have properties intermediate between reactant and product. An intermediate will have transition states for its formation and breakdown; the intermediate obviously having a structure that possesses features of both transition states. Thus a tetrahedral intermediate in ester hydrolysis will have much transition state character.

The Hammond postulate is generally, although less reliably, extended to encompass the relationship between the overall thermodynamics of a reaction and the extent to which the transition state structure can be compared with that of the reactant or product. Reactions that are strongly favoured thermodynamically are expected to have reactant-like or "early" transition states while thermodynamically unfavourable reactions are expected to have product-like or "late" transition states. Since the absolute reaction rate theory is not concerned with events after the transition state the above expectations fit in well with this theory and seem intuitively reasonable. The extreme case of a diffusion-limited encounter, thermodynamically favourable reaction will have a very early transition state as shown in figure 2.5.

A bimolecular reaction, taken in isolation, must be treated with caution when an attempt is made to apply the Hammond postulate due to the importance of the entropic terms involved in approximation but this does not apply to the interpretation of structure–reactivity relations. If a

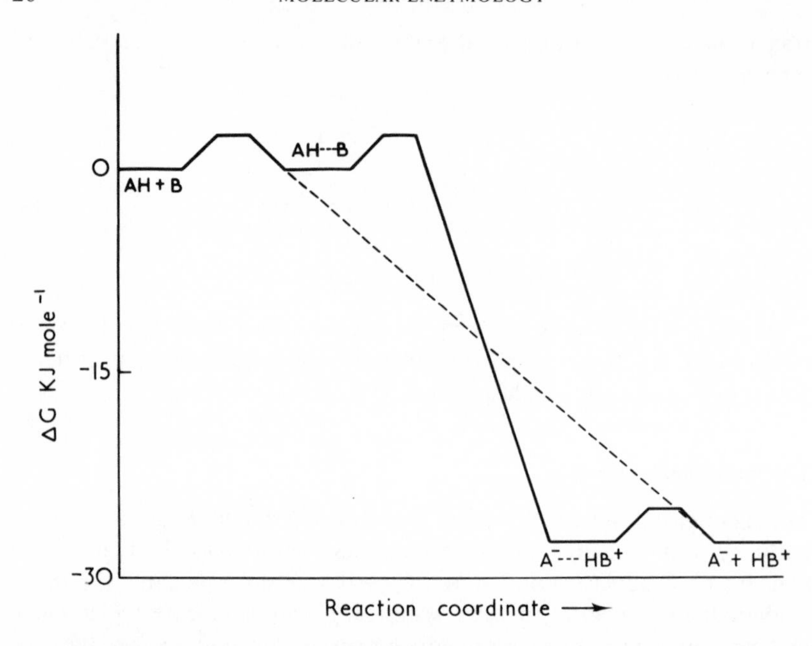

Figure 2.5 Free energy diagram for a diffusion-controlled proton transfer $AH+B \rightarrow A^- + HB^+$. The dotted line joins the transition state to the products in a simplified form of the diagram which clearly demonstrates the "early" transition state. The solid line indicates encounter complexes for both reactants and products. The overall *rate* in the reverse direction $A^- + HB^+ \rightarrow AH + B$ may be limited by the diffusive dissociation of AH . . . B since it is preceded by an unfavourable equilibrium and will have a low concentration. According to transition state theory a rate constant will have a value of one-third of the theoretical maximum (kT/h) if a 2.5 kJ mole^{-1} barrier is present between reactants and products (dotted line, 300°K).

Brønsted plot is linear the degree of charge transfer in the transition state is constant over the whole range of acidity covered by the experimental data. The entropic terms are thus the same for each reaction and cannot interfere with an application of the Hammond postulate in a relative sense. It is apparent that a general base-catalysed reaction will become progressively more thermodynamically favourable the stronger the base catalyst. Thus as the basicity of the catalyst is strengthened the reaction proceeds more rapidly, the thermodynamics become more favourable and the transition state, which does not change its degree of charge transfer, occurs earlier on the reaction coordinate and is more reactant-like. The above rationalization is entirely consistent with the concept of base strength or electron availability—the stronger the base the more it wishes to share its electrons, and so the earlier on the reaction coordinate it reaches a particular stage of attack.

The Hammett equation

Another important structure-reactivity relationship that has been used to good effect in both organic chemistry and in enzymology is that based upon the Hammett equation (2.15),

$$\log(k/k_0) = \rho\sigma \tag{2.15}$$

which may be written as equation (2.16).

$$\log k = \log k_0 + \rho\sigma \tag{2.16}$$

If values of $\log k$ are plotted against values of σ, the Hammett substituent constant, a straight line of intercept $\log k_0$ and slope ρ results which is characteristic of the reaction, provided a linear free energy relationship exists between the thermodynamic and kinetic properties of the reaction. The Hammett equation applies to many reactions of substituted aromatic compounds, where k_0 is the rate constant for the unsubstituted compound. The substituent values relate to the electron-donating or -withdrawing effect of a particular substituent. The σ values are the pK differences between appropriately *m*- and *p*-substituted benzole acids or phenols and benzoic acid or phenol. Hammett plots can be used to predict the likely structure of transition states by making use of the ρ value which is equivalent to the α and β values of Brønsted plots. Substituents capable of strong resonant interaction with the reaction centre frequently give rise to large deviations from the Hammett equation and special (σ^+ and σ^-) values have been devised to take the resonant interaction into account. Values of σ are large and positive for strongly electron-withdrawing substituents and small or negative for electron-releasing groups. The Hammett plot for the hydrolysis of phenyl acetates has a ρ value of 0.8, the hydrolytic rate being dependent upon the electron-withdrawing power of the substituent. This is expected since both tetrahedral intermediate formation and breakdown will be enhanced by electron withdrawal in the leaving group. The ρ value of 0.8 indicates some negative charge accumulation in the transition state which is presumably located mainly on the carbonyl oxygen, but will be shared to some extent with the carbonyl carbon and leaving group oxygen atoms. Values of ρ can be negative indicating cationic character in the transition state, and overall can have values which range from ca. -8 to 8.

Curved free energy relationships

Non-linear free energy relationships may result from a change in the rate-limiting step of the reaction as the reactant properties are changed. Compensation between two (or more) steps of a reaction, both of which are partially rate-limiting, can lead to a zero or small slope which can be mistaken for diffusion limitation. Clearly diffusion limitation will lead to a zero slope since the rate is independent of the chemical nature of the reactants, and so care must be exercised when the interpretation of non-linear free energy plots is undertaken.

Application of structure–reactivity correlations in enzyme chemistry

It must be apparent to the reader that the *direct* experimental applicability of the Brønsted catalysis laws to the study of enzyme catalysis is essentially nil. The enzyme whose catalytic properties are under study is not a variable of the system, in that it has a unique structure with specific catalytic groups at its active centre. The object of the study is the elucidation of the nature and reactivity of these groups and their involvement in events upon the reaction coordinate. Thus in order to make use of the methods of structure–reactivity correlation the structure and reactivity must be varied. This can only be achieved in the case of enzymes that have sufficiently general specificity requirements that they will tolerate considerable interference with the structure and chemical nature of the *substrates*. The proteolytic enzymes represent good examples of this in that they have both esterase and amidase catalytic activity. As a result the structure of substrates that can be hydrolysed by these enzymes may be varied widely provided some basic rules are obeyed (see chapter 9).

Thus it is possible, at least in principle, to use the Hammett equation in order to gain information concerning the nature of the transition state for the rate-limiting step of a protease-catalysed hydrolysis reaction. The slopes of such plots must be interpreted with extreme caution since the substrate-binding properties of the enzyme are likely to influence the relative rates at which the enzyme catalyses the hydrolysis of the series of substrates. An important factor in enzyme-substrate interaction is the hydrophobicity of the substrate molecule and it is thus important to ensure that the Hammett σ values do not correlate with the π (hydrophobicity) values since the observed results could be ambiguous.

The reader may well wonder why so much space has been devoted to structure–reactivity correlations when they appear to have rather limited applicability in studies of enzyme catalysis. The answer is that a clear knowledge of the form and interpretation of these correlations forms an essential basis of physical organic chemistry upon which is based our understanding of the mechanisms of organic catalysis. Our analysis of the mechanisms of enzyme-catalysed reactions must be based upon our knowledge of the physical organic chemistry of simple molecules. It is an important article of faith that the principles of physical organic chemistry are completely valid in the enzyme active site environment provided the environment is properly taken into account.

Kinetic ambiguity

It is sometimes tempting to think that the mechanism of a reaction has been solved when the kinetic rate equation has been determined. For example the rate equation (2.17)

$$\frac{dp}{dt} = k(\mathrm{HA})(\mathrm{S}) \tag{2.17}$$

for the acid-catalysed hydrolysis of S could relate to the mechanism

H_2O: S----H—A

which is general acid catalysis. However consideration of the rate equation will reveal that it must also apply to the mechanism

$A^{\ominus}$ H—O(—H) $SH^{\oplus}$

which is general base-specific acid catalysis. The true value of the rate constant will be different in each case since the pH-dependence of the species involved in the reaction is effectively part of the rate constant. This difference cannot be resolved by simple kinetic measurements since the rate law is the same for both. The true value of the rate constant is, of course, unknown since it can only be calculated from the observed rate constant once the mechanisms have been distinguished. It is important to note that the observed pH-dependence of the two mechanisms is identical and that the pH-dependence cannot therefore be used to distinguish between these mechanisms.

It is sometimes possible to eliminate one mechanism by means of calculation. The concentrations of the species proposed to be involved in a particular mechanism are calculated for a particular pH value using the appropriate pK_a values. The value of the rate constant needed to give the observed rate may be determined. If the value of the rate constant is greater than that for diffusion-limited proton transfer of ca. 10^{11} M^{-1} sec^{-1} then clearly the proposed mechanism cannot be correct. Another useful method for the distinction of such equivalent mechanisms requires that the catalytic species (or substrate) be confined to a single protonic form. This can be achieved quite easily if one of the reactants has a pK_a near neutrality. If, in the examples above, HA has a pK_a near neutrality it will be essentially completely protonated in 1 M HCl and unable to act as a general base. If the reaction proceeds in 1 M HCl it cannot involve general base catalysis but must proceed by specific and/or general acid catalysis. Structure–reactivity correlations can sometimes be used to distinguish between mechanisms of this type, by yielding information concerning the effectiveness of electron-withdrawing or donating substituents in the substrate. The ambiguity that can arise when attempting to "place the proton" in a particular mechanism sometimes arises in studies of enzyme catalysis (see chapter 6). Acid-base catalysis usually leads* to bell-shaped pH-dependence of the reactivity; the above ambiguity means that it is not possible to state which of the two groups (each characterized by a particular apparent pK_a value) acts as an acid and which acts as a base (see chapter 9).

Isotope effects

At a temperature of absolute zero the energy of a diatomic molecule, as distinct from a single atom, is not zero but has so-called zero point energy of $\frac{1}{2}h\nu$ where ν is the vibrational stretching frequency. In the case of hydrogen or deuterium attached to C, O, N or S the heavy atom can be regarded as stationary while the proton or deuteron vibrates in relation to the heavy atom. Transition state theory, mentioned in chapter 1, requires vibrational $\rightarrow$ translational conversion along the reaction coordinate. Accordingly, the vibration that provides the zero point energy is lost in the

* Not always so, since α-chymotrypsin-catalysed hydrolyses have sigmoid pH-dependence for acylation and deacylation.

transition state of a reaction which involves proton transfer between two heavy atoms. The frequency of the vibration that provides the zero point energy is dependent upon the mass of the atoms involved and so bonds to hydrogen and deuterium will have different zero point energies. Thus a reaction in which a proton or deuteron is transferred will involve the loss of different amounts of zero point energy on transition state formation. Since ν is inversely proportional to the atomic mass, a bond to deuterium will have a lower zero point energy than a bond to hydrogen. Thus reactions which involve deuteron transfer will be expected to proceed more slowly than a reaction that involves proton transfer. The zero point energy difference corresponds to a rate difference of 6–10-fold depending upon the atom to which the proton or deuteron is attached but this range of values is sometimes exceeded.

The study of deuterium isotope effects in catalytic reactions has been widespread, and in many cases useful deductions have been made from the results. If a "normal" primary isotope effect is found, for example $k_H/k_D = 5–10$, proton transfer is involved either in the rate-limiting step of the reaction or in a fast pre-equilibrium step. The absence of an isotope effect does not provide evidence for the absence of proton transfer since compensating effects may be present.

Isotopic substitution at a carbon atom results in a non-exchangeable isotopic atom while substitution at O, N, or S gives rise to an atom that will rapidly exchange with solvent. The effect, if any, of the isotopic solvent upon the general structural properties of enzymes is generally neglected in studies of deuterium kinetic isotope effects in the hope that this is a valid assumption. The small isotope effects seen in the deacylation steps of protease-catalysed hydrolyses have been ascribed to general base catalysis and more recent "proton inventory" studies support this view (see chapter 9).

Heavy atom isotope effects have also been measured in some instances. N^{14}/N^{15}, O^{16}/O^{18} and C^{12}/C^{13} isotope effects can be measured using mass spectrophotometric techniques. These measurements must be performed with great precision since the numerical values of these isotope effects, even when fully expressed, are close to unity (e.g. ~ 1.03 in the case of N^{14}/N^{15}). Measurement of nitrogen isotope effects in the acylation of α-chymotrypsin and papain by amide substrates has indicated that C–N bond breaking is partially rate-limiting in α-chymotrypsin acylation and completely rate-limiting in papain acylation. The substrates are usually used at their natural isotopic abundance so expensive enrichment procedures are not required: an expensive isotope ratio spectrometer, is however, required. The isotope effect is determined by measurement of the fractional isotopic content of product relative to reactant. This method, although technically demanding, is very useful for the detection of rate-limiting cleavage of bonds between heavy atoms.

Approximation, internal rotation and strain

Each of the reactants in a bimolecular reaction has three degrees of translational and three degrees of rotational freedom. Upon transition state formation three degrees of translational and rotational freedom are lost since the transition state is a single molecule, but six new vibrational degrees of freedom and one new internal rotation are present in the transition state due to its increased complexity. Intramolecular reactions involve no nett change in translation, vibrational or rotational degrees of freedom since a single molecule is involved at all stages.

The changes in degrees of freedom that result on transition state formation in a bimolecular reaction are expressed in the entropy term for the free energy of activation. The loss of three degrees of freedom of translation and overall rotation serve to decrease the entropy of the transition state by large (40 kJ at 25°) and approximately equal quantities. Neither of these factors is very sensitive to molecular size since the numerical value of the entropy is largely determined by the values of the physical constants of the entropy equations under standard state conditions (1 M, 25°). The loss of entropy that results from loss of translational degrees of freedom is inversely proportional to concentration and is thus larger at lower concentrations. The extent of the loss of entropy that occurs on transition state formation depends upon how well-defined or "tight" is the structure of the transition state.

A loose transition state will possess several partly formed or broken bonds, the reaction coordinate being extended over several atoms and thus the structure will be characterized by the presence of low frequency vibrations. Low frequency vibrations make a significant contribution to the total entropy of a structure in which they are present and so a transition state of this type will have a significant residual entropic contribution to the free energy. Thus reactions which are characterized by loose transition states will benefit less in rate enhancement terms from ground state effects (absorptive interactions) that render the reaction more intramolecular. Indeed in the extreme case they will benefit less from chemical bonding in the ground state that ensures intramolecularity. A comparison of nucleophilic and general basic catalysis shows this effect particularly clearly and is discussed at the end of this chapter together with examples.

Calculations that include all of the above factors indicate that the rate of an intramolecular reaction should be favoured over that of an inter-molecular reaction by a factor of ca. 10^8-fold if the transition state is "tight". Thus rate accelerations of the order of 10^8-fold can be expected simply as a result of approximation *if* the energy required for this approximation can be supplied. This energy may be supplied by providing strong positive interaction between the reactants either in the form of chemical bonds or strong specific absorptive interactions. Thus the provision of favourable enthalpic forces can be used to overcome the unfavourable entropic requirements of an intermolecular bimolecular reaction and render the reaction effectively intramolecular. It should now be apparent to the reader how the above applies to catalytic reactions and enzymes; the desirability of binding as an intimate component of catalysis has been made plain. The binding must be tight enough to achieve the maximum loss of entropy consistent with the structures of the reactants and the *mode* of binding must be as close as possible to that which provides a complex that is on the minimum energy route (i.e. will pass through the saddle point) of the potential energy diagram of the reaction. It is clear that strong non-productive "wrong way" binding can inhibit a reaction by tying up reactant in a form that cannot react and so the *mode* is as important as the strength of binding.

We state above that a 10^8-fold rate advantage could be achieved by inter → intra-molecular conversion, but no mention has been made of units. An intramolecular reaction is first order, and the rate constant has units sec^{-1} while an intermolecular (bimolecular) reaction is second order and its rate constant has units $M^{-1}\ sec^{-1}$. In order to compare these we divide the former by the latter giving a rate constant enhancement of 10^8 M. This means that the reactive components of an intramolecular reaction are capable of achieving an "effective concentration" of 10^8 M one with respect to the other. This figure comes as something of a surprise since one usually thinks in terms of 55 M as being the maximum concentration that can be achieved in aqueous systems, this value being the concentration of pure water. 55 M is a good estimate of the rate constant enhancement expected simply as a result of placing two molecules next to each other in aqueous solution but it must be remembered that under these circumstances they retain nearly all of the entropy that may be lost when the interaction between them becomes strong as in a "tight" transition state. Thus although at first sight this large figure of 10^8 M may seem extraordinary, it arises because we do not usually think in terms of concentrations when species are strongly interacting: we think in terms of the overall concentration of the complex of interacting species, i.e. the concentration of the transition state.

Rotamer distribution

Intramolecular reactions involve cyclization which in turn implies the elimination of internal rotation. This may be illustrated as shown in (2.18).

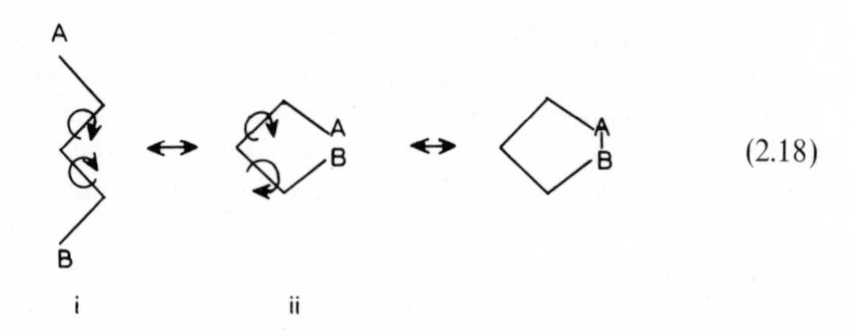

Rotamer (ii) resembles the product ring more closely than does rotamer (i) and so rotamer (ii) will be entropically closer to the ring closure conformation than rotamer (i) which cannot ring close in this conformation. An extensive range of rotamer structures like (i) gives a high entropy that must be lost to give a rotamer like (ii) before reaction can take place. Thus any influence that is capable of "freezing out" internal rotation in such a way as to lead to an enhanced population of rotamer (ii) will lead to increased reactivity, as the entropy loss required for transition state formation is decreased. Factors which affect rotamer distribution represent a special aspect of the more general entropic approximation effect.

Ground state strain

If it were possible to introduce a factor into rotamer (ii) above which would push A and B together by a "nut cracker" effect, then the cyclization reactivity would be increased. The introduction of productive strain in the ground state will lower the free energy gap between ground and transition states and increase the concentration of transition state. The strain must be relieved upon transition state formation otherwise no change in rate will result. The "nutcracker" effect applied to rotamer (ii) is shown diagrammatically in (2.19).

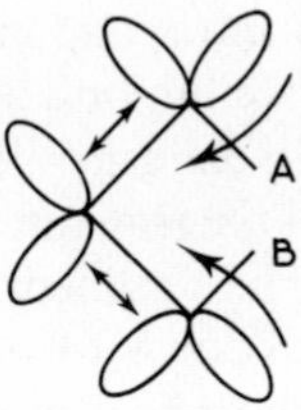

(2.19)

This type of effect is mainly *enthalpic* in that it contributes *directly* towards bond formation in contrast to the entropic factors discussed earlier.

Some examples of simple compounds that demonstrate the approximation, internal rotation and strain effects

The specific acid-catalysed esterification of hydrocoumarinic acids is believed to proceed according to (2.20).

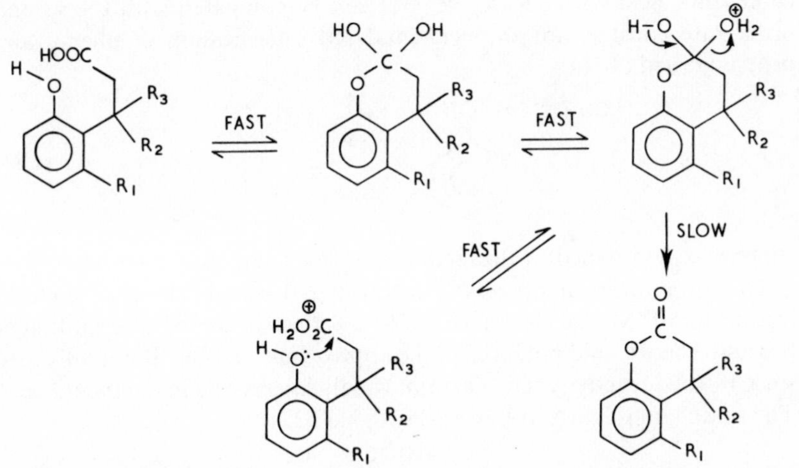

(2.20)

The precise mechanistic pathway has not been determined but it has been established that at neutral pH tetrahedral intermediate formation is fast and that breakdown is rate-limiting. Intermolecular esterification is generally supposed to be characterized by rate-limiting attack of the alcoholic component upon the protonated carboxylic acid in contrast to the rate-limiting tetrahedral intermediate breakdown that has been proposed in the case of the lactonisation reaction above. Thus if we wish to compare the rate constants for intra- and intermolecular esterification we must bear in mind that the rate-limiting steps are probably different in each case. Such a comparison must generally be regarded as bad practice and will yield a low estimate of the true rate enhancement of the slow step in the bimolecular reaction. The tetrahedral intermediate must be substantially protonated for breakdown to occur and the involvement of $-CO_2H_2^+$, the protonated carboxylic acid group, is not considered to be essential for tetrahedral intermediate formation in lactonisation reactions. Some hydrocoumarinic acids (e.g. R, R_2, R_3 = Me) cyclize rapidly at neutral pH where the concentration of $-CO_2H_2^+$ will be extremely small. The parent acid, R, R_2, R_3 = H does not cyclize in 1 M HCl at a measurable rate and the equilibrium for cyclization is very unfavourable. Accordingly the specific acid-catalysed lactonisation rate constant was determined by calculation, using values of the reaction equilibrium constant and the lactone acidic hydrolysis rate constant which were easily measurable. The rate constant for the intramolecular cyclization of the parent hydrocoumarinic acid where R, R_2, R_3 = H can be compared with the second order rate constant for the acid-catalysed esterification of phenol and propionic acid (2.21),

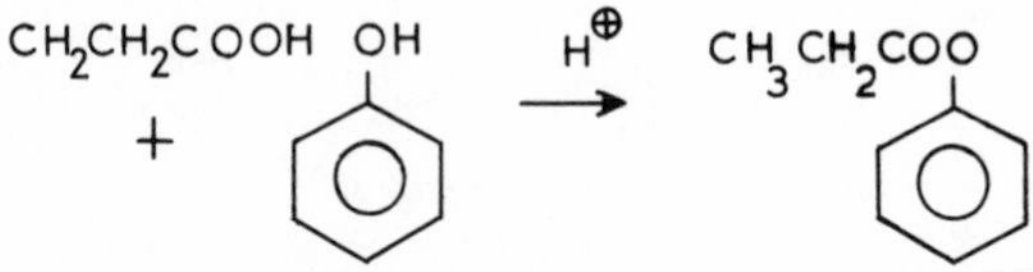

$$CH_2CH_2COOH + C_6H_5OH \xrightarrow{H^{\oplus}} CH_3CH_2COO{-}C_6H_5 \quad (2.21)$$

which was determined by the method described above.

The intramolecular process is characterized by an "effective" concentration of 10^5 M and is thus 10^5-fold as effective as the intermolecular process. The 10^8-fold potential of the approximation effect is not fully seen since the two reactive centres are not rigidly interacting in the parent acid. The side chain is free to rotate as shown in (2.22).

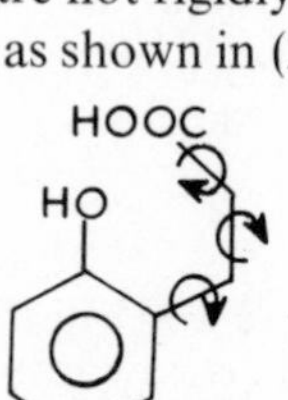

HOOC, HO (2.22)

Thus the volume available to the carboxyl group is still considerable and has not been reduced to an extent that produces permanent stationary (non-rotating) contact with the hydroxyl group. Table 2.1 illustrates the consequences of substitution of methyl groups for hydrogen atoms in

Table 2.1 Comparison of specific acid-catalysed inter- and intramolecular esterification*

Intermolecular reaction

$$C_6H_5OH + CH_3CH_2CO_2H \xrightarrow{[H^+]} CH_3CH_2CO.OC_6H_5$$

Intramolecular reactions

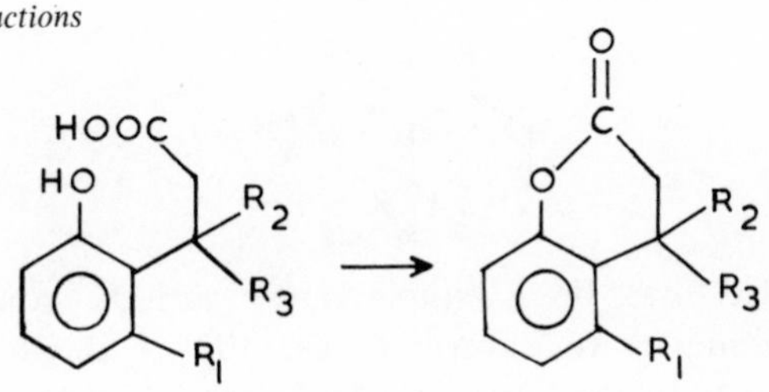

R_1, R_2, R_3	Ratio $\frac{k_{intra}}{k_{inter}}$ (M)	Rel. k_{intra}
H, H, H	10^5	1
Me, H, H	1.05×10^5	1.05
H, Me, Me	4.4×10^8	4400
Me, Me, Me	10^{16}	10^{11}

positions R_1, R_2, R_3, of (2.20). The remarkable rate enhancement that results from methyl group substitution must be explained, at least in part, in terms of rotational restriction. Indeed model-building studies and quantum mechanical calculations indicate that rotation about the bond adjacent to the benzene ring will be severely restricted. The conformational restriction resulting from the interaction of the three methyl groups has been described as the "trialkyl" lock. The reaction mechanism as shown in equation (2.20) proceeds via a tetrahedral intermediate which is formed rapidly, in a pre-equilibrium before rate-limiting breakdown with loss of water. The rate-enhancing effect of methyl substitution thus arises partly as a result of tetrahedral intermediate stabilization relative to the ground state as well as an increased rate of tetrahedral intermediate breakdown.

Clearly the overall rate enhancement factor of 10^{16}(M) is much greater than the factor of ca. 10^8(M) that can be explained on the basis of the approximation effect. An enthalpic strain effect must be present in the ground state which is at least partly relieved in the transition state. The large improvement in the favourability of the equilibrium with respect to

* See note added in proof, p. 43.

lactone formation that results from methyl substitution indicates that extensive strain is relieved in the product relative to the reactant. The strain effect may, in part, be visualized as a variation of the "nutcracker" effect (2.23).

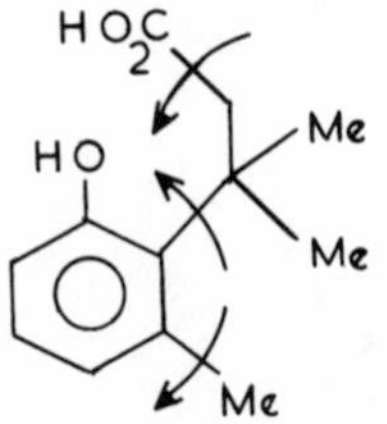

(2.23)

The strain effect has been estimated by various experimental and theoretical techniques to contribute ca. 10^7 of the total 10^{11} rate enhancement resulting from methyl substitution. This leaves ca. 10^4 for the rotational restriction entropic effect. Thus of the overall 10^{16} (M) ca. 10^9 is seen as arising from the approximation entropic effect and ca. 10^7 from the enthalpic strain effect.

It is readily apparent that a combination of entropic and enthalpic effects as seen in the model reactions described above is amply capable of producing rate enhancements of the order of those found in enzyme catalysed reactions. We can confidently expect both these factors to be important in enzyme-catalysed reactions but we may expect to have some difficulty in resolving the contributions of each.

Intramolecular hydrolysis of maleamic acids

Substituted maleamic acids are hydrolysed according to the scheme (2.24):

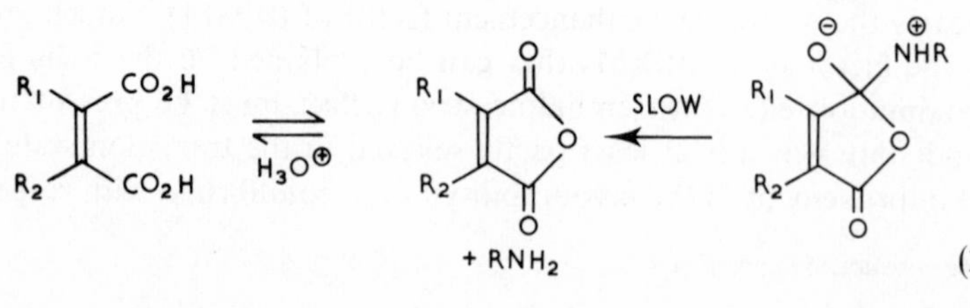

(2.24)

Table 2.2 Maleamic acid hydrolyses

R_1	R_2	R_3	*Rate enhancement M "effective concentration"*
H	H	Me	2.0×10^6
H	H	Ph	1.2×10^7
H	^{t}Bu	Me	1.3×10^8
Me	Me	^{n}Pr	4.0×10^{10}

As the formic acid-catalysed hydrolysis of simple amides is not detectable, the rate constants of maleamic acid hydrolyses are compared with that of the specific acid-catalysed hydrolysis of *N*-methylacrylamide. The rate enhancement factors are given in table 2.2.

The "effective concentration" of 2×10^6 M is large in the case of the unsubstituted acid but assumes the remarkable figure of 4×10^{10} M for the doubly methyl-substituted molecule. The nature of the leaving group will have most effect on the C–N bond cleavage step which is rate-limiting and this is reflected in the six-fold increase in the rate enhancement on replacing methylamine with aniline as the leaving group. Alkyl substitution upon the maleic acid moiety results in more favourable cyclization giving a higher concentration of ring-closed species prior to C–N bond cleavage. Thus the large rate enhancement seen in this series is due to an effect upon a pre-equilibrium as was the case in hydrocoumarinic acid lactonisation. In the present series there is no question of rotation about the double bond so the rate enhancement must be explained in terms of the introduction of ground state strain producing a "nutcracker" effect. Substitution by t-butyl in the R_2 position increases the rate enhancement ca. 100-fold, and the magnitude of the increase is related to the size of the alkyl substituent as far as t-butyl. Dialkyl substitution gives a nearly maximal rate enhancement when R_1 and R_2 are methyl groups; little further rate enhancement results when these groups become as large as, for example, isopropyl. Thus there appears to be a maximum degree of steric compression that can be expressed in terms of an increased rate enhancement. Presumably any further compression produced by substitution of larger alkyl groups leads to compensating unfavourable torsional strain which decreases the favourability of the nucleophilic insertion reaction and hence decreases the concentration of cyclized intermediates. It is worth noting that for effective reaction the carboxyl group, whose lone pair electrons are in the O–C–O plane, must attack the amide group on a line perpendicular to the N–C–O plane of the amide group. Thus the amide group must rotate out of the plane of the maleamic acid double bond for reaction to take place. Consideration of the figures given in table 2.2 shows that about a 10^4-fold rate enhancement is provided by steric compression which perhaps incorporates a component representing improvement of orbital alignment as a result of double bond twisting. The remaining 10^6-fold factor results from the conversion from inter → intramolecularity, although the numerical value may be an underestimate since general acid catalysis of amide hydrolysis could not be measured. It is again apparent that intramolecular nucleophilic catalysis with a tight transition state can enhance rates very greatly, and, when combined with steric compression, can be responsible for the rate enhancements characteristic of enzyme reactions.

The hydrolysis of monoaryl malonates

Monoaryl malonates (2.25) are susceptible to intramolecular general base catalysis as shown in equation (2.26).

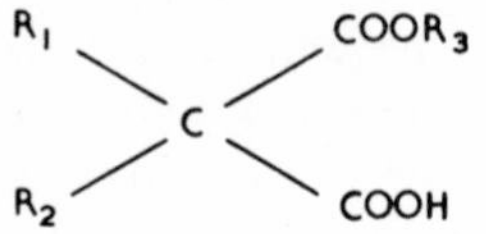

(2.25)

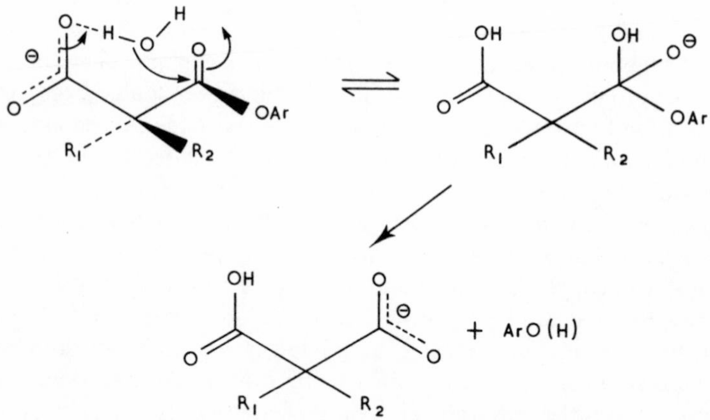

(2.26)

(Protonic equilibria of intermediates are omitted in equation (2.26) for the sake of clarity, arbitrary ionic forms being given.)

Anhydride formation is precluded by the energetically unfavourable

Table 2.3 Intramolecular general base catalysis by the carboxyl group in monoaryl malonates and monoaryl cyclopropanedicarboxylate

Structure	*Angle*	*Effective concentration of carboxylate*
cyclopropane-$C(COOAr)(COO^{\ominus})$	118.4°	60 M
$H_2C(COOAr)(COO^{\ominus})$	110°	25 M
$Me_2C(COOAr)(COO^{\ominus})$	106.2°	11 M
$Et_2C(COOAr)(COO^{\ominus})$	—	0.3 M

nature of the four-membered ring and thus general base catalysis is seen. The effective concentration of the carboxyl group is much lower than in nucleophilic catalysis as shown by the data in table 2.3.

Intramolecular general base catalysis is characterized by "effective concentration" values up to ca. 100 M and in this series the largest value is associated with an effect that is opposite to the nutcracker effect. This could be termed the "decompression" effect, where the bond angle α is opened beyond its normal value of 109°. (By induction it may be presumed that the diethyl compound has an angle α less than 106° and so apparently compression of the groups has a deleterious effect.) As in the examples of intramolecular nucleophilic catalysis discussed earlier, the plane of the O–C–O atoms of the carboxyl group must be approximately orthogonal to the plane of the O–C–O atoms of the ester group and a water molecule must be fitted in between the two. The figures in table 2.3 suggest that the water molecule fits in with less strain and more favourable geometry when the angle α is deformed above 109°. The deuterium solvent kinetic isotope effect of 2.2 for this reaction is characteristic of general base catalysis and the Hammett ρ of 0.97 is the same as that for aspirin hydrolysis which is known to proceed via intramolecular general base catalysis. The low ρ value indicates relative insensitivity to the nature of the leaving group and so is characteristic of a reaction that involves rate-limiting tetrahedral intermediate formation.

It seems reasonable to suppose that the rather modest effectiveness of general base catalysis might be improved if the "water" moiety could be made part of the ground state structure by means of strong chemical bonding. Clearly this cannot be achieved in the case of water itself but an alcohol could be so incorporated. An example of this is shown by a comparison of (2.27) with (2.28).

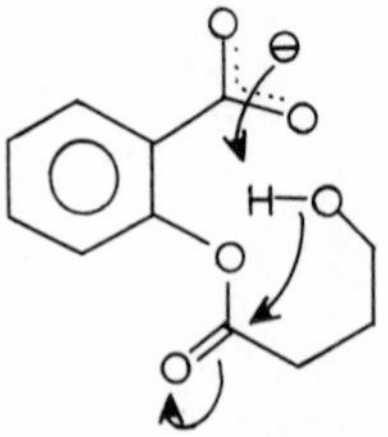

(2.27)

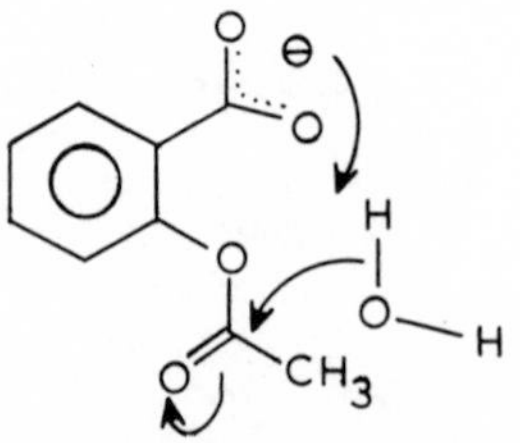

(2.28)

Compound (2.27) hydrolyses via a general base mechanism 150-fold faster than (2.28) presumably because the hydroxyl group is included as a formal part of the structure of the molecule in (2.27). The rate would be further enhanced by structural variation which might optimize the geometry of a reaction of the type shown in (2.27). It is important to note that the "effective concentration" of the carboxylate groups in (2.27) and (2.28) is the same, so the rate enhancement is not due to an improvement in the geometric alignment of the general basic group but to an increase in the "effective concentration" of the nucleophilic (water) component of the reaction system. We shall see in later chapters that enzymes have evolved to exploit exactly this feature of general base catalysis and are thus able to promote extremely effective general base catalysis despite the seemingly absolute limitation of 100 M upon the magnitude of the "effective concentration" of a general base.

Conclusions

Some readers may query the desirability of such an extended treatment of chemical catalysis in a book supposedly devoted to enzyme catalysis. At the end of chapter 1 we argued that a sound understanding of the bases of chemical dynamics was essential to an understanding of catalysis. Here again we argue that a very sound understanding of the principles of chemical catalysis is absolutely essential before a serious approach can be made to the understanding of enzyme catalysis. The aim of the molecular enzymologist is to ensure that the latest advances in our understanding of chemical catalysis are promptly and correctly applied to the study of enzymic catalysis and it is the aim of this book to assist in the realization of this process.

Note added in proof
The equilibria and rate constants of hydrocoumarinic acids have recently been remeasured (Caswell, M. and Schmir, G. L. (1980) *J. Amer. Chem. Soc.*, 102, 4815–4821). The trimethyl derivative has been found to have a rate constant for lactonisation some 10^{-4}-fold less than the previously measured value. Thus the relative rate constant (Rel. k_{intra}, table 2.1) for the "trialkyl lock" has a value of approximately 4×10^5 rather than 10^{11} as shown in the table. The overall rate enhancement is of the order of 5×10^{10}, as compared with 10^{16}. This reduction in the rate enhancement requires a reappraisal of the contribution of ground state strain and/or entropic factors which have been invoked to explain the overall effect.

CHAPTER THREE

PROTEIN STRUCTURE

THE STANDARD ASPECTS OF PROTEIN STRUCTURE ARE WELL DESCRIBED IN MANY biochemistry textbooks. Accordingly it seems appropriate to devote much of this chapter to a discussion of the underlying factors which, for any given protein, contribute to the formation of a singular structure. The degree of singularity of such structures will also be considered, since this area, which includes such postulated phenomena as internal structural mobility, is currently attracting much experimental attention and theoretical speculation.

We start with a consideration of the structure and functional properties of the peptide bond. The lone pair electrons of the nitrogen atom are partially donated to the carbon—nitrogen bond which in turn implies electron accumulation on oxygen. Thus the peptide C—N bond has about 50% double bond character, is somewhat shorter than an amino C—N bond, and the atoms attached to the C—N pair lie in a plane (3.1).

Note that this is a resonance stabilization, not a tautomeric effect since the proton is not transferred to the carbonyl oxygen atom. The degree of resonance stabilization is considerable, about 80 kJ mole^{-1}. The double bond character of the C—N bond will be lost upon twisting if the overlap of the *p*-electrons is reduced and this results in loss of the resonance stabilization energy. Thus the C—N axis of the peptide bond will be resistant to twisting although rotation can occur in the bonds to the C and N atoms as shown in (3.1).

The planarity and transoid nature of the peptide bond favours the formation of two types of structure in the interior of proteins. These are the α-helix and the β-pleated sheet. The regular presentation of

$$>N{-}H \quad \text{and} \quad >C{=}O$$

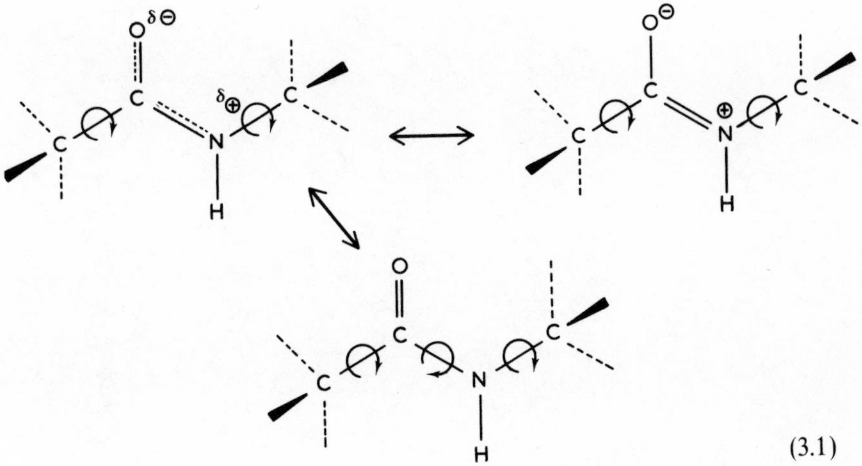

(3.1)

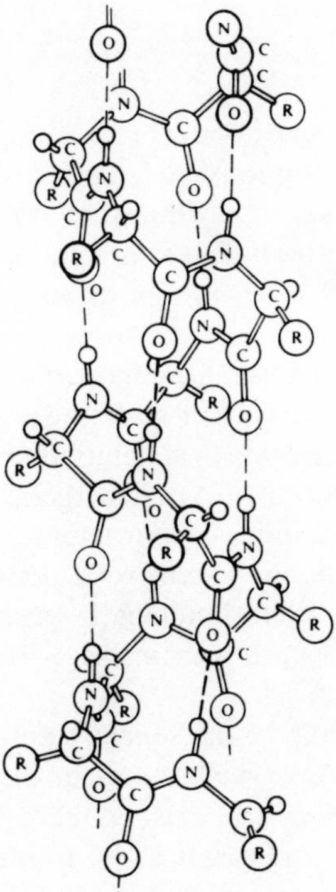

(3.2)

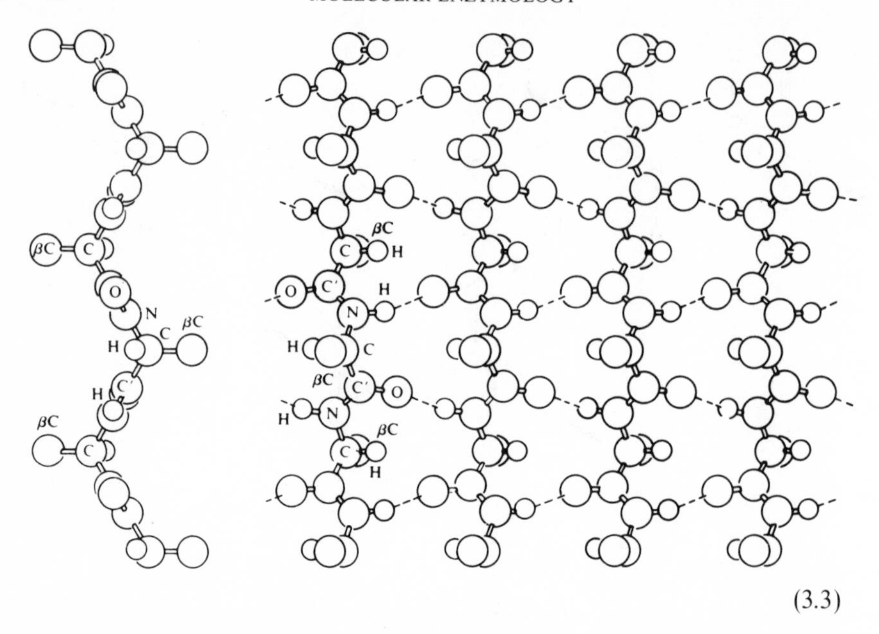

(3.3)

groups for potential hydrogen bonding interaction is an important factor in the stabilization of the established structures. The degree to which hydrogen bond formation contributes to the driving force for the formation of these structures in aqueous solution will be considered later. The α-helix and β-pleated sheets are shown diagrammatically in (3.2) and (3.3).

Hydrogen bond formation is highly cooperative in such structures since they will "zip up" when the correct structure is energetically favourable. Amino acid homo- or co-polymers in solution undergo sharp helix ↔ coil transitions dependent upon the overall energetic favourability of the final structure. The so-called random coil structure of a popypeptide chain in solution does not possess any extensive regions of organized structure other than the normal covalent bonding. In particular, repeating units of hydrogen bonding such as are seen in the α-helix or β-pleated sheet of proteins are strictly absent.

We note in passing some of the salient features of the α-helix and β-pleated sheet. The R groups, which represent the amino acid side chains, are directed outwards from the axis of the α-helix and the hydrogen bonding between amino acid residues 3.6 residues apart in the helix is highly directional. In the case of the β-pleated sheet the R-groups are disposed in a plane approximately perpendicular to that of the hydrogen bonding mesh and thus in both cases the hydrogen bonding system is not subject to direct interference by the R groups. Clearly such factors as electrostatic interaction between neighbouring R groups can have an indirect effect upon the stability of the hydrogen bonding network. A β-pleated sheet can be made up from parallel chains (i.e. $C \rightarrow N$ bonds directed in an approximately parallel direction) or from antiparallel chains. The structure is not restricted to two-chain interaction but can be built up from multi-chain interaction to give a two dimensional "bed" structure. The fibrous protein silk fibroin is entirely composed of parallel β-structure and the R-groups are small and uncharged thus allowing close interaction between two-dimensional arrays of protein structure. As we shall see, most protein structures are composed of multiple regions of α-helix and β-pleated sheet with small linking regions between these two types of structure.

Hydrogen bonds

The peptide link provides the possibility of regular hydrogen bonding. Since this is clearly of great importance in protein structures it will be discussed first.

The strongest known hydrogen bond has a free energy of formation of $-155\,\text{kJ mole}^{-1}$ in the crystalline state. This bond is present in the bifluoride ion $[\text{F—H—F}]^-$ which is a symmetrical single potential well system (see figure 3.1). The free energy of formation of the bond is reduced to $-3.4\,\text{kJ mole}^{-1}$ in water; a dramatic reduction in the strength of this bond. The reason for the reduction of the apparent bond strength in water is that competitive hydrogen bonding with water enormously reduces the tendency for the bifluoride ion to form. The equation for the formation of the bifluoride ion is given in (3.4).

$$\text{M}^{\oplus}\ \text{F}^{\ominus}\text{---H—OH} + \text{HF---H—OH} \rightleftharpoons \text{M}^{\oplus} + [\text{F---H---F}]^{\ominus} + \text{H—}\underset{}{\overset{\text{H}}{\overset{|}{\text{O}}}}\text{---H—}\overset{\text{H}}{\overset{|}{\text{O}}} \qquad (3.4)$$

The free energy change that results from bifluoride ion formation represents the difference between these alternatives, and due to the multifarious hydrogen bonding ability of water, the difference is small. The very large reduction in the free energy of formation of the hydrogen bond

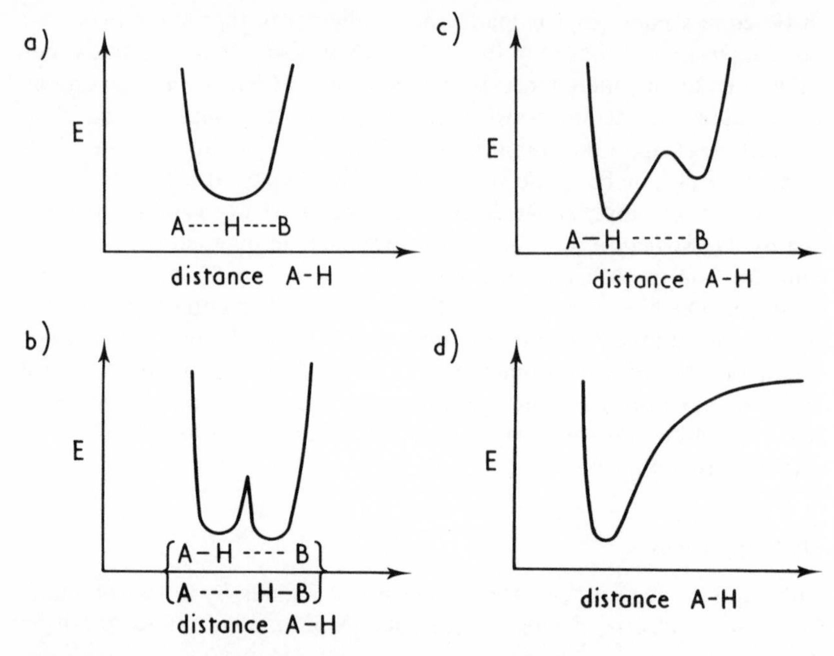

Figure 3.1 Energy diagrams for various types of hydrogen bonds.
(a) Single potential well. Likely if A and B (which may be identical) are close and if H-bonding is strong.
(b) Double potential well. A and B have similar (or identical) electronegativity. Proton may transfer from one well to the other by tunnelling.
(c) Asymmetric double well. The electronegativity of the atoms is different; the proton resides closer to one atom than the other virtually all the time.
(d) The dissociation curve for A–H.

on going from the crystalline state to aqueous solution suggests that the bifluoride ion is likely to be the only intermolecular hydrogen bonded species to exist in observable quantity in aqueous solution. Hydrogen bonds (having typically -20 kJ to -40 kJ free energy of formation) form quite readily in solvents where competitive hydrogen bonding is much reduced or absent.

Intramolecular hydrogen bonds probably exist in aqueous solution because the "effective" concentration of the participating species can be large. Evidence suggestive of this type of hydrogen bonding is provided by the difference in the pK values of the acidic groups of dibasic acids (3.5). The pK difference of dibasic acids can be up to 8 pH units in cases where steric effects force the acidic groups into close proximity.

Fumaric acid	Maleic acid
HOOC–CH=CH–COOH (trans)	HOOC–CH=CH–COOH (cis)
⇌ $pKa_1 = 3{\cdot}0$	⇌ $pKa_1 = 1{\cdot}9$
$HOOC–CH=CH–COO^{\ominus}$	$HOOC–CH=CH–COO^{\ominus}$
⇌ $pKa_2 = 4{\cdot}5$	⇌ $pKa_2 = 6{\cdot}3$
$^{\ominus}OOC–CH=CH–COO^{\ominus}$	$^{\ominus}OOC–CH=CH–COO^{\ominus}$
$\Delta pKa = 1{\cdot}5$	$\Delta pKa = 4{\cdot}4$

(3.5)

It thus seems unlikely that stable hydrogen bonds can form in aqueous solution unless the donor and acceptor are well oriented and located with respect to each other and the local dielectric constant is reduced by hydrophobic shielding.

The various forms of hydrogen bond that might be supposed to exist in biochemical systems are shown in figure 3.1. In fact, hydrogen bonds found in biochemical systems other than transition states approximate to type (c) in figure 3.1, with some being more nearly symmetrical than others. The interaction energy is not high enough for single potential well formation. The hydrogen bonds that occur in proteins are primarily of the $>N—H\ldots O{=}C<$ type although other types, e.g. $\geqslant N\ldots H—O—$ may be found in special circumstances. Hydrogen bonds are commonly regarded as forming primarily as a result of favourable electrostatic interaction, for example as shown in (3.6).

$$^{+}A-\overset{\delta+}{H}\cdots B^{-} \quad \text{or} \quad A-\overset{\delta-}{H}\cdots\overset{\delta-}{B} \qquad (3.6)$$

The electrostatic contribution is most important when the A–B distance is long, but a covalent contribution as shown in (3.7) becomes progressively more important as the A–B distance is shortened.

A H --- B

← increase in covalent character (3.7)

Hydrophobic interactions

The source of the cohesive energy of a hydrocarbon liquid is provided by van der Waals–London dispersion forces. These result from mutually induced dipolar interaction. It is clear that these dispersion forces are not very strong since small hydrocarbons have low boiling points and high volatility. Introduction of polar atoms such as chlorine into a hydrocarbon causes a dramatic increase in the boiling point, partly as a result of a decrease in the mean free path (due to increased size) and partly as a result of permanent dipole–dipole interaction (which increases the cohesive energy). Introduction of a substituent capable of hydrogen-bonding interaction (e.g. an hydroxyl group) causes an even larger increase in the cohesive energy. This is illustrated by the boiling points of methane, methyl chloride and methanol which are −164°, −24·2° and 65° respectively.

Water has a remarkably high cohesive energy density having a boiling point of 100° compared with that of hydrogen sulphide (−61°) and hydrogen telluride, a much larger molecule (−2°). The reason for this high cohesive energy density is provided by the existence of extensive hydrogen bonding in liquid water, together with the strongly dipolar nature of the water molecule.

Introduction of a hydrophobic molecule into an aqueous solution requires the formation of a "cavity" in the aqueous solution. In this cavity the cohesive energy density of water has been dispersed and so energy will be required for the creation of such cavities. The water molecules at the periphery of a cavity will organize themselves in such a way as to maximize dipolar interaction with their neighbours. This implies ordering of the water molecules since they cannot form favourable dipolar interaction in a direction towards the cavity.

The process of cavity formation is energetically unfavourable but to a degree depending upon the surface to volume ratio. There is a strong tendency for this to be optimized with respect to the quantity (volume) of hydrophobic material to be introduced, and hence for a cavity to be formed. This is so because optimal reduction of the surface to volume ratio will decrease the quantity of water that must undergo a decrease in entropy.

Consider a quantity of hydrophobic material that is introduced into an aqueous solution in the form of two spheres of unit radius. A cavity must be created for each and the surface to volume ratio is 6. Now we allow the spheres to coalesce to form a single sphere, the surface to volume ratio of which is 4.74. The surface to volume ratio is reduced by 20% and this reduction provides a strong driving force for the coalescence of hydrophobic moieties in aqueous solution. In this way, during protein folding, hydrophobic groups (the amino acid side chains, Phe, Trp, Tyr, Val, Leu, Ileu, etc.) will tend to coalesce, depending on their sequential positions and the nature of the local structure. The ideas introduced above are shown diagrammatically in (3.8).

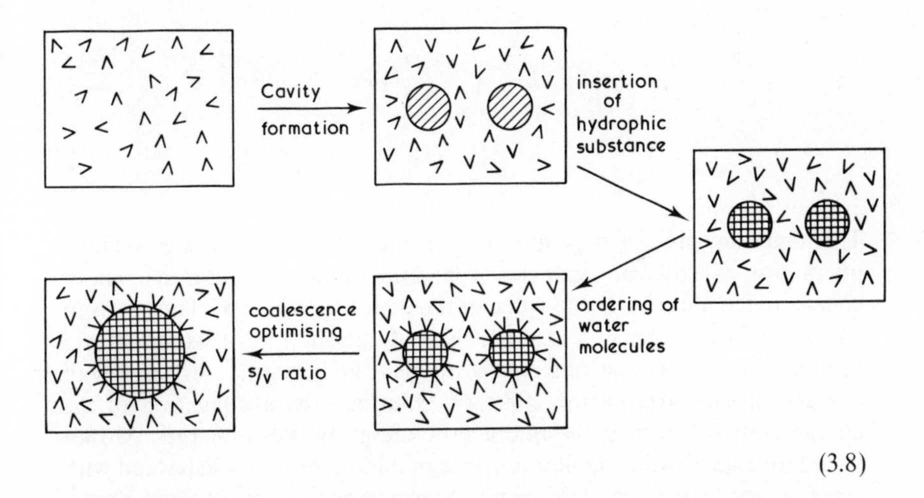

(3.8)

It is apparent from the discussion above that the concept of the "hydrophobic bond" in aqueous systems is, in essence, a negative one. Hydrophobic bonding results primarily from a minimization of unfavourable interactions rather than from any positive driving force. The energetics of hydrophobic interaction can be assessed by measurement of the partition coefficient of a particular molecule between a reference hydrophobic phase (*n*-octanol is often used) and water. The free energy of transfer can be calculated from the partition coefficient and gives a value of

ΔG of ca. 3.4 kJ mole^{-1} per methylene unit of a hydrocarbon chain transferred from a non-polar solvent to water.

Electrostatic interactions

Charged species interact with each other according to the equation

$$F = \frac{Q_1 \, . \, Q_2}{D \, . \, r^2}$$

where F is the interactive force, r is the distance of separation of point charges Q_1 and Q_2 and D is the dielectric constant of the medium separating the charges. The force will be attractive if Q_1 and Q_2 have opposite sign and repulsive if the sign is common. The work required to bring the charges Q_1 and Q_2 from infinite separation to a distance of separation equal to r is equivalent to the electrostatic interaction energy ΔE,

$$\Delta E = \int_{\infty}^{r} \frac{Q_1 Q_2}{D r^2} \, dr$$

$$= \frac{Q_1 \, . \, Q_2}{D \, . \, r}$$

This equation gives a dependence of $1/r$ for the energy of electrostatic interaction which means that this type of interaction is long-range but, of course, much affected by the value of the dielectric constant. Thus a given energy of interaction in water will be only about 1% of its value in a vacuum. Water has a strong dipole moment and interacts strongly with charged species, having the effect of "diluting" the propagation of the charge and so lowering the interaction energy between charges. Monovalent ions have rather similar energies of interaction with water and with other charged species, and this means a weak tendency for the formation of ion pairs in aqueous solution. The free energy of monovalent ion pair formation in water is approximately zero kJ mole^{-1}. This compares with a value of 200–800 kJ mole^{-1} for interactions between water and ionic species as well as ion–ion interaction in a vacuum. Thus the driving force for ion pair formation in aqueous solution is very weak, being dependent upon a small difference between large numbers. This concept can be expressed as in (3.9) which is conceptually closely related to (3.4) where hydrogen bonding interaction is described.

$$\oplus\!\sim\!H_2O + \ominus\!\sim\!H_2O \rightleftharpoons \oplus\!\sim\!\ominus + H_2O\text{---}H_2O \quad (3.9)$$

Divalency increases the favourable energy of ion pair formation considerably and some di-divalent ion pairs are relatively stable in aqueous solution.

Since electrostatic interaction is dependent upon the distance of separation of the charges it might be supposed that small ions might bind to charged structures (such as ion exchange resins) more effectively than larger ions whose distance of closest approach would be larger. Alkyl quaternary ammonium ion exchangers bind the halide anions in the order $I^- > Br^- > Cl^- > F^-$, which is the opposite of that expected from simple electrostatic considerations. Similarly strongly acidic cation exchangers bind the alkali metal cations in the order $Cs^+ > Rb^+ > K^+ > Na^+ > Li^+$; again an order contrary to expectation. The ionic mobilities of these ions show a similar order. The free energies of hydration of the anions and cations mentioned above have a reversed order. It is eminently reasonable that a small ion with a very high charge density at its surface should bind water molecules much more tightly than larger ions. Small ions such as Li^+ and F^- bind water molecules so tightly as to decrease the degrees of freedom of the water and so have a "structure making" effect upon the water system. In addition the "cavity" that must be created for insertion into the aqueous medium will be relatively small. The cohesive energy density per unit volume of the solution will be increased in the presence of such ions. Ions such as K^+ or Cl^-, of intermediate size, require a larger cavity for their insertion into the aqueous system and interact less strongly with the water molecules immediately surrounding them. The net result is that the cohesive energy density per unit volume is decreased by the presence of an ion of this type and such an ion may be regarded as "structure-breaking". Very large ions such as I^- and tetraethylammonium ion have a "structure-making" effect. This process is different from that which is predominant in small ion–water interactions. The ions have a relatively large surface area and their large size results in effective shielding of the ionic charge from solvent interaction. The large rather hydrophobic surface imposes order upon the solvent in the same way as do ordinary hydrophobic interactions as described earlier in this chapter.

The ions at the N— and C— terminal ends of proteins and in amino acid side chains are mainly of the COO^- and NH_3^+ types. The ammonium group has an ionic radius comparable to that of potassium and may therefore be regarded as "structure-breaking" although its hydrogen bonding potential may somewhat reduce this tendency. The carboxylate anion is larger and the charge more delocalized. These factors place the carboxylate anion somewhere between obviously "structure-making" and "structure-breaking" ions but it is probably mildly "structure-breaking". The guanidino group of arginine falls into the latter category.

Ion pair formation between ammonium and carboxylate anions attached to separate small molecules in aqueous solution is no more likely than ion pair formation in dilute NaCl, which is of the order of hydrated ion diffusional encounter. However when these groups form part of a protein structure a further effect comes into play. This factor, known as the "chelate" effect, is an important consideration, as is demonstrated by the extremely effective complexation of ions by multidentate ligands.

The chelate effect may be seen as a multivalency effect in which the valency of one ion is increased. Thus an interaction of a carboxylate anion with several amino groups in the same molecule may be seen in these terms as a mono-multivalent interaction. It will not be as effective as a simple increase in valency of a single ion since the charge will be more widely dispersed. Clearly all such effects will be greatly enhanced by the exclusion of water molecules. The interaction of a closely mated ion pair will be very strong in a region of low dielectric constant and it is only in regions of this nature that ion pair formation can be predicted with any certainty. It is most important to appreciate, however, that the energy required to transfer an individual ion from an aqueous environment to a hydrophobic non-polar environment will be very unfavourable as it involves reversal of the hydration process. Transfer of an intimate ion pair will, in contrast, be relatively favourable since extensive dehydration must have occurred for ion pair formation to take place.

Unpaired charge is rarely, if ever, found in the non-polar interior of proteins, but paired charges are found, owing to their favourable energy of mutual interaction. For example, α-chymotrypsin is stabilized in its active conformation as a result of a buried charged-pair interaction between Asp-194 and N-terminal Ileu-16. At high pH where the Ileu-16 amino group is partially deprotonated the enzyme equilibrates between two conformations, one of which has a buried charge pair and the other has Ileu-16 exposed to solvent. Only the conformer with the buried charge pair is enzymically active since the other conformer is unable to bind the substrate. Estimates of the equilibrium concentrations of the two conformers yield a value of $12\,\mathrm{kJ\,mol^{-1}}$ for the electrostatic stabilization energy of the buried salt bridge.

Summary

The three types of force that contribute to conformational stability in proteins have been described in some detail. The contribution of each type of interaction to the driving force for folding may now be estimated. It is well known that regions of α-helix and/or β-structure can form in solution from the random configuration of a long polypeptide chain, and the

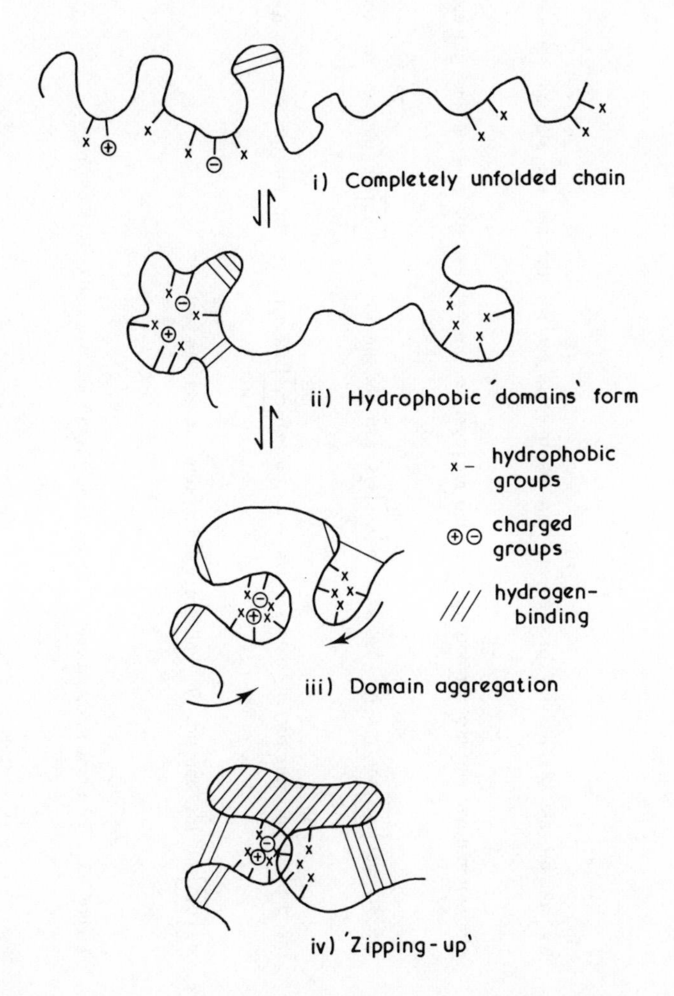

Figure 3.2 Schematic diagram of the folding of an arbitrary globular protein.

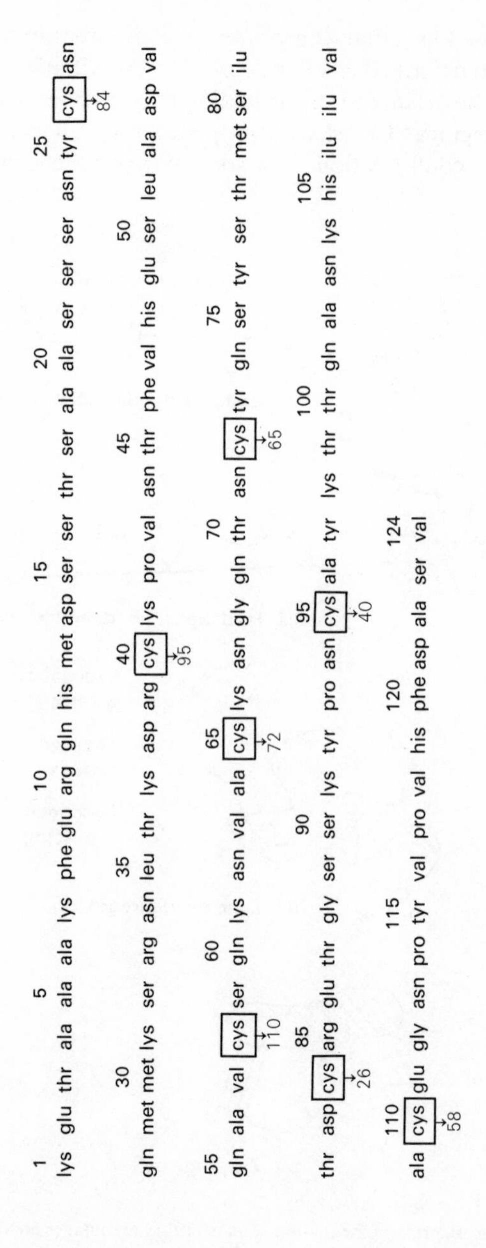

Figure 3.3 The amino acid sequence of ribonuclease. From Dickerson and Geis (1969), p. 79.

nucleation structure which starts the protein folding process may often be such an α-helix or β-structure. The formation of local structure, in turn, depends primarily upon nearest-neighbour short-range forces. The highly cooperative nature of hydrogen bond formation in these structures is an important factor in the driving force, while the shielding provided by the R-groups, which represents a form of hydrophobic effect, is clearly also of great importance. Thus cooperative hydrogen bond formation in the presence of shielding represents the driving force during the early stages of folding. The nucleation sites subsequently grow into nucleation domains which have hydrophobic cores, and long-range interactions (i.e. between residues far apart in the sequence) are clearly important in the aggregation of these nucleation domains. This aggregation has the additional effect of decreasing the effective surface to volume ratio of the hydrophobic core. Thus it is apparent that a delicate interplay of forces is involved in protein folding since it is not possible to single out any of the component forces as being completely dominant.

To describe in detail the exact sequence of events during the folding of a protein represents a difficult task since there may be many pathways of similar energy that lead to the same final conformation. Despite this, the main features of the folding process are likely to occur in the same sequence as each individual molecule folds. In addition, the further folding proceeds, the less variation will be possible as a result of the cooperative nature of the process. The folding of an arbitrary globular protein is shown diagrammatically in figure 3.2.

The structure of pancreatic ribonuclease (RNase)

Pancreatic ribonuclease is a small globular protein of molecular weight 13 700. The single polypeptide chain is comprised of 124 amino acids, and there are five disulphide bridges giving 105 possible combinations for the internally cross-linked structure. The sequence of the 124 amino acids is given in figure 3.3, in which the disulphide bonds of the native structure are shown. The S protein is produced by proteolytic cleavage between residues 20–21 followed by dissociation of the N-terminal S-peptide fragment. The S protein is enzymically inactive, but activity is recovered when S-peptide is mixed with S-protein. The hydrogen bonding network in the RNase S protein–S peptide complex is shown in figure 3.4. Thus the native structure can be achieved despite the lack of a peptide bond at the original cleavage position. The active site of the enzyme (figure 3.5) is created by chain folding; His-12 and His-119 provide acid–base catalysis of the hydrolysis

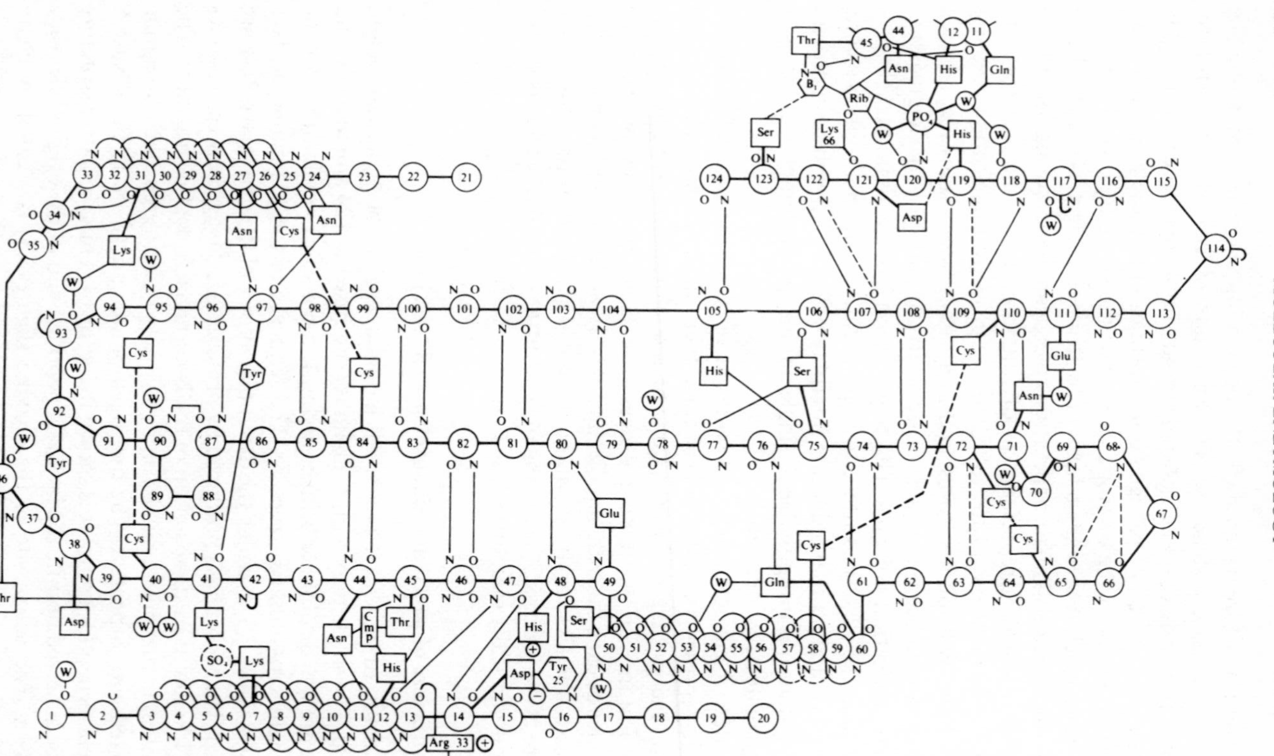

Figure 3.4 Complete hydrogen bonding matrix of RNase S + S-peptide which is fully active enzymically. From "Atlas of Molecular Structure in Biology", Vol. I, 1973, p. 16.

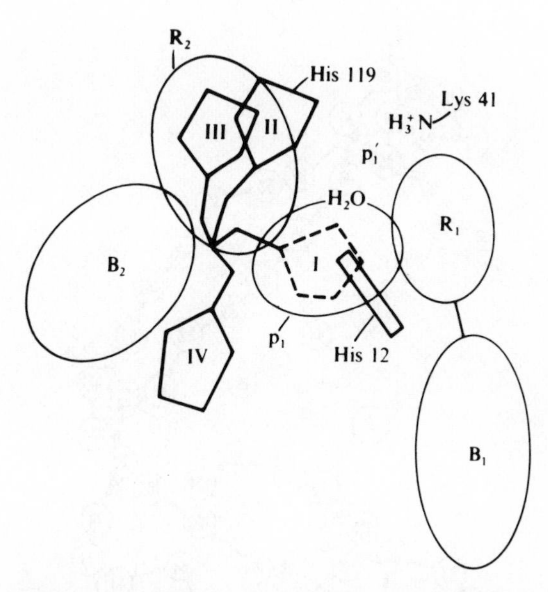

Figure 3.5 Outline diagram of the putative binding mode of a dinucleoside phosphate to the active site of ribonuclease. B_1 and R_1 represent the base and ribose components of the 3′-linked nucleotide and B_2 and R_2 the relevant components of the 5′-linked nucleotide. 3′CMP and 3′UMP occupy positions B_1 and R_1 and 5′AMP occupies B_2 and R_2 when these are diffused into the crystals. His-119 may occupy any of the positions I–IV. The phosphate group is p_1. His-119 is forced into position IV when 5′-AMP is present. From "Atlas of Molecular Structure in Biology", Vol. I, 1973, p. 3.

of RNA. One of the two histidine residues that contribute to the acid–base catalysis is included in the S-peptide fragment, the other being part of the S-protein. Thus non-covalent interactions are sufficiently strong and specific to determine the relative location of the histidine residues to a degree that yields full enzymic activity. The polypeptide chain conformation is shown in figure 3.6. The α-helical content of the protein is ca. 25 %, comprising three regions. A rather unusual feature is the inclusion of His-12 in an α-helical region since catalytically active residues are not usually incorporated in α-helical segments. It is, however, at the end of a helical segment. The amount of α-helix in this protein is rather less than is often found ($\sim$ 40 %) but nonetheless the structure is compact, there being only a small number of water molecules in the crystals of the protein. As seen in figure 3.6 there are four regions of β-pleated sheet all of the anti-parallel form.

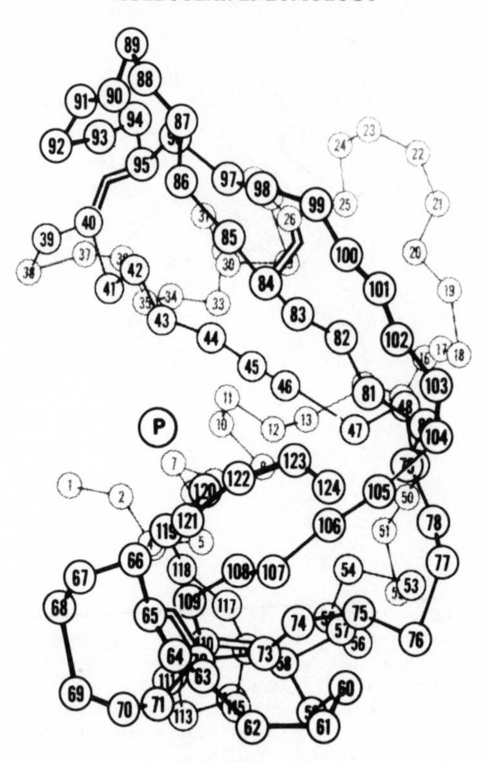

Figure 3.6 The polypeptide chain conformation of RNase A determined by X-ray crystallography. From Dickerson and Geis (1969), p. 80.

The contribution of X-ray crystallography to the determination of structure of RNase and other enzymes

The complete structure of RNase has been determined by X-ray crystallography. This technique (which has now been employed for the solution of the complete structure of a large number of enzymes) has had an enormous impact upon enzymology and protein chemistry. Furthermore, the crystalline forms of several enzymes have been shown to be enzymically active. This was achieved by bathing the enzyme crystals in substrate solution under conditions in which the crystals would not dissolve. Calculations which take into account the limited diffusive access of the substrate to the enzyme indicate that in some cases the crystals then behave as fully active enzyme.

Important deductions regarding the mechanism of action of many enzymes including RNase have been made on the basis of the X-ray structure. This has been greatly facilitated by difference Fourier analysis of crystals which have substrate-relevant compounds (i.e. analogues, products, etc.) bound at the enzymic active sites. In the case of ribonuclease, 3′ uridine monophosphate (3′-UMP) can be diffused into the crystals and is bound at the active site.

Most proteins have some regions which yield "fuzzy" electron density maps and such "fuzzy" regions may be interpreted as resulting from multiple conformations. A study of the temperature-dependence of the electron density in these regions (temperature factor) can provide important information concerning the conformational mobility in these regions. In RNase, residues 21–22 are poorly defined in the electron density maps. It is of interest to note that residue 21 is the N-terminal amino acid of RNase-S, and perhaps this explains why a covalent link is not necessary between residues 20–21 since it would not be expected to provide much conformational stability.

The technique of X-ray crystallography, although capable of providing vital information for the enzymologist, is not of much interest to him *per se*. All enzymologists are, however, very interested in obtaining the most refined structure available for the particular enzyme protein upon which they perform kinetic and other experiments. The present-day limit of resolution in protein crystallography is around 1.5 Å (0.15 nm) which enables atoms to be placed within ± 0.1 Å. Such a resolution enables a sequence to be directly fitted to the electron density maps and obviates the need for sequencing of the protein. The best available structures for enzymes are generally at lower resolution than this but many may be refined to this level in the course of (computer) time.

Despite the enormously important contribution of X-ray crystallography it does suffer several important disadvantages. Perhaps the most important of these is the near impossibility of measuring dynamic events. Many ligands may be diffused into protein crystals and the structure of the enzyme-ligand complex determined, but it is not currently possible to perform this with the substrate(s) itself (unless the equilibrium of the overall reaction is extreme). Thus in general it is not possible to observe directly the fully productive enzyme–substrate complex. In the future, however, cryogenic techniques together with holography may allow the rapid direct determination of the structure of fully productive complexes. Another important disadvantage of crystallography is the artefactual nature of the protein during the structural determination. A protein molecule in solution may have more conformational mobility as compared with the crystal structure. That some enzymes have been shown to be fully active in the crystalline state somewhat allays our fears in this quarter. NMR spectra of the small enzyme lysozyme in solution show that some of the benzene rings of phenylalanine residues buried in the enzyme are capable of rapid rotation. Also, buried tryptophan residues undergo a flapping motion of up to 60°. Temperature factor measurement of the electron density maps in the appropriate regions would confirm whether this is also the case in the crystalline enzyme.

The conformational dynamics of ribonuclease

Ribonuclease was the first enzyme shown to be capable of reversible unfolding. The enzyme can be denatured and the disulphide bonds reduced to give a fully unfolded random coil configuration. On removal of the denaturing agent and in the presence of a mixture of oxidized and reduced glutathione, the protein spontaneously folds into its native configuration in a period of a few hours (50 % enzymic activity in 90 min). The number of reduced cysteine residues in the protein decreases at a notably more rapid rate than the activity appears. Thus a randomly cross-linked chain must occur early on the overall folding pathway, followed by re-shuffling of the disulphide bonds until the native set is formed. Another important observation is that the *fully* reduced enzyme has been shown to have 0.04 % of the full enzymic activity. This has been interpreted to mean that a very small fraction of the total protein is in the native configuration, at least in so far as activity is concerned in the absence of any disulphide bond formation. This in turn strongly suggests that the pathways of folding are the same or similar whether or not disulphide bonds are involved. It is notable that enzymes which have disulphide bonds usually perform their catalyses in surroundings which are unlikely to allow disulphide exchange, e.g. RNase in the small intestine. The enzyme unfolded in the absence of disulphide reduction refolds to the native conformation very much more rapidly than does the fully reduced enzyme.

A most interesting finding is that RNase-S protein is incapable of folding to *its* correct conformation under conditions which allow the complete protein to fold to its native conformation. However, addition of the S-peptide allows the mixture to form active RNase. Similarly, neither insulin nor chymotrypsin will fold to the native conformation following reduction and unfolding. Both proinsulin and chymotrypsinogen readily fold to their respective native conformations. This makes good sense since it is in these forms that the proteins are originally folded consequent upon or immediately after synthesis.

A variety of experimental techniques have been employed in the study of ribonuclease folding and unfolding. NMR and proteolytic digestion

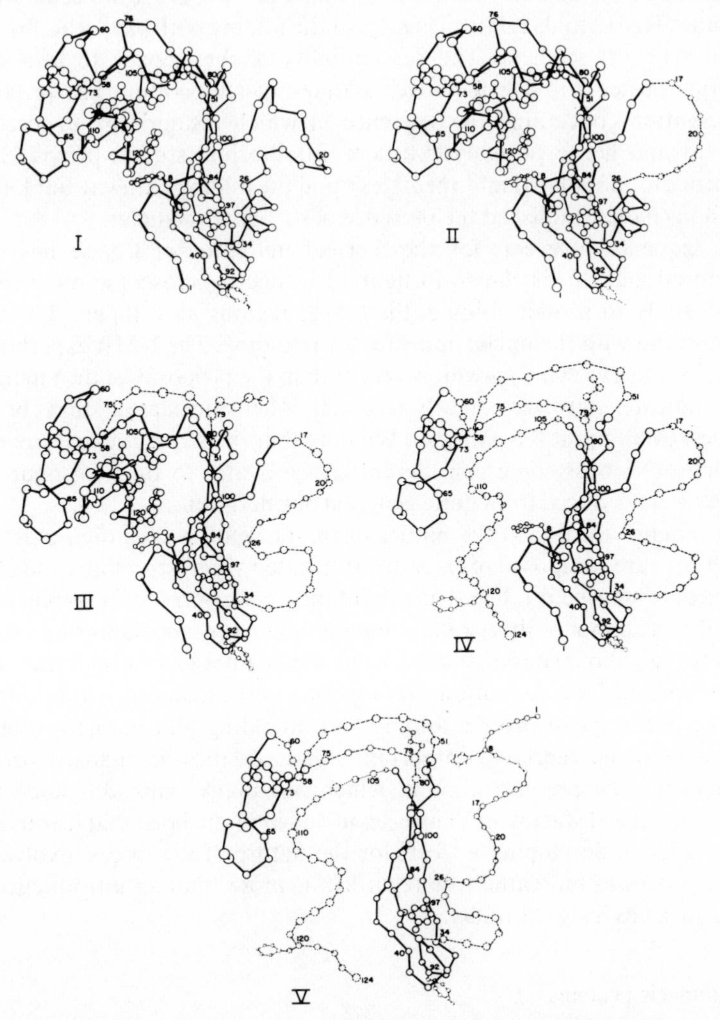

Figure 3.7 Schematic pathway for the thermal unfolding of ribonuclease A. Disulphide bonds join residues 26–84, 40–95, 58–110 and 65–72. Sequence proceeds from I through V to the fully unfolded form. Note that NMR evidence suggests that the *N*-terminal (1 → 20) region folds first in the folding sequence. From Anfinsen and Scheraga (1975).

studies have proved particularly useful. NMR observation of the C-2 protons of the four histidine residues in RNase indicates that an intermediate occurs on the folding pathway in which His-12 is in a folded environment while the remaining three histidine residues are in an unfolded environment. Thus the N-terminal portion of the molecule which includes His-12 folds relatively early on the folding pathway—this portion is an α-helical segment. The susceptibility of the enzyme to attack by various proteolytic enzymes over a range of temperature has provided information concerning the sequence in which residues are exposed to solvent and hence proteolytic attack. Equilibrium studies provide little evidence for populated intermediates upon the folding pathway but kinetic methods have established the presence of such intermediates.

A sequence of events for the thermal unfolding of RNase has been proposed and this is shown in figure 3.7. Since microscopic reversibility must apply to protein folding, the folded regions in V (figure 3.7) must correspond with the nucleation sites for refolding. The NMR experiments described above, which are more recent than the proposals shown in figure 3.7, indicate, however, that N-terminal α-helix formation may be the earliest event upon the pathway. We note that the structural features seen in figure 3.7 bear some considerable resemblance to those of figure 3.2 which were deduced from quite general considerations.

It remains to discuss the nature of the nucleation sites (figure 3.7, V). Both sites are composed of extended (β-pleated sheet) structures and chain reversals (*not* α-helix). The structure of the nucleation sites is stabilized by nearest neighbour short-range interactions and medium-range (next nearest neighbour) interactions. Longer-range interactions (in terms of the sequence) "tie" the nucleation sites together in the folded structure.

The pathways of protein folding and unfolding and the nature of the forces involved represent important areas of study in modern protein chemistry. The present account, which has many omissions and some overgeneralized statements, has been included in the hope that it will assist the reader in developing a "feel" for the nature of the forces involved in these processes but cannot be regarded as more than an introduction to the subject as it stands today.

Oligomeric proteins

Most enzymes which act catalytically in an intracellular environment have a multi-subunit structure. Dimeric and tetrameric structures are by far the most common oligomeric forms. Haemoglobin (which is tetrameric, having two α and two β chains) is the classical non-enzymic example. Each of the four chains has an oxygen binding site and the sites interact cooperatively during oxygen binding. In other words the four sites do not operate independently but are able to exchange information concerning the occupancy (by oxygen) of the other sites. This gives rise to the familiar sigmoid curve for oxygen binding to haemoglobin (see chapter 7).

The subunits of oligomeric enzymes are held together by non-covalent interactions. For this reason they tend to be much more delicate than the extracellular enzymes, which are monomeric and stabilized by intrachain disulphide bonds. The interfacial regions of subunit interaction are characterized by the presence of predominantly hydrophobic amino acid side chains. The sequence of the amino acids in the interfacial regions tends to be highly conserved when a given oligomeric enzyme from several species is compared. This indicates that a "best way" of ensuring optimal subunit interaction has evolved for any particular oligomeric enzyme. Most oligomeric enzymes have similar or identical subunits, each of which has an active site, but in many cases there is no observable cooperativity between the catalytic sites. This raises the question of why they are oligomeric if the possibility of cooperativity (see chapter 7) is not exploited.

The properties of oligomeric enzymes will be discussed in terms of a specific example, lactate dehydrogenase. Lactate dehydrogenase has a molecular weight of 140 000 and is composed of four subunits. The dogfish enzyme whose complete sequence is known has 329 amino acids per subunit, the enzymes from other species being similar in size. In mammals two types of chain are present, H-chains being synthesized predominantly in heart muscle and M-chains in skeletal muscle. For instance, in rat heart tissue 78 % of the popypeptide chains are of the H-type, while in rat leg muscle only 11 % of the chains are of the H-type. The size, amino acid composition and sequence, where known, of the H and M chains are similar but the chains may readily be distinguished by electrophoretic migration. Enzyme is isolated from a particular tissue as a mixture of five isoenzymes H_4, H_3M, H_2M_2, HM_3, and M_4, the relative proportion of each isoenzyme being dependent upon the ratio of H : M chain synthesis in that tissue. If an equimolar mixture of H and M chains, e.g. $[H_4]=[M_4]$ is combined in conditions where hybridization can occur, a binomial distribution (1 : 4 : 6 : 4 : 1) of the isoenzyme results. Hybridization which results from subunit interchange may be induced by repeated freezing and thawing, mild denaturation and dilution.

Each subunit of the enzyme has an active site which acts independently of the three other sites in the tetramer. Dissociated subunits are inactive, which implies that conformational changes take place upon subunit association. This is not surprising, since the hydrophobic region of the monomer which is responsible for intersubunit interaction will be exposed to solvent. The solitary subunit in solution will be expected to assume a conformation which minimizes such an unfavourable interaction. Association of subunits will be accompanied by a decrease in the surface area to volume ratio which will result in stabilization of the tetramer. The

Figure 3.8 6 Å low resolution structure of tetrameric lactate dehydrogenase. Taken from Adams *et al.* (1970).

enzyme, which may be regarded as being assembled from two dimers, has three orthogonal two-fold axes of symmetry that meet at the centre of the tetramer. A low resolution structure of the enzyme is shown in figure 3.8. Approximately 40 % of the structure is α-helix and some 23 % in the form of β-structure. Figure 3.9 is a schematic diagram of the folding of an LDH subunit showing the relative dispositions of the α-helices and β-structure.

The subunit structure may be analysed in terms of three domains. These may be described as the N-terminal arm, the coenzyme binding domain and the catalytic domain. The coenzyme binding and catalytic domains may be further divided into two sub-domains each. The *coenzyme binding domain* has one subdomain for adenine binding and one for nicotinamide binding. The general form of the coenzyme binding domain (which is the region of the six stranded β-structure) is conserved among the dehydrogenases, and this suggests that a common precursor protein is responsible for this domain. The *catalytic domain* which contains the essential His-195 may also be divided into two parts. The domains are shown in exploded form in figure 3.10 in which the catalytic domain corresponds to the top pair and the coenzyme binding domain the bottom pair. The adenosyl portion of NAD binds to the enzyme initially, which results in the creation of a binding site for the nicotinamide moiety. Diffusion of coenzyme into apoenzyme crystals leads to disintegration of the crystals, suggesting a major conformational change. On formation of the NAD : pyruvate abortive ternary complex (3.10) the α-helix αD moves 11 Å and a loop of polypeptide chain moves to enclose substrate and coenzyme. Arg-101 forms

R, N, ⊕, C=O, NH_2, CH_2, C–O–H, $CO_2^{\ominus}$, :B → R, N, C=O, NH_2, CH_2, C=O, H–B, $CO_2^{\ominus}$

NAD + enol-pyruvate

(3.10)

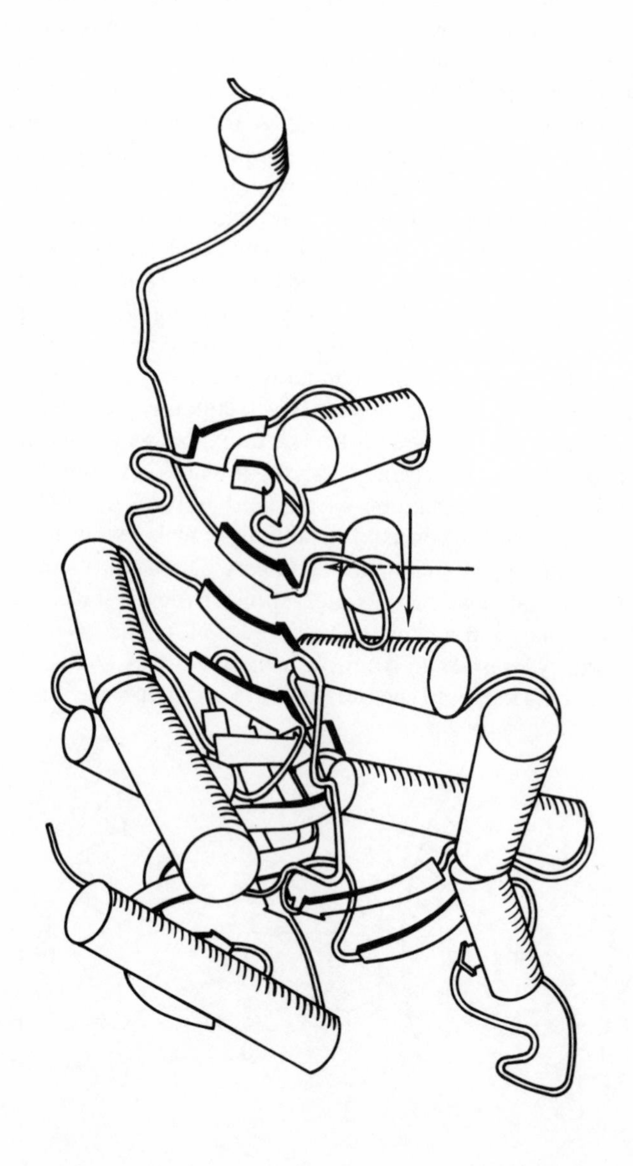

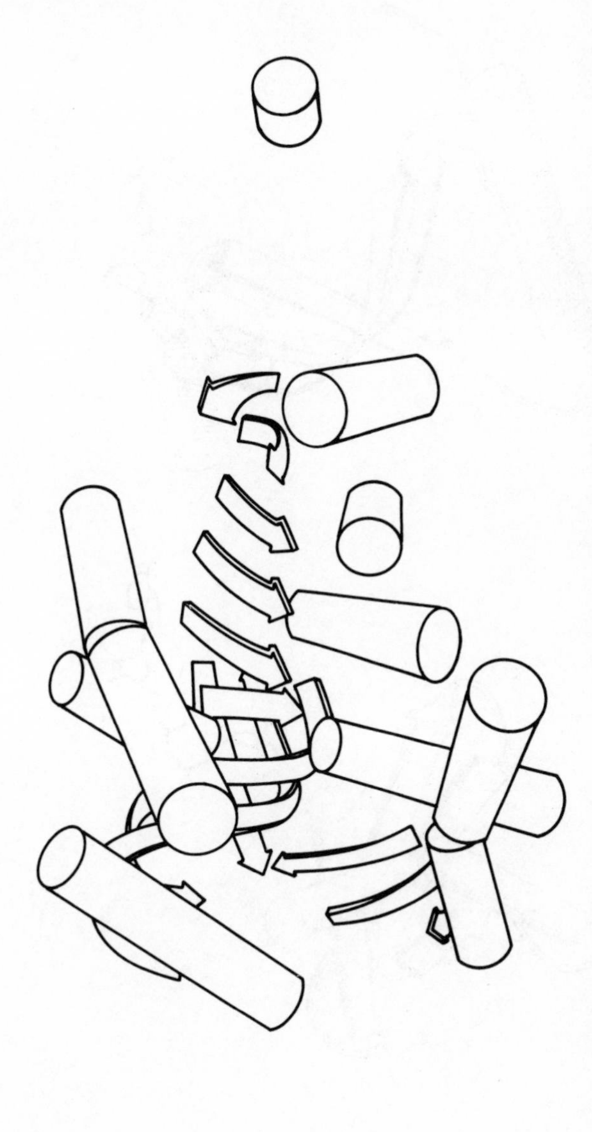

Figure 3.9 Schematic diagram of the folding of a single subunit of lactate dehydrogenase. α-helices are shown as cylinders and β-structures as strips. Taken from *The Enzymes*, Vol. XI, 1975, p. 226.

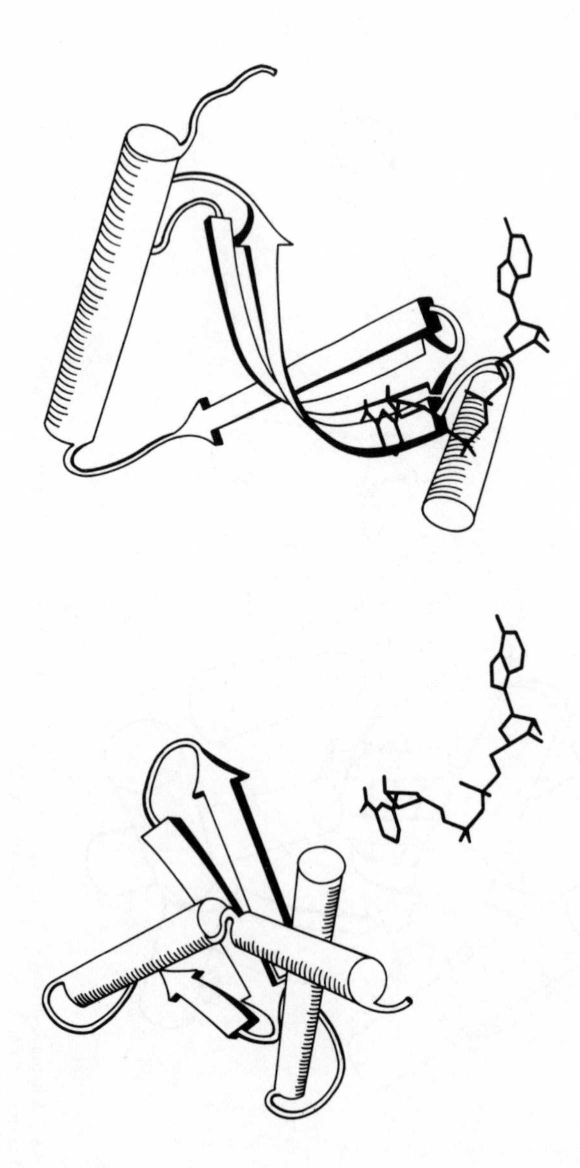

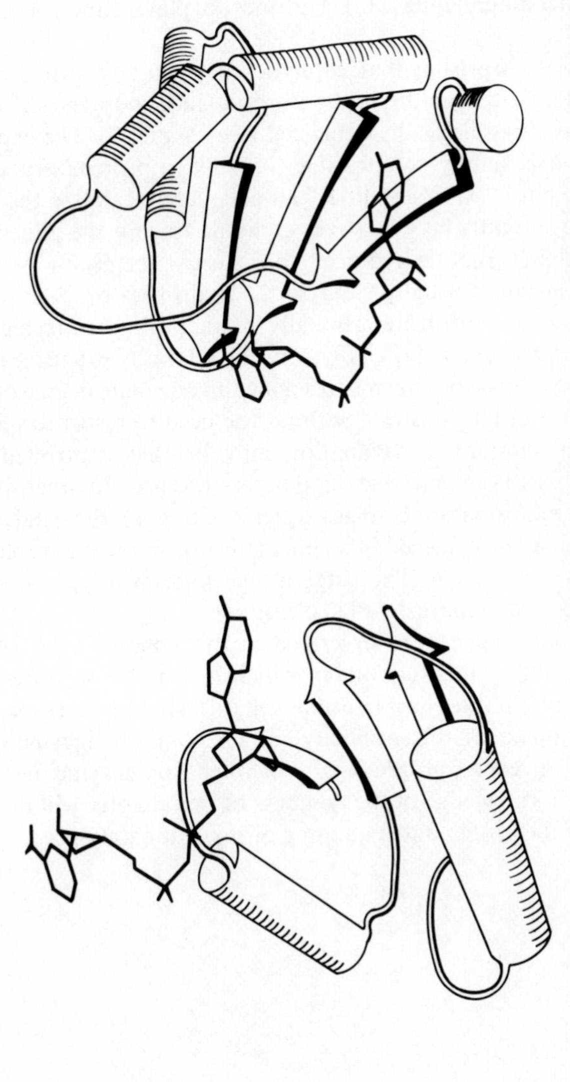

Figure 3.10 The domain structure of lactate dehydrogenase, see text for details. Taken from *The Enzymes*, Vol. XI, 1975, p. 220.

an ion pair with the coenzyme pyrophosphate grouping and this may act as a trigger for the conformational changes. Arg-109, whose side chain is exposed to solvent in the apo-enzyme, moves 14 Å to form a buried ion pair with the substrate carboxyl group. The domain created by the *N-terminal arm* is involved in intersubunit interaction and does not play a direct role in binding or catalysis.

It is perhaps surprising that catalytic cooperativity is not observed, given the tetrameric structure of the enzyme, since cooperativity might be expected to be physiologically beneficial (see chapter 7). The explanation may be that it is simply not possible to evolve a protein structure with molecular weight of ca. 35 000 that can adequately catalyse the required reaction. The subunits may however combine to gain the added conformational stability that is required for effective catalysis. Monomeric extracellular enzymes usually catalyse the hydrolysis of large polymeric substrates, and these substrates generally interact with the enzyme through several monomeric units of the polymer. The relatively extensive nature of the interaction allows the enzyme to achieve an adequate degree of binding and orientation of the substrate without the need to resort to oligomeric structure. An alternative explanation may be that controlled conformational mobility is enhanced in oligomeric structures. In other words, the considerable conformation changes which occur upon the catalytic pathway of LDH may be achieved in an energetically more favourable fashion by an oligomeric protein. The catalytic mechanism of lactate dehydrogenase is described in more detail in chapter 9.

The forces that have been described in this chapter have, of course, general applicability to situations in which intermolecular interaction is involved. Thus all the considerations of this chapter may be used to provide a rationale for the specificity and energetics of enzyme-substrate interaction. The reader, progressing to chapter 9 on enzyme mechanisms, will find that a knowledge of the contents of this chapter will prove most important for the proper understanding of enzymic catalysis.

CHAPTER FOUR

SIMPLE ENZYME KINETICS

THE MOST IMPRESSIVE FEATURES OF ENZYMES ARE THE SPECIFICITY AND efficiency of their catalytic function. For this reason, most studies of enzymes have been concerned with characterizing their kinetic properties. Such studies not only lead to an increased understanding of the mechanisms of enzymic catalysis but are also useful in defining the role of enzymes in the context of the functions of living organisms.

Historically, the development of enzyme kinetics has been based on the derivation of equations describing the rates of conversion of a single reactant, termed the substrate (S), to one or more products (P). In fact, very few enzyme-catalysed reactions involve only one substrate, but for many reactions, e.g. hydrolyses, the second substrate is the solvent, and its concentration is so large (in the case of water $\sim$55M) that it varies negligibly with respect to the variation in the concentration of S. Moreover, it will be seen (chapter 7) that even where two or more substrates are involved with concentrations of similar magnitude, the experimental conditions can be designed so as to make the single-substrate case applicable to more complex systems.

Let us then consider the irreversible uncatalysed reaction

$$S \rightarrow P$$

This is a first-order reaction and its *time course* or *progress curve* will have the appearance of figure 4.1 where the concentration of P is plotted against time t. The slope of the tangent to the progress curve $d(P)/dt$ (the *rate* at time t) decreases from a finite value, which depends on S_0, the concentration of S at $t = 0$, to zero when the reaction reaches completion. The shape of the curve is exponential and the equation describing the variation of (P) with t can be derived by integration of the empirical first-order rate

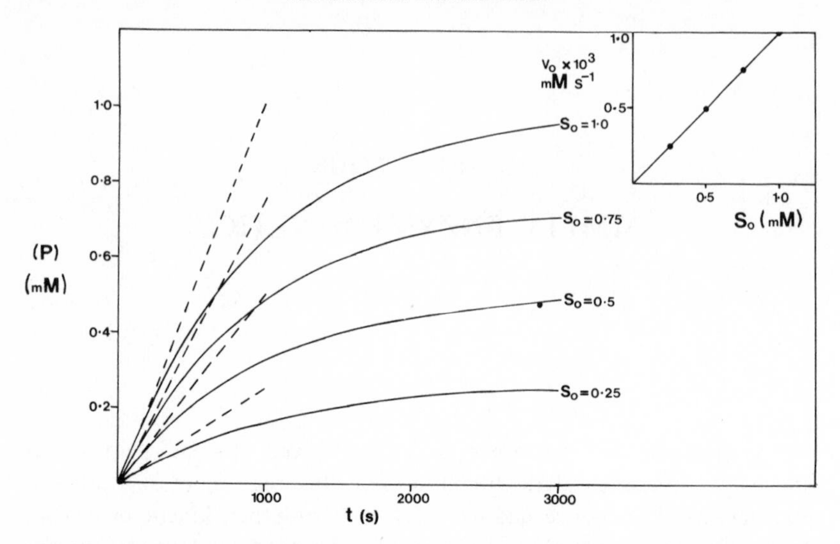

Figure 4.1 Progress curves for a first order reaction $S \rightarrow P$ with rate constant $k = 10^{-3}\ s^{-1}$ for different initial reactant concentrations, S_0 (mM), as given on each curve. The dashed lines represent the initial rates, v_0, drawn as tangents to the progress curves at $t = 0$. Inset: plot of v_0 against corresponding S_0. The slope of this plot gives the value of the rate constant, k.

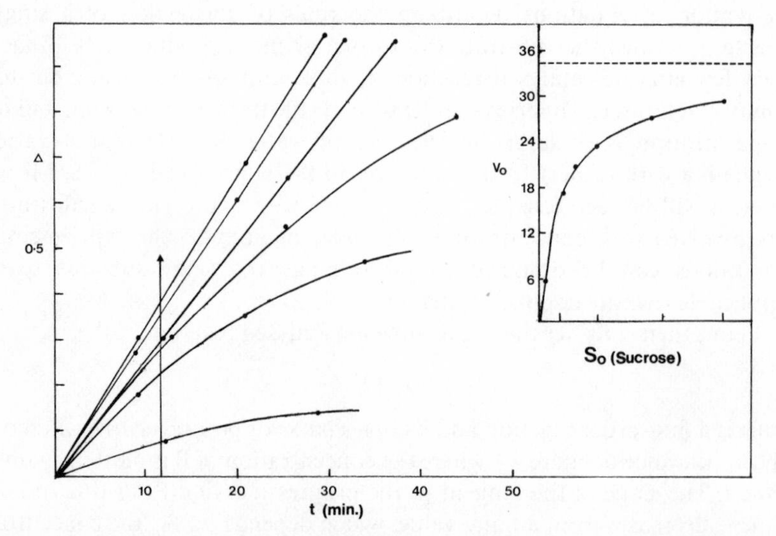

Figure 4.2 Progress curves for the invertase-catalysed hydrolysis of sucrose. Taken from Haldane (1930) after Kuhn (1923). Arrow indicates increasing initial sucrose concentration. Inset: plot of initial rate, v_0, against sucrose concentration drawn using data of Kuhn (1923). The dashed line represents the limiting velocity, V.

equation

$$\frac{d(\mathrm{P})}{dt} = k(\mathrm{S}_0 - (\mathrm{P})) = -\frac{d(\mathrm{S})}{dt}$$

where k is the first-order rate constant. The equality $d(\mathrm{P})/dt = -d(\mathrm{S})/dt$ follows from the fact that no intermediates occur in the reaction mechanism, so the disappearance of one molecule of S corresponds to the appearance of one molecule of P. It can be seen that when $t = 0$ (and $(\mathrm{P}) = 0$) the slope of the tangent to the progress curve (the initial rate, $d\mathrm{S}_0/dt$ or v_0) will be directly proportional to S_0 and a plot of v_0 against S_0 is a straight line through the origin with slope k (figure 4.1, inset). This is characteristic of first-order reactions. In the presence of a chemical catalyst, such as H^+, the linear relationship between v_0 and S_0 would hold but the slope of the line would be steeper. The value of the apparent first-order rate constant will be proportional to the concentration of the catalyst, but the same rate law would apply.

Michaelis–Menten equation

One of the first enzymes to be studied by kineticists was invertase, which catalyses the hydrolysis of sucrose to fructose and glucose. This can be regarded as effectively a one-reactant system. In the presence of a constant amount of enzyme and at varying initial sucrose concentrations, progress curves are obtained which appear similar to those described above (though they are not, in fact, the same shape) in that the rate depends on the initial sucrose concentration and decreases with time (figure 4.2). However, it is found that the initial rate v_0 varies linearly with S_0 only at low values of S_0; as S_0 increases, v_0 approaches a limiting value independent of S_0 (figure 4.2, inset). Thus the catalytic effect of the enzyme involves more than simply increasing the first-order rate constant.

At the beginning of this century Brown and Henri proposed mechanisms to account for this kinetic behaviour; these were further developed by Michaelis and Menten and later by Briggs and Haldane. The mechanisms assumed that formation of a complex between enzyme and substrate was an essential part of the catalytic process. O'Sullivan and Thompson had previously observed that the stability of invertase to thermal denaturation was enhanced in the presence of sucrose, and this provided independent evidence for the existence of an enzyme-substrate complex.

The mechanism used below to derive the Michaelis–Menten equation assumes that the substrate combines with the enzyme E, in a reversible manner to give complex ES, which can then dissociate to enzyme and substrate or react further (irreversibly) to give enzyme and product.

$$E+S \underset{k_{-1}}{\overset{k_1}{\rightleftharpoons}} ES \xrightarrow{k_2} E+P$$

This mechanism represents three reactions, each characterized by a rate constant. Note that E appears as a reactant in the first step and as a product in the final step, a condition demanded by its function as a catalyst.

The rate of production or consumption of any component of a reaction mechanism will be the algebraic sum of the rates of all the individual steps involving that component as reactant or product. The overall velocity of the reaction, v, is defined as the rate of appearance of product

$$v = \frac{d(\mathrm{P})}{dt} = k_2(\mathrm{ES}) \tag{4.1}$$

or the rate of disappearance of substrate

$$-\frac{d(\mathrm{S})}{dt} = k_1(\mathrm{E})(\mathrm{S}) - k_{-1}(\mathrm{ES}) \tag{4.2}$$

Because of the high catalytic efficiency of enzymes,* enzyme assays are usually run under conditions such that $S_0 \gg E_0$ and the total enzyme concentration is itself small relative to the changes observed in (S) or (P). It follows that the concentration of bound substrate, (ES), is negligible. Therefore we may equate the rate of increase in (P) with the rate of decrease in (S).

$$-\frac{d(\mathrm{S})}{dt} = \frac{d(\mathrm{P})}{dt} \tag{4.3}$$

and so

$$k_1(\mathrm{E})(\mathrm{S}) - k_{-1}(\mathrm{ES}) = k_2(\mathrm{ES}) \tag{4.4}$$

or

$$\frac{(\mathrm{E})(\mathrm{S})}{(\mathrm{ES})} = \frac{k_{-1}+k_2}{k_1} \tag{4.5}$$

What is required is an expression for v in terms of the known quantities E_0 and (S). To obtain this we invoke the conservation equations for substrate

* By this is meant that enzymes enhance reaction rates at concentrations which can be many orders of magnitude smaller than those of the reactants.

and enzyme. For substrate

$$S_0 = (S)+(P)+(ES) \tag{4.6}$$

The experimental conditions allow the (ES) term in equation (4.6) to be dropped because (ES) $\ll$ (S). A further simplification can be made by considering only the initial slope, v_0, of the progress curve so that (S) $\gg$ (P) in which case the (P) term can be dropped leaving

$$(S) = S_0 \tag{4.6a}$$

There are also several practical advantages, peculiar to enzyme-catalysed reactions, for considering only the initial rate. These will be discussed later.

For enzyme the conservation equation is

$$E_0 = (E)+(ES) \tag{4.7}$$

Solving equation (4.5) for (E) using (4.6a) and substituting into (4.7) gives

$$E_0 = \frac{(ES)(k_{-1}+k_2)}{k_1 S_0} + (ES)$$

Solving for (ES)

$$(ES) = \frac{E_0}{1+(k_{-1}+k_{-2})/k_1 S_0} \tag{4.8}$$

which can be substituted for (ES) in equation (4.1) giving

$$v_0 = \frac{k_2 E_0}{1+(k_{-1}+k_2)/k_1 S_0} \tag{4.9}$$

The constant term $(k_{-1}+k_2)/k_1$ in the denominator is usually defined by the symbol K_m and is called the *Michaelis constant*. It can be seen that if $S_0 \gg K_m$ the second term in the denominator of equation (4.9) becomes small and the value of the denominator approaches one. Under these conditions

$$v_0 = k_2 E_0 = V$$

and the rate is independent of S_0. This corresponds to the limiting velocity seen in figure 4.2 (inset) and therefore is defined as the *maximum velocity*, termed V, with units the same as those of v_0, i.e. concentration/unit time.

Equation (4.9) is usually written in the form

$$v_0 = \frac{V}{1+K_m/S_0} = \frac{VS_0}{K_m+S_0} \tag{4.10}$$

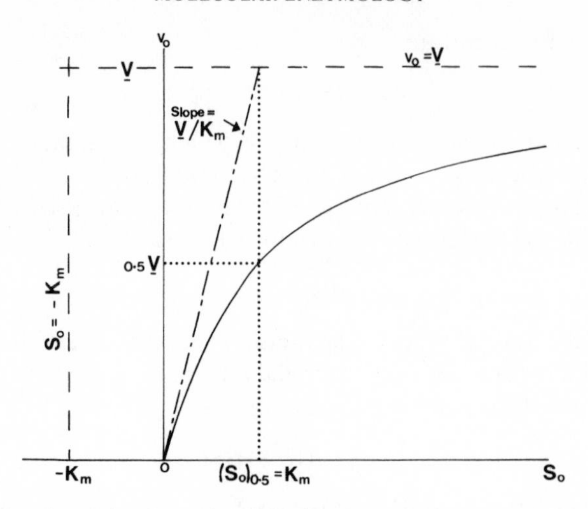

Figure 4.3 Plot of v_0 against S_0 for the Michaelis–Menten equation showing the asymptotes (dashed lines) to the rectangular hyperbola.

and is known as the *Michaelis–Menten equation.* Equation (4.10) describes a rectangular hyperbola (i.e. one with orthogonal asymptotes) in v_0, S_0 space passing through the origin with asymptotes $v_0 = V$ and $S_0 = -K_m$ (figure 4.3).

When $S_0 \ll K_m$ equation (4.10) reduces to

$$v_0 = \frac{VS_0}{K_m} \tag{4.11}$$

and as V and K_m are constants at fixed E_0, equation (4.11) predicts that v_0 will be directly proportional to S_0 indicating a first-order reaction. The rectangular hyperbola will thus approach linearity at low S_0, the tangent at $S_0 = 0$ being V/K_m. The deviation of the slope of the tangent from V/K_m is less than 1% at $S_0 = 0.01K_m$, 5% at $S_0 = 0.05K_m$ and about 9% at $S_0 = 0.1K_m$. This region of the hyperbola is sometimes termed the *first-order region.* The proportionality constant V/K_m has the units of reciprocal time and is the first-order rate constant for the enzyme-catalysed transformation

$$S \xrightarrow{E} P$$

Equations (4.9) and (4.10) also predict a linear relationship between v_0 and E_0 at any fixed S_0. The direct proportionality of v_0 and E_0 is often utilized in estimating enzyme concentration in tissue extracts and should always be verified when rate assays are used. Although the relationship holds for any fixed S_0, it is advantageous to choose S_0 carefully when assaying enzyme activity. Inspection of figure 4.3 shows that errors in S_0 will have a much smaller effect on v_0 if S_0 is in the plateau region of the rectangular hyperbola, i.e. if $S_0 \gg K_m$. This is discussed in chapter 10.

Maximum velocity and k_{cat}

As defined above, the maximum velocity, $V = k_2E_0$, and is the overall reaction rate when $S_0 \gg K_m$. Under these conditions, $(ES) \simeq E_0$ and the enzyme is said to be *saturated* with its substrate, i.e. all the catalytic sites are occupied. The ratio v/V can be interpreted as the fractional saturation of catalytic sites, and its value varies from 0 to 1. The V asymptote is reached only at infinite value of S_0. When $S_0 = 9K_m$, $v_0 = 0.9V$; at $S_0 = 19K_m$, $v_0 = 0.95V$; in this region of S_0 the rectangular hyperbola is nearly parallel to the S_0 axis and v_0 changes little. This is the *zero-order region*, and its significance in the determination of enzyme concentration by rate assay has been described above.

As $V = k_2E_0$, its value varies linearly with the total enzyme concentration. However, V/E_0 is an important *constant* characteristic of a given enzyme and substrate under defined conditions. For the mechanisms discussed above involving a single intermediate $V/E_0 = k_2$, but this will not be true in general. For more complex mechanisms V/E_0 may be a function of several rate constants and is generally denoted k_{cat} and called the *catalytic constant*. As it is calculated by dividing a rate by a concentration, it has the units of reciprocal time and represents the number of substrate molecules transformed to product per enzyme molecule per unit time. Thus k_{cat} reflects the catalytic effectiveness of an enzyme. Although k_{cat} is a more useful parameter than V it can be calculated only if the purity of the enzyme preparation, the molecular weight, and the number of catalytic sites per enzyme molecule are known. Many enzymes are composed of subunits and have more than one catalytic site per molecule. For such a system the catalytic constant is given by V/nE_0 where n is the number of catalytic sites per enzyme molecule. k_{cat} is sometimes referred to as the *turnover number*.

Michaelis constant

It will be shown that the Michaelis–Menten equation (4.10) can be applied to many kinetic mechanisms more complex than the case discussed above, and the Michaelis constant, K_m, will represent various combinations of rate constants depending on the mechanism. For this reason K_m is best defined operationally as the *substrate concentration at which the initial rate is one-half the maximum velocity* and therefore is sometimes termed $(S_0)_{0.5}$. This is easily shown, for if $K_m = S_0$ then substitution of S_0 for K_m in equation (4.10) gives $v_0 = V/(1+S_0/S_0) = V/2$. For the simple irreversible mechanism, K_m is the sum of two first-order rate constants divided by a second-order rate constant $(k_{-1}+k_2)/k_1$. Its units are therefore those of concentration and its value is independent of E_0 (but see below). The Michaelis constant is characteristic of a given enzyme and substrate but the conditions under which it is determined must be specified as its value can vary with, e.g. temperature, pH or ionic strength.

The lower the value of K_m, the higher will be the rate for a given non-saturating substrate concentration. For this reason K_m is sometimes interpreted as being inversely proportional to the affinity of the enzyme for its substrate, i.e. as the dissociation constant, K_S, for the equilibrium

$$ES \underset{k_1}{\overset{k_{-1}}{\rightleftharpoons}} E+S$$

At equilibrium, the rates of association and dissociation are equal

$$k_1(E)(S) = k_{-1}(ES)$$

or

$$\frac{(E)(S)}{(ES)} = \frac{k_{-1}}{k_1} = K_S$$

Thus $K_S = k_{-1}/k_1$ whereas $K_m = (k_{-1}+k_2)/k_1$. The physical significance of K_m for any mechanism will depend on the relative magnitudes of the rate constants involved. In this case $K_m \simeq K_S$ only if $k_2 \ll k_{-1}$. In their original paper, Michaelis and Menten made the (unnecessary) assumption that equilibrium between E, S and ES is established very rapidly compared to the breakdown of ES to E and P. The rate equation derived from the single-substrate, one-intermediate mechanism using this equilibrium assumption is identical in form to equation (4.10) but the meaning of K_m $(=k_{-1}/k_1)$ is different. Another mechanism that may be suitable for the description of some enzyme-catalysed reactions assumes that the formation of ES from E and S is irreversible

$$E+S \xrightarrow{k_1} ES \xrightarrow{k_2} E+P$$

This mechanism implies that $k_2 \gg k_{-1}$. Derivation of the rate equation again leads to a relationship identical in form to the Michaelis–Menten equation but the K_m term is now k_2/k_1. So the significance of K_m in terms of rate constants will depend on the mechanism from which the rate equation is derived. For this reason it is generally unwise to interpret K_m as a measure of the affinity of an enzyme for its substrate unless independent evidence can be obtained that $K_m = k_{-1}/k_1$, i.e. that the k_2 term is negligible.

A final point concerning K_m should be made. It has been pointed out that the operational definition of K_m as $(S_0)_{0.5}$ contains the assumption, implicit in the derivation of equation (4.10), that $S_0 \gg E_0$. If this assumption cannot be made then it can be shown that $(k_{-1}+k_2)/k_1 = (S_0)_{0.5}-(E_0/2)$, and the v_0, S_0 curve will not be a simple rectangular hyperbola. Thus the K_m operationally defined as $S_{0.5}$ will be larger, by $E_0/2$, than the "true" K_m as defined by the Michaelis–Menten equation (4.9) (i.e. $(k_{-1}+k_2)/k_1$). Under the conditions normally used in steady-state kinetic measurements where E_0 is small this will make little difference. However, in the cell the concentrations of enzymes are often of the same order of magnitude as their substrates (and their K_m). This has led to the view that the Michaelis–Menten equation may not adequately describe the behaviour of enzymes *in vivo*. But this results from a flaw in the operational definition of K_m rather than in the Michaelis–Menten equation. If the concentration of *free* substrate, (S), is substituted for S_0 in equation (4.10) the Michaelis–Menten equation will hold under these conditions. Thus the problem comes down to the important and difficult one of determining the true concentration of free substrate *in vivo* and not on the validity of the Michaelis–Menten equation.

Units of enzyme activity

Enzymologists have usually described enzyme activity in terms of "international" units as defined by the Enzyme Commission of the International Union of Biochemistry. One *enzyme unit* (U) is the amount of enzyme which will catalyse the transformation of substrate to product at a rate of *one μmole per minute* under defined conditions. Note that the enzyme unit is defined in terms of the *amount* of substrate converted, *not* the resulting concentration change. *Specific activity* is a useful term for describing the purity of an enzyme preparation and is defined as U/mg protein. If the molecular weight of the enzyme is known, *molecular activity* may be defined as the number of *molecules of substrate transformed per molecule of enzyme per minute* or U/μmole enzyme. The *catalytic centre activity* is the molecular activity divided by the number of catalytic sites per enzyme molecule and is thus the number of substrate molecules transformed per active site per minute. At saturating substrate concentration this is equivalent to the catalytic constant defined earlier.

With the introduction of SI units, the *katal* has been recommended as a replacement for the enzyme unit. The *katal* is defined as the amount of activity catalysing the conversion of *one mole of substrate per second* and is abbreviated kat. Specific activity is defined as katals per kilogram of protein; similarly molecular activity is kat/mole enzyme. The conversion factors for units and katals are thus 1 kat $= 6 \times 10^7$ U or 1 U $= 16.67$ nkat. Both kat and U are in current use.

Presentation of kinetic data

The Michaelis–Menten equation (4.10) can be rearranged into the following form:

$$\frac{S_0}{v_0} = \frac{S_0}{V} + \frac{K_m}{V} \tag{4.12}$$

Equation (4.12) predicts that a plot of S_0/v_0 against S_0 (the "half-reciprocal plot") will be a straight line of slope $1/V$, y-axis intercept of K_m/V and x-axis intercept of $-K_m$ (figure 4.4). This is not the only linear transformation of the Michaelis–Menten equation but, for reasons discussed in chapter 10, is the one that we recommend for presenting the results of measurements of v_0 at different S_0.

We do not advocate using the half-reciprocal plot to *determine values* of

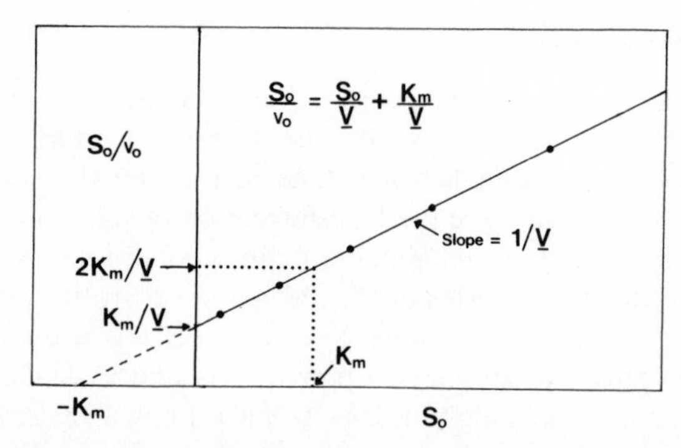

Figure 4.4 Half-reciprocal plot of S_0/v_0 against S_0 for data obeying the Michaelis–Menten equation.

K_m and V from kinetic data. The *direct linear plot* is recommended for this purpose. This plot is based on equation (4.13) which is also a rearranged form of the Michaelis–Menten equation:

$$\frac{V}{v_0} - \frac{K_m}{S_0} = 1 \tag{4.13}$$

Equation (4.13) has the form

$$\frac{x}{a} + \frac{y}{b} = 1$$

which is the equation of a straight line in xy space with x-axis intercept a and y-axis intercept b. In order to apply equation (4.13) to kinetic data, rectangular axes are set up and labelled K_m and V for abcissa and ordinate respectively. The value of $-S_0$ is marked off on the K_m axis and the corresponding v_0 on the V axis. Any single observation S_0, v_0 can be represented by a *line* (rather than a point) passing through $-S_0$ on the K_m-axis and the corresponding v_0 on the V-axis. The coordinates of any point on this line will give values of K_m and V which satisfy the Michaelis–Menten equation *for that observation.* If such a line is drawn for each observation a series of lines will be obtained which intersect at a common point in the first quadrant as shown in figure 4.5. The coordinates of the common intersection point define the value of K_m and V which satisfy the Michaelis–Menten equation for *all* observations.

In practice, the observations will be subject to error and a cluster of intersections will be obtained. The *best estimate* of K_m and V is then taken

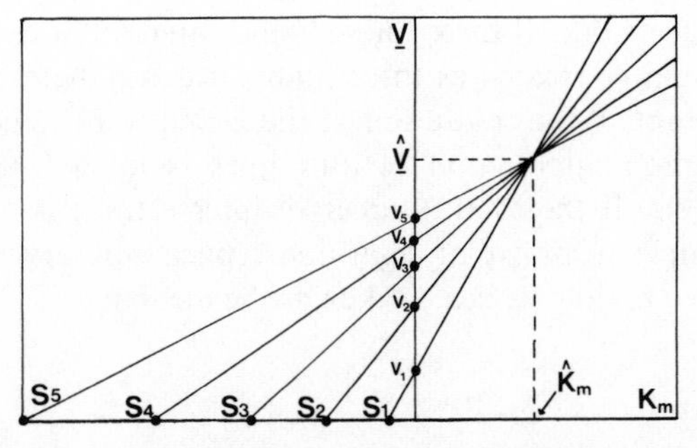

Figure 4.5 Direct linear plot. Each observation is represented by a straight line drawn through v_0 plotted on the ordinate axis and the corresponding $-S_0$ plotted on the abscissa axis. The values of v_0 and S_0 correspond to those plotted as points on the half-reciprocal plot in figure 4.4. The coordinates of the common intersection point gives the estimates $\hat{V}$ and $\hat{K}_m$.

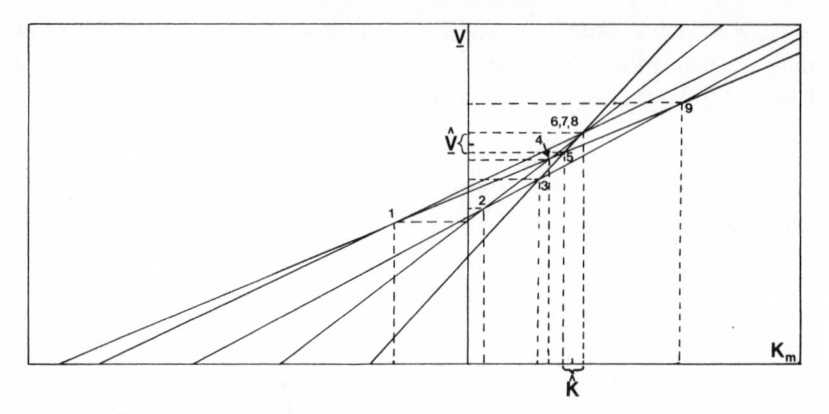

Figure 4.6 Effect of experimental error on the direct linear plot. The ranking of intersections is shown. There are five observations and therefore ten intersections although the 10th (largest) intersection is off-scale. Note that the 6th, 7th and 8th intersection coincide. The estimates of the kinetic parameters $\hat{V}$ and $\hat{K}_m$ are taken as the mean of the coordinates of the 5th and 6th intersections.

as the middle point or *median* of the intersection cluster. For real data the total number of intersection points will be $n(n-1)/2$ where n is the number of observations. If the data are poor some intersections may be off scale, usually at large positive values of K_m and V. The *values* of the coordinates of such intersections are unimportant but their number should be taken into account when ranking the intersections to find the median. The procedure is illustrated in figure 4.6. Intersections may occasionally occur in the third quadrant; in this case the numerical value of such intersections is taken as being large and positive. Intersections in the second quadrant are taken at face value. If three lines should intersect at a common point this point should be treated as three points because there would be three intersection points if the resolution of the graph were sufficient to reveal them. A common intersection of four lines is treated similarly as six intersections, etc. If the total number of intersections is even (as would arise, for example, from five or eight non-replicate observations) then the mean of the two middle values is taken as the median.

The steady state

We have seen that the single-substrate irreversible model gives rise to the same form of rate equation regardless of mechanistic differences (e.g. the first step is irreversible, or E and S are at equilibrium). It is evident that the treatment developed above can tell us little about the kinetic behaviour of (E) and (ES) during the catalytic process. Examination of equation (4.4) shows why this is so. Rearrangement gives

$$k_1(\mathrm{E})(\mathrm{S}) - (k_{-1} + k_2)(\mathrm{ES}) = 0$$

which is also the rate equation for $d(\mathrm{ES})/dt$. So it can be seen that the condition that $d(\mathrm{ES})/dt = 0$ (and $d(\mathrm{E})/dt = 0$) is implicit in the derivation of equation (4.10). This is known as the *steady-state approximation*, and the kinetic relationships obtained under conditions where the approximation is valid are termed *steady-state kinetics*. During the time over which v_0 is determined the concentration of E and ES remain constant. (This can

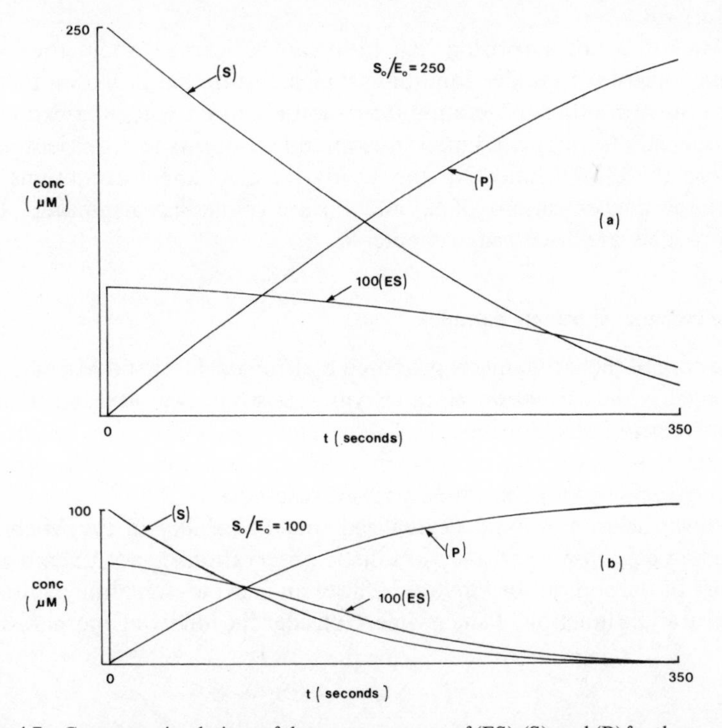

Figure 4.7 Computer simulations of the progress curves of (ES), (S), and (P) for the reaction $\mathrm{E+S \rightleftharpoons ES \rightarrow E+P}$ to illustrate the effect of the ratio $\mathrm{S_0/E_0}$ on $d(\mathrm{ES})/dt$. The progress curves for (E) are not shown; they will be the inverse of those for (ES). (a) $\mathrm{S_0/E_0} = 250$, $\mathrm{E_0} = 1\ \mu\mathrm{M}$, $K_\mathrm{m} = 50\ \mu\mathrm{M}$, $k_\mathrm{cat} = 1\ \mathrm{s}^{-1}$; (b) $\mathrm{S_0/E_0} = 100$, other conditions as in (a). The simulation uses the equilibrium assumption ($K_\mathrm{m} = K_\mathrm{S}$) but over the time-scale shown virtually identical curves would be obtained using the steady-state assumption if realistic values were assumed for k_1 and k_{-1}. (Computer simulation kindly provided by Professor H. Gutfreund.)

never be exactly true, as (S) must change as the reaction proceeds and therefore (ES) = (E) (S)/K_m must also change. It would be more accurate to say that (ES) changes little compared to the change in (S).) This is illustrated in figure 4.7 which shows the effect of the ratio of S_0 to E_0 on the variation of the concentration of the components of the single-substrate irreversible mechanism with time. Expressed as a fraction of the overall reaction, S → P, (ES) changes by 15 % over the first two-thirds of the reaction when $S_0/E_0 = 250$ but by 25 % when $S_0/E_0 = 100$ in the example given. On a time basis (ES) changes by 55 % over the first 200 seconds when $S_0/E_0 = 100$ but only by 22 % when $S_0/E_0 = 250$. In steady-state kinetic studies the S_0 to E_0 ratio is usually at least 10^3 and often very much greater so that (ES) will remain virtually constant over the period of observation.

Thus it is not surprising that little can be learned about the kinetic behaviour of (ES) under conditions which ensure that it does not change with time. In order to obtain information about enzyme intermediates, it is necessary to carry out kinetic measurements during the transient period before the establishment of the steady state or under conditions such that the concentrations of S_0 and E_0 are of similar magnitude. These approaches are discussed in chapter 8.

The enzyme–substrate complex

The core of the mechanisms proposed by Brown, Henri and Michaelis and Menten is the formation of an enzyme–substrate complex and its direct involvement in the conversion of substrates to products. This has become a central dogma in enzymology and forms the basis for all current theories of the mechanisms of enzyme-catalysed reactions.

Nevertheless it should be realized that the ability of the Michaelis–Menten equation to fit a set of kinetic observations is not unambiguous proof of the compulsory involvement of an enzyme–substrate complex in the catalytic function of an enzyme. Consider the following mechanism:

$$E + S \overset{K_S}{\rightleftharpoons} ES$$

$$E + S \xrightarrow{k} E + P$$

In this mechanism substrate is converted to product in a bimolecular step with a second-order rate constant k. The enzyme thus acts "at a distance". An enzyme–substrate complex is also formed, with a dissociation constant

K_S, but it is assumed to be inert and plays no part in the catalytic step. The velocity of the overall reaction is given by

$$v = \frac{d(\mathrm{P})}{dt} = k(\mathrm{E})(\mathrm{S}) \tag{4.14}$$

Again we require an expression relating v to the known quantities E_0 and S_0. The conservation equation for substrate is $S_0 = (S)+(ES)+(P)$, and if only initial velocities are considered, and if $S_0 \gg E_0$, (P) and (ES) can be ignored so that

$$\mathrm{S}_0 = (\mathrm{S}) \tag{4.15}$$

as before. The conservation equation for enzyme is

$$\mathrm{E}_0 = (\mathrm{E})+(\mathrm{ES}) \tag{4.16}$$

As the catalytic step does not involve a change in the concentration of E we may assume that E and S are in equilibrium with ES. Thus,

$$\frac{(\mathrm{E})(\mathrm{S})}{(\mathrm{ES})} = K_\mathrm{S}$$

and

$$(\mathrm{ES}) = \frac{(\mathrm{E})(\mathrm{S})}{K_\mathrm{S}} \tag{4.17}$$

Substituting equation (4.17) into (4.16) we obtain

$$\mathrm{E}_0 = (\mathrm{E}) + \frac{(\mathrm{E})(\mathrm{S})}{K_\mathrm{S}} \tag{4.18}$$

and solving equation (4.18) for (E) gives

$$\mathrm{E} = \frac{\mathrm{E}_0}{1+(\mathrm{S})/K_\mathrm{S}} \tag{4.19}$$

Substitution of equations (4.19) and (4.15) into (4.14) gives

$$v_0 = \frac{k\mathrm{E}_0\mathrm{S}_0}{1+\mathrm{S}_0/K_\mathrm{S}} = \frac{kK_\mathrm{S}\mathrm{E}_0\mathrm{S}_0}{K_\mathrm{S}+\mathrm{S}_0} \tag{4.20}$$

Note that equation (4.20) is of the same form as the Michaelis–Menten equation with $V = kK_S E_0$ and $K_m = K_S$. Thus $k_{cat} = kK_S$ and not k_2 as before, but as k is a second-order rate constant, the dimensions of k_{cat} are the same, i.e. reciprocal time.

So it can be seen that two quite different mechanisms can give rise to the same rate equation. This ambiguity, generally true for any rate equation, is sometimes termed *homomorphism.* In fact, nearly thirty years elapsed between the publication of Michaelis and Menten's paper and the provision, by Britton Chance, of independent evidence of the essential role of the enzyme–substrate complex.

Reversible reactions

The irreversible one-substrate model used to derive the Michaelis–Menten equation is unsatisfactory for two reasons. First, all chemical reactions are, in theory, reversible. Even for reactions whose equilibrium position lies very far to the right it is usually possible to detect formation of S from P provided that high concentrations of P can be obtained and sensitive methods of detection of S, e.g. radioactive tracers, are available. Second, as substrate and product generally have similar structures, it would be unreasonable to expect the substrate but not the product to form a complex with enzyme.

Let us then consider the reversible model

$$\mathrm{E+S} \underset{k_{-1}}{\overset{k_1}{\rightleftharpoons}} \mathrm{X} \underset{k_{-2}}{\overset{k_2}{\rightleftharpoons}} \mathrm{E+P}$$

where (X) represents the sum of the concentrations of all the enzyme–substrate and enzyme–product complexes. If the overall reaction S → P is considered, the velocity of the overall reaction is the rate of formation of P, which is given by

$$v = \frac{d(\mathrm{P})}{dt} = k_2(\mathrm{X}) - k_{-2}(\mathrm{E})(\mathrm{P}) \tag{4.21}$$

The rate of disappearance of S is

$$-\frac{d(\mathrm{S})}{dt} = k_1(\mathrm{E})(\mathrm{S}) - k_{-1}(\mathrm{X}) \tag{4.22}$$

Under the condition that (S) and (P) $\gg \mathrm{E}_0$, $-d(\mathrm{S})/dt = d(\mathrm{P})/dt$ and therefore

$$k_1(\mathrm{E})(\mathrm{S}) - k_{-1}(\mathrm{X}) = k_2(\mathrm{X}) - k_{-2}(\mathrm{E})(\mathrm{P}) \tag{4.23}$$

Invoking the conservation equation for enzyme, $\mathrm{E}_0 = (\mathrm{E}) + (\mathrm{X})$, we have

$$(\mathrm{E}) = \mathrm{E}_0 - (\mathrm{X}) \tag{4.24}$$

and substitution of equation (4.24) in (4.23) gives

$$k_1(\mathrm{E}_0-\mathrm{X})(\mathrm{S})-k_{-1}(\mathrm{X}) = k_2(\mathrm{X})-k_{-2}(\mathrm{E}_0-\mathrm{X})(\mathrm{P}) \tag{4.25}$$

Rearrangement of equation (4.25) gives

$$(\mathrm{X}) = \frac{(k_1(\mathrm{S})+k_{-2}(\mathrm{P}))\mathrm{E}_0}{k_1(\mathrm{S})+k_{-2}(\mathrm{P})+k_{-1}+k_2} \tag{4.26}$$

Similar substitution of equation (4.24) in (4.21) leads to

$$v = k_2(\mathrm{X})-k_{-2}(\mathrm{E}_0-(\mathrm{X}))(\mathrm{P}) \tag{4.27}$$

Now substituting equation (4.26) for (X) in (4.27) and clearing the fraction gives the rate equation

$$v = \frac{k_1k_2(\mathrm{S})\mathrm{E}_0-k_{-1}k_{-2}(\mathrm{P})\mathrm{E}_0}{k_1(\mathrm{S})+k_{-2}(\mathrm{P})+k_{-1}+k_2} \tag{4.28}$$

Note that if initial rates are used, $(\mathrm{S}) = \mathrm{S}_0$ and $(\mathrm{P}) = 0$ so that the second term in the numerator and denominator disappears giving

$$v_0 = \frac{k_1k_2\mathrm{S}_0\mathrm{E}_0}{k_1\mathrm{S}_0+k_{-1}+k_2} \tag{4.29}$$

Dividing top and bottom of equation (4.29) by $k_1\mathrm{S}_0$ gives

$$v_0^{\mathrm{S}} = \frac{k_2\mathrm{E}_0}{1+(k_{-1}+k_2)/k_1\mathrm{S}_0}$$

which is equation (4.9), the rate equation for the irreversible reaction $\mathrm{S} \rightarrow \mathrm{P}$. Defining $k_2\mathrm{E}_0 = V^{\mathrm{S}}$ and $(k_{-1}+k_2)/k_1 = K_{\mathrm{m}}^{\mathrm{S}}$ (the superscript referring to the fact that S is the substrate) we obtain $v_0^{\mathrm{S}} = V^{\mathrm{S}}/(1+K_{\mathrm{m}}^{\mathrm{S}}/\mathrm{S}_0)$ which is the Michaelis–Menten equation for the reaction $\mathrm{S} \rightarrow \mathrm{P}$.

A similar argument can be applied to the reverse reaction $\mathrm{P} \rightarrow \mathrm{S}$. If only the initial rate of formation of S from P is considered, $(\mathrm{S}) = 0$ and $(\mathrm{P}) = \mathrm{P}_0$, the first terms in the numerator and denominator of equation (4.28) disappear, and

$$v_0^{\mathrm{P}} = \frac{k_{-1}k_{-2}\mathrm{P}_0\mathrm{E}_0}{k_{-2}\mathrm{P}_0+k_{-1}+k_2} \tag{4.30}$$

(The minus sign in the numerator has been dropped as we are now reversing the direction of the overall reaction.) Dividing numerator and denominator of equation (4.30) by $k_{-2}\mathrm{P}_0$, we obtain

$$v_0^{\mathrm{P}} = \frac{k_{-1}\mathrm{E}_0}{1+(k_{-1}+k_2)/k_{-2}\mathrm{P}_0} \tag{4.31}$$

Note that equation (4.31) is of the same form as equation (4.9); indeed it is the Michaelis–Menten equation for the reverse reaction where the Michaelis constant for P, $K_m^P = (k_{-1}+k_2)/k_{-2}$, and the maximum velocity for $P \rightarrow S$, $V^P = k_{-1}E_0$. Dividing numerator and denominator of equation (4.28) by $(k_{-1}+k_2)$, and substituting K_m^S, V^S, K_m^P and V^P gives

$$v = \frac{V^S(S)/K_m^S - V^P(P)/K_m^P}{1+(S)/K_m^S+(P)/K_m^P} \tag{4.32}$$

which is the rate equation for the reversible reaction in the direction $S \rightarrow P$.

Equation (4.32) has two terms containing (P). Increase in the magnitude of either term will reduce the rate. The negative numerator term also contains the maximum velocity for the reverse reaction and thus reflects the reduction in rate due to the increasing importance of the reverse reaction relative to the forward reaction as (P) increases. The significance of the $(P)/K_m^P$ term in the denominator is not so obvious. To understand the significance of this term a more realistic mechanism must be considered.

Such a mechanism would be one which explicitly assumes that the complexes formed by the enzyme with the substrate and the product are different; thus

$$E+S \underset{k_{-1}}{\overset{k_1}{\rightleftharpoons}} ES \underset{k_{-2}}{\overset{k_2}{\rightleftharpoons}} EP \underset{k_{-3}}{\overset{k_3}{\rightleftharpoons}} E+P$$

This mechanism, which implies that conversion of substrate to product occurs in the enzyme–substrate complex, is in accord with the accepted theories of enzyme catalysis.

The overall rate

$$v = \frac{d(P)}{dt} = k_3(EP) - k_{-3}(E)(P) \tag{4.33}$$

and

$$-\frac{d(S)}{dt} = k_1(E)(S) - k_{-1}(ES)$$

Under steady-state conditions, i.e. $S_0 \gg E_0$, $-d(S)/dt = d(P)/dt$ and

$$k_1(E)(S) - k_{-1}(ES) = k_3(EP) - k_{-3}(E)(P) \tag{4.34}$$

As two enzyme complexes are involved, the steady state must be assumed explicitly for (either) one of them by setting the rate equation for the complex equal to zero.

For (ES),

$$\frac{d(\mathrm{ES})}{dt} = k_1(\mathrm{E})(\mathrm{S}) - k_{-1}(\mathrm{ES}) + k_{-2}(\mathrm{EP}) - k_2(\mathrm{ES}) = 0 \qquad (4.35)$$

Solving equation (4.35) for (ES) gives

$$(\mathrm{ES}) = \frac{k_1(\mathrm{E})(\mathrm{S}) + k_{-2}(\mathrm{EP})}{k_{-1} + k_2} \qquad (4.36)$$

The conservation equation for enzyme is

$$\mathrm{E}_0 = (\mathrm{E}) + (\mathrm{ES}) + (\mathrm{EP}) \qquad (4.37)$$

Equation (4.36) can now be substituted for (ES) in (4.34) and (4.37) respectively giving

$$k_1(\mathrm{E})(\mathrm{S}) - \frac{k_{-1}(k_1(\mathrm{E})(\mathrm{S}) + k_{-2}(\mathrm{EP}))}{k_{-1} + k_2} = k_3(\mathrm{EP}) - k_{-3}(\mathrm{E})(\mathrm{P}) \qquad (4.38)$$

and

$$\mathrm{E}_0 = (\mathrm{E}) + \frac{k_1(\mathrm{E})(\mathrm{S}) + k_{-2}(\mathrm{EP})}{k_{-1} + k_2} + (\mathrm{EP}) \qquad (4.39)$$

Equation (4.39) can be solved for (E) and the resulting equation substituted into equation (4.38) giving an expression for (EP) in terms of rate constants, (S) and (P). Similar substitution into equation (4.33) gives an expression for v in terms of (EP), (S), (P) and constants. Combination of the expressions for (EP) and v involves some tedious algebra and will not be dealt with here as the principles are identical to those already described. A convenient method for obtaining rate equations for complex mechanisms is described in chapter 10.

The resulting rate equation appears to be highly complicated.

$$v = \frac{k_1 k_2 k_3(\mathrm{S})\mathrm{E}_0 - k_{-1}k_{-2}k_{-3}(\mathrm{P})\mathrm{E}_0}{k_1(\mathrm{S})(k_2 + k_{-2} + k_3) + k_{-3}(\mathrm{P})(k_{-1} + k_2 + k_{-2}) + k_{-1}k_2 + k_{-1}k_3 + k_2 k_3} \qquad (4.40)$$

However, examination of equation (4.40) shows that it is of the same form as equation (4.28) for the simple reversible case. The numerator of both equations contains two terms, a positive term involving $(\mathrm{S})\mathrm{E}_0$ and a negative one involving $(\mathrm{P})\mathrm{E}_0$. In the denominator the first two terms contain (S) and (P) respectively and the three constant terms of equation (4.40) correspond to the two constant terms of (4.28).

Dividing top and bottom of equation (4.40) by the constant terms in the denominator and defining

$$K_m^S = (k_{-1}k_{-2}+k_{-1}k_3+k_2k_3)/k_1(k_2+k_{-3}+k_3),$$
$$V^S = k_2k_3E_0/(k_2+k_{-2}+k_3),$$
$$K_m^P = (k_{-1}k_{-2}+k_{-1}k_3+k_2k_3)/k_{-3}(k_{-1}+k_2+k_{-2}),$$
$$V^P = k_{-1}k_{-2}E_0/(k_{-1}+k_2+k_{-2})$$

gives equation (4.32)

$$v = \frac{V^S(S)/K_m^S - V^P(P)/K_m^P}{1+(S)/K_m^S+(P)/K_m^P}$$

So both reversible mechanisms give parametrically identical rate equations which are indistinguishable by steady-state kinetics.

However, the second mechanism provides some useful insight into reversible enzyme-catalysed reactions. In any reaction mechanism the occurrence of one irreversible step means that the overall reaction must be irreversible. If we assume the final step $EP \rightarrow E+P$ of the above mechanism to be irreversible, i.e. $k_{-3} = 0$, K_m^P becomes infinite and the (P) terms in both the numerator and denominator of equation (4.32) disappear giving

$$v = \frac{V^S(S)/K_m^S}{1+(S)/K_m^S} = \frac{V^S(S)}{K_m^S+(S)}$$

the Michaelis–Menten equation for $S \rightarrow P$. K_m^S and V^S retain their significance as defined above.

But if the overall reaction is made irreversible by making the conversion of ES to EP irreversible, i.e. $k_{-2} = 0$, $V^P = 0$ and the (P) term in the numerator disappears. However, the K_m^P term does *not* become zero when $k_{-2} = 0$ (although its meaning is altered), and the $(P)/K_m^P$ term remains in the denominator and the rate equation becomes

$$v = \frac{V^{S'}(S)/K_m^{S'}}{1+(S)/K_m^{S'}+(P)/K_m^{P'}} \tag{4.41}$$

(Primed symbols are used for the kinetic parameters to indicate that their meaning has been altered by the deletion of the k_{-2} terms.) Equation (4.41) is important because it shows how the presence of product can lower the rate of an enzyme-catalysed reaction even if the overall conversion of S to P is effectively irreversible. The reason for this can be seen by inspecting the mechanism. Making the second step irreversible does not prevent the product from combining with the enzyme. Thus the product competes with substrate for binding sites and reduces the concentration of E available for reaction with substrate. This phenomenon is known as *product inhibition*, and is a form of competitive inhibition, which is discussed in chapter 6.

Equilibrium and the Haldane relationship

At equilibrium the rates of the reactions S → P and P → S are equal and the net rate, v, is zero. If $v = 0$, the numerator of equation (4.32) must be zero so

$$V^{S}(S)_{eq}/K_{m}^{S} - V^{P}(P)_{eq}/K_{m}^{P} = 0$$

where the subscript "eq" refers to the equilibrium condition. Rearranging gives

$$\frac{(P)_{eq}}{(S)_{eq}} = \frac{V^{S}K_{m}^{P}}{V^{P}K_{m}^{S}} = \frac{k_1 k_2}{k_{-1}k_{-2}} = K_{eq}$$

The relationship for the two-intermediate reversible case is also $K_{eq} = V^{S}K_{m}^{P}/V^{P}K_{m}^{S}$ but in terms of rate constants this is $k_1k_2k_3/k_{-1}k_{-2}k_{-3}$. Relationships of the steady-state parameters to the equilibrium constant, K_{eq}, are known as *Haldane relationships* and can be derived for any reversible enzyme mechanism which follows Michaelis–Menten kinetics.

The apparent equilibrium constant of a chemical reaction is normally determined by measuring the concentrations of reactants and products when the net rate is zero. The Haldane relationships provide an independent check on this. V^{S} and K_{m}^{S} can be readily determined from initial rate measurements of S → P in the absence of P, and V^{P} and K_{m}^{P} from similar measurements of the reverse reaction in the absence of S. The Haldane ratio of these parameters should agree with the equilibrium ratio $(P)_{eq}/(S)_{eq}$.

Substrate inhibition

The high specificity of enzymes towards their substrates is thought to result, at least in part, from the ability of the protein to provide a multidentate binding site for the ligand. But this feature of enzymes can also result in inhibition of catalysis. This is illustrated schematically in figure 4.8a which shows a hypothetical molecule X–Y and its binding site containing complementary subsites $\underline{x}$ and $\underline{y}$. Subsite $\underline{z}$ is a group which is assumed to play an essential role in the disruption of the X–Y bond. In

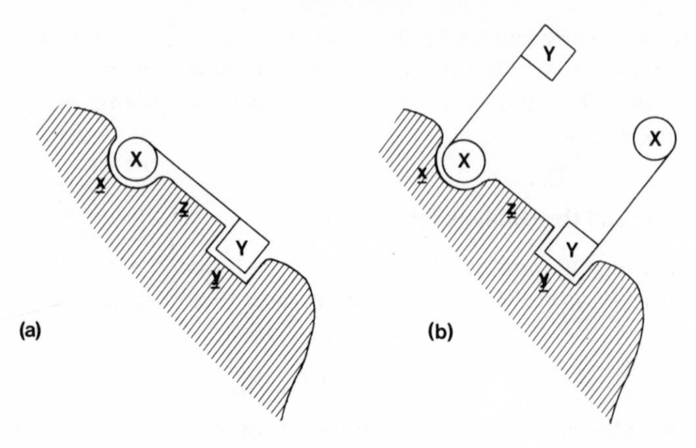

Figure 4.8 Schematic illustration of a molecular mechanism for substrate inhibition. For details see text.

figure 4.8b is shown the situation which might arise if X–Y is present at high concentration. Here the attachment of two substrate molecules prevents the occurrence of catalysis.

This situation can be expressed by the following mechanism:

$$\mathrm{E+S} \underset{k_{-1}}{\overset{k_1}{\rightleftharpoons}} \mathrm{ES} \xrightarrow{k_2} \mathrm{E+P}$$

$$\mathrm{ES+S} \underset{k_{-3}}{\overset{k_3}{\rightleftharpoons}} \mathrm{ES_2}$$

ES_2 can be taken to represent the complex illustrated in figure 4.8b and is a *dead-end complex*, i.e. it does not react further. The rate is

$$v = \frac{d(\mathrm{P})}{dt} = k_2(\mathrm{ES}) \tag{4.42}$$

As E is regenerated in the overall reaction $-d(\mathrm{E})/dt = d(\mathrm{E})/dt$ and $k_1(\mathrm{E})(\mathrm{S}) = (k_{-1}+k_2)(\mathrm{ES})$. Rearranging gives

$$\frac{(\mathrm{E})(\mathrm{S})}{(\mathrm{ES})} = \frac{k_{-1}+k_2}{k_1} = K_\mathrm{m}$$

and

$$(\mathrm{E}) = K_\mathrm{m}(\mathrm{ES})/(\mathrm{S}) \tag{4.43}$$

Defining the dissociation constant for ES_2 as $K' = k_{-3}/k_3$, we have

$$(ES_2) = k_3(S)(ES)/k_{-3} = (S)(ES)/K' \tag{4.44}$$

The conservation equation for enzyme is

$$E_0 = (E)+(ES)+(ES_2) \tag{4.45}$$

Substituting equations (4.43) and (4.44) in (4.45) gives

$$E_0 = (K_m/(S)+1+(S)/K')(ES) \tag{4.46}$$

Solving equation (4.46) for (ES) and substituting into equation (4.42) under initial rate conditions gives the rate equation

$$v_0 = \frac{k_2E_0}{1+K_m/S_0+S_0/K'} = \frac{V}{1+K_m/S_0+S_0/K'} \tag{4.47}$$

When S_0 is sufficiently small so that S_0/K' is insignificant compared to $1+K_m/S_0$, equation (4.47) becomes identical to the Michaelis–Menten equation. As S_0 increases, the value of S_0/K' will increase until it becomes the significant term in the denominator, and the value of v_0 will decrease. As substrate is present at all times the contribution of S_0/K' to the rate will depend on the relative values of K_m/S_0 and S_0/K' which will depend in

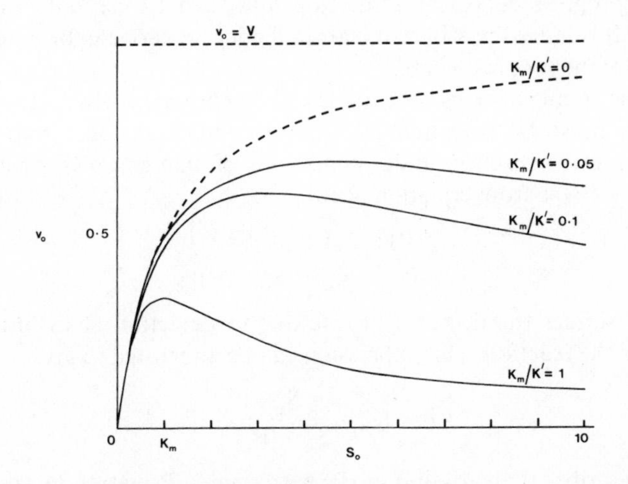

Figure 4.9 Effect of substrate inhibition on the v_0, S_0 profile. Solid curves calculated from $v_0 = V/(1+K_m/S_0+S_0/K')$ for different values of K' assuming $V = K_m = 1$ (arbitrary units). Dashed curve represents the uninhibited reaction and is calculated from the Michaelis–Menten equation assuming $V = K_m = 1$.

turn on the ratio K_m/K'. The smaller this ratio the less will be the inhibiting effect of the substrate at high concentration.

Figure 4.9 shows the effect of different values of K_m/K' on the shape of the v_0 versus S_0 profile. The most obvious indication of substrate inhibition is a decrease in v_0 as S_0 is raised. However, limitations of solubility, availability or interference with the assay may not allow sufficiently high concentrations of S_0 to be used to reach the point where v_0 begins to decrease. Substrate inhibition may thus be present at low values of S_0 and its effect on this region of the v_0, S_0 curve will not necessarily be obvious.

If it is not realized that substrate inhibition is present, the effect will be to cause V, and hence K_m, to be underestimated. The highest value of v_0 attainable, designated V_{opt}, will be reached when $S_0 = (K_m K')^{1/2}$. For values of K_m/K' of 1, 0.1 and 0.05, V_{opt} will be 33%, 61% and 69% respectively of the true V. Substrate inhibition may be more prevalent than is commonly realized and the possibility of its presence should always be considered.

Analysis of the progress curve

The analysis of steady-state kinetics described thus far has been restricted to the use of *initial velocities*. It may appear wasteful to determine the time course (progress curve) of a reaction and then to use only the initial portion. It is indeed possible, at least in theory, to derive information from analysis of the progress curve.

For the simple irreversible case described above the Michaelis–Menten equation must be rewritten in order to reflect the fact that we are considering the reaction under conditions of changing (S). At any time $(S) = S_0 - (P)$ so from equation (4.10)

$$\frac{d(P)}{dt} = \frac{V(S_0 - (P))}{K_m + (S_0 - (P))}$$

which describes the decrease in rate due to depletion of (S) during the course of the reaction. This expression can be integrated to give

$$Vt = K_m \ln \frac{S_0}{S_0 - (P)} + (P) \tag{4.48}$$

which describes the variation of (P) with time, t. Equation (4.48) is not a simple first-order exponential rate law and this explains why the progress curves of enzyme-catalysed reactions differ in shape from first-order reactions as described at the beginning of this chapter.

A number of graphical procedures have been devised for the analysis of progress curves following equation (4.48). For example, rearranging (4.48) gives

$$(\mathrm{P})/t = V - K_m \ln \frac{\mathrm{S}_0}{(\mathrm{S}_0 - (\mathrm{P}))} \bigg/ t$$

and a plot of $(\mathrm{P})/t$ against $\ln[\mathrm{S}_0/(\mathrm{S}_0 - (\mathrm{P})]/t$ will have slope $-K_m$ and intercept V.

If $\mathrm{S}_0 < 0.1K_m$, the reaction will approximate to first-order kinetics because $d(\mathrm{P})/dt \simeq V(\mathrm{S})/K_m$. Integration gives $\ln[\mathrm{S}_0 - (\mathrm{P})/\mathrm{S}_0] = -(V/K_m)t$ so that a plot of $\ln[\mathrm{S}_0 - (\mathrm{P})/\mathrm{S}_0]$ against t will be a straight line of slope $-V/K_m$.

It must be pointed out that this treatment applies only to the simple one-substrate irreversible case. If the reaction is reversible, or if product or substrate inhibition is present, or if the enzyme becomes progressively inactivated during the experiment, equation (4.48) is inapplicable. It is possible to derive equations which allow for these more realistic situations but these are much more complex and the experimental results are more difficult to analyse. For these reasons, most steady-state kinetic studies are restricted to consideration of initial rates only.

Summary

1. For most enzymes the variation of v_0 with S_0 follows a rectangular hyperbola through the origin which is described by the *Michaelis–Menten equation*. The *Michaelis constant* is the substrate concentration giving one-half the maximum velocity. The initial rate of an enzyme-catalysed reaction, v_0, is proportional to the substrate concentration, S_0, when S_0 is small. As S_0 is increased, v_0 approaches a limiting value known as the *maximum velocity*, V. The maximum velocity and v_0 are directly proportional to the enzyme concentration.

2. The *Michaelis constant* is characteristic of a given enzyme and substrate under defined conditions. Dividing the maximum velocity by the concentration of catalytic sites gives the *catalytic constant* which is also characteristic of an enzyme under defined conditions and is a measure of catalytic efficiency.

3. The mechanism of enzyme-catalysed reactions involves the formation of one or more *enzyme–substrate complexes*. If the concentration of substrate greatly exceeds the concentration of enzyme, the concentration

of these complexes will remain constant, i.e. in a *steady state* for a significant portion of the time course of the overall reaction.

4. The presence of product can reduce the rate of an enzyme-catalysed reaction either by reversal of the overall reaction or through *product inhibition* resulting from the competition of product with substrate for complex formation with free enzyme molecules and the formation of enzyme–product complexes.

5. The equilibrium constant of a reversible enzyme-catalysed reaction can be expressed as a function of the maximum velocities and Michaelis constants of the forward and reverse reaction through the use of *Haldane relationships.*

6. *Substrate inhibition* can arise by the formation of abortive complexes of substrate with enzyme.

7. The Michaelis–Menten equation can be integrated to give an expression which allows the kinetic parameters to be evaluated from progress curves. This approach is difficult to apply to realistic cases involving product inhibition, reversibility and enzyme inactivation so that steady-state kinetic studies are generally restricted to consideration of *initial rates* only.

CHAPTER FIVE

COENZYMES

COENZYMES MAY BE DESCRIBED AS THE COMMON SECOND SUBSTRATE OR product of a multiplicity of enzymes. Those biochemists who are primarily interested in the metabolic aspects of the subject tend to think of coenzymes in terms of their ability to act as buffer stores of chemical potential. Thus such factors as the ratio of the concentration of the different forms of the coenzymes, their localization and their transport assume central importance when considering metabolic status. Most enzymologists think of coenzymes simply as substrates for the particular enzymes in which they are interested. In this chapter we shall deal mainly with coenzymes of wide versatility.

1. Adenosine triphosphate

Adenosine triphosphate (ATP) takes pride of place amongst the coenzymes primarily because of the large number of different ways in which its chemical potential is utilized in enzyme reactions. As we shall see, almost every conceivable mode of reaction is exploited. In biochemical circles there has been sharp argument for many years regarding the proper interpretation of the so-called "high energy" nature of ATP. We shall consider the question in terms of "available free energy" since this is most convenient from the chemical standpoint. In fact ATP is a compound of moderate free energy, near the middle of the range of available free energies found in biochemical compounds. It is useful to remember that one molecule of NADH $(+H^+)$ gives rise to three molecules of ATP during oxidative phosphorylation.

The remarkable versatility of ATP is best illustrated by a list of the reaction types in which it may participate (table 5.1). Bear in mind the structure (5.1) of $Mg:ATP^{2-}$ (the substrate-active species) during consideration of this list.

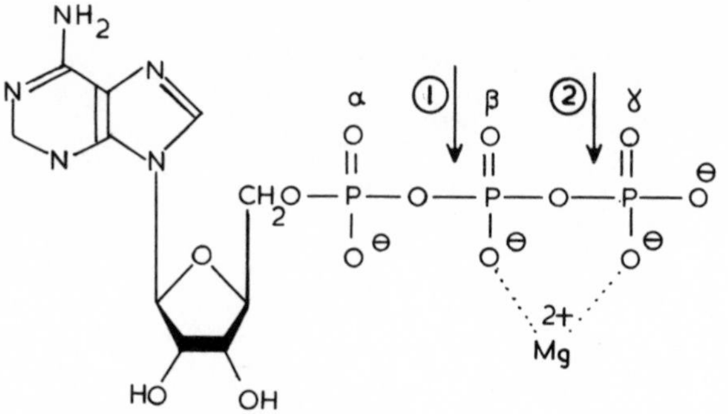

(5.1)

Note that in all cases where ΔG is positive the reaction is unfavourable in that direction, and that the product (other than ADP or AMP) has a higher available free energy than does ATP. From a consideration of the ΔG values in table 5.1 it is simple to work out a "league table" of "available free energy" at least as far as transfer to ADP and hydrolysis are concerned. ATP is almost universally cleaved by nucleophilic attack upon the phosphorus atoms adjacent to the arrows labelled (1) and (2) in the diagram of the structure of $Mg:ATP^{2-}$ (5.1), bond breaking occurring at the sites of these arrows. Reactions of class (1) involve splitting ATP to AMP and PP_i—2, 4, 5 and 6 in the table—4 and 5 are activation reactions in biosynthetic pathways, and 6 is an activation reaction for both biosynthetic and catabolic pathways. Reactions of class (2) (such as 1, 3, 6, 8, 9, 10 and 12) involve attack at the γ-phosphate. The remarkable versatility of ATP as a coenzyme must be expressed in terms of binding selectivity by the enzymes involved in the many different reactions involving ATP. The selectivity must arise from correct orientation of the ATP molecule with respect to the substrate and placement of the catalytic groupings in appropriate juxtaposition with the bound ATP and substrate. This is borne out by the high degree of selectivity often found for ATP relative to GTP, UTP, etc. For example, hexokinase, although able to catalyse the phosphorylation of a range of hexose sugars with similar ease will accept ATP and none else as phosphate group donor. Yeast phosphofructokinase on the other hand utilizes all nucleotides with equal facility.

The catalytic requirements of an enzyme which catalyses a reaction involving ATP are in principle rather simple, in contrast with the highly

Table 5.1 The versatility of ATP as a coenzyme

	Examples
1. $ATP + H_2O \rightleftharpoons ADP + P_i$ $\Delta G'_0 \simeq -30\,kJ/mole$*	ATPase, e.g. muscle contraction
2. $ATP + H_2O \rightleftharpoons AMP + PP_i$ $\Delta G'_0 \simeq -30.5\,kJ/mole$	This is not a natural enzymic process but is included to indicate partitioning of free energy
3. $R{-}OH + ATP \rightleftharpoons ROP + ADP$ $\Delta G'_0 \simeq -17\,kJ/mole$	Glucokinase $Glucose + ATP \rightleftharpoons Glucose\text{-}6\text{-}P + ADP$
4. $R{-}OP + ATP \rightleftharpoons ROPP + ADP$ $\Delta G'_0 \simeq 0$	Phosphomevalonate kinase
5. $RCOOH + ATP \rightleftharpoons RCO \cdot AMP + PP_i$ $\Delta G'_0 \simeq +29\,kJ/mole$	Aminoacyl t-RNA synthetase; intermediate in thiokinase reaction
6. $RCOOH + ATP \rightleftharpoons RCO \cdot P + ADP$ $\Delta G'_0 \simeq +21\,kJ/mole$	Phosphoglycerate kinase; acetyl CoA synthesis
7. $ATP + AMP \rightleftharpoons 2ADP$ $\Delta G'_0 \simeq 0\,kJ/mole$	Adenylate kinase
8. $enol + ATP \rightleftharpoons enol\text{-}P + ADP$ $\Delta G'_0 \simeq +31\,kJ/mole$	Phosphoenol pyruvate kinase
9. $R{-}N(R'){-}C({=}NH_2^+){-}NH_2 + ATP \rightleftharpoons R{-}N(R'){-}C({=}NH_2^+){-}NH{-}P + ADP$ $\Delta G'_0 \simeq +12\,kJ/mole$	Creatine kinase
10. $NH_3 + CO_2 + ATP \rightleftharpoons NH_2CO{-}P + ADP$ $\Delta G'_0 \simeq +8\,kJ/mole$ pH 9.5, 10°	Carbamyl kinase
11. $Methionine + ATP \rightleftharpoons S\text{-}adenosyl\text{-}methionine + PPP_i$ $\Delta G'_0 \simeq -2.5\,kJ/mole$	
12. $ATP \rightleftharpoons 3',5'\text{-}cyclic\ AMP + PP_i$ $\Delta G'_0 \simeq +7\,kJ/mole$	Adenyl cyclase
13. $PP_i \rightleftharpoons 2P_i + H_2O$ $\Delta G'_0 \simeq -35\,kJ/mole$	Pyrophosphatase (included for purposes of comparison)
14. $R{-}OP + N\dagger TP \rightleftharpoons ROPPN + PP_i$ $\Delta G'_0 \simeq +3\,kJ/mole$	Uridine diphosphoglucose pyrophosphorylase

* Note that at intracellular concentration $\Delta G'_0$ will be more negative $\simeq -40\,kJ/mole$.
† N = U, A or C.

specific binding requirements. The most important, and the only absolute, requirement is for a general base group or system of groups that is able to remove a proton from the attacking nucleophilic group. In addition a general acid group or groups may provide assistance by protonation of the leaving group(s). A very elegant mechanism which employs these concepts was proposed for creatine kinase as long ago as 1962. A slightly modified version of this mechanism is shown below:

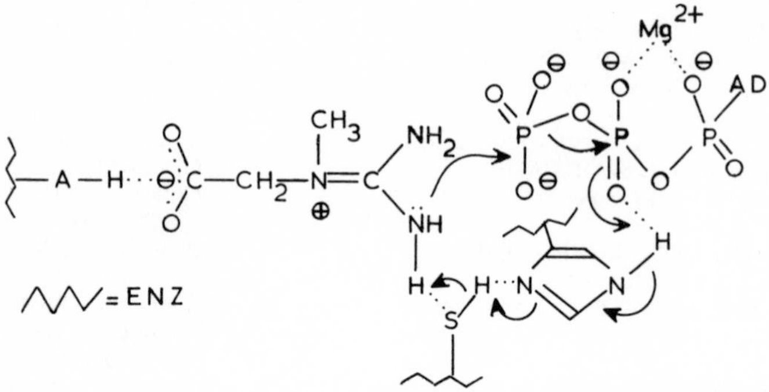

(5.2)

The cyclic nature of the reaction is interesting; general acid assistance is provided by the same (His...Cys) system that provides general base assistance for the nucleophilic attack of the creatine nitrogen on the γ-phosphate of ATP. The reactive form of the His...Cys pair must be reviewed in the light of evidence that suggests that papain, which also has a His...Cys pair at the active site (although sulphur reacts as a nucleophile in papain-catalysed reactions) may catalyse its reactions by employing this assembly in the form Cys^{-}...His^{+} (see chapter 9).

We feel little need to "explain" the "high energy" nature of ATP since, as we have seen, it has a medium level of available free energy; this level being an effective compromise between relative ease of synthesis and efficient utilization. Factors that contribute to the medium level of available free energy are: (1) resonance stabilization of products ($ADP + P_i$) relative to ATP; (2) charge repulsion between the four negative charges on ATP^{4-} (presumably considerably reduced when in the form of $Mg^{2+}:ATP^{4-}$); and (3) the moderately strong electrophilic character of the phosphorus atoms, in particular the β atom. Note that phosphorus is less electropositive with respect to oxygen than is carbon. Thus a comparison of ATP with acetic anhydride would suggest that the latter should have the larger free energy of hydrolysis, which indeed it does, to the extent of $\simeq 420$ kJ/mole.

An important and flourishing area of current research concerns the stereochemical investigation of phosphoryl transfer reactions. Such studies involve introduction of chirality into one or more of the phosphate groups of ATP or into the phosphate group(s) of sugar phosphates, etc. These stereochemical techniques allow some of the detailed mechanisms possible for phosphoryl transfer reactions to be distinguished, i.e. S_N1, S_N2, etc.

2. Nicotinamide adenine dinucleotide (NAD)

NAD (5.3) and the derivative phosphorylated on the 2 position of the ribose ring of the adenyl moiety (NADP) represent a buffer storage system for reducing power. The structure is given below:

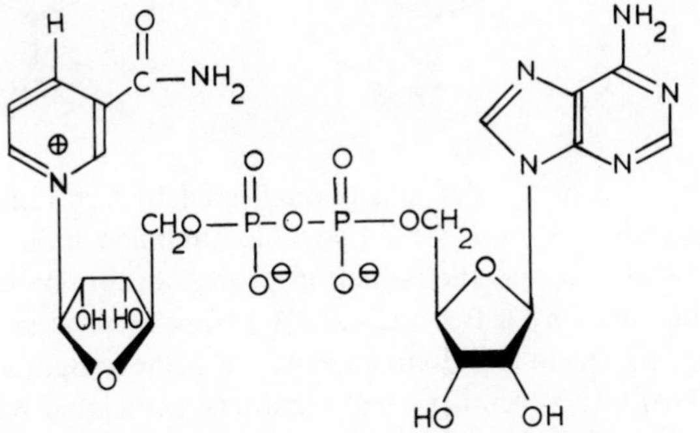

(5.3)

The enzyme-catalysed reactions in which NAD^+ participates have the general form:

$$AH_2 + NAD^+ \underset{}{\overset{\text{dehydrogenase } (\Delta)}{\rightleftharpoons}} A + NADH^+ + H^+$$

or more specifically, for example

$$HO-\overset{|}{\underset{|}{C}}-H + NAD^+ \overset{\Delta}{\rightleftharpoons} O{=}C\langle \; + NADH + H^+$$

The reduction of NAD and NADP involves hydride addition to position 4 of the nicotinamide ring with loss of the positive charge on the pyridinium nitrogen. As we have already mentioned, oxidation by electron transport of one molecule of NADH allows the phosphorylation of three ADP molecules, this being equivalent to an exchange of free energy of approximately $91\,kJ\,mol^{-1}$. The total free energy available from NADH calculated from the redox potential is $\simeq 167\,kJ\,mol^{-1}$ and NADH can rightly be regarded as a compound of moderately high energy (cf. acetic anhydride, above).

Model studies, using analogues of the nicotinamide moiety of NAD^+, show that as far as the chemical reactivity is concerned, this is the important part of the molecule. A good example of a model reaction is given by the reaction of thiobenzophenone and 1-benzyl-4-hydronicotinamide.

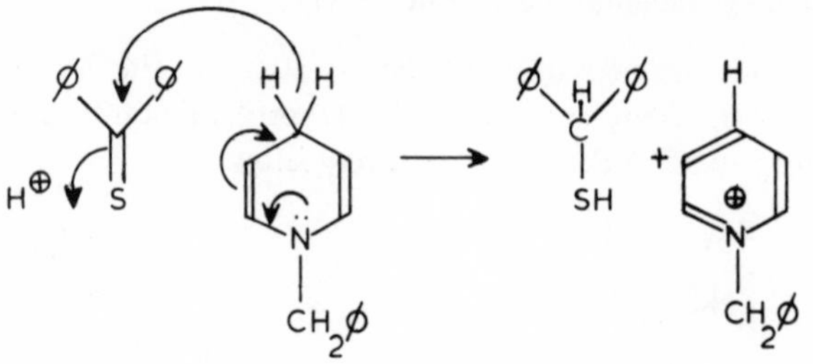

(5.4)

This reaction is second order and is unaffected by free radical trapping agents. This indicates that the 4-hydronicotinamide moiety is a good reducing reagent and that the reduction process occurs by hydride (H^-) transfer rather than by a free radical ($H^{\cdot}$) transfer process. The adenyl moiety of the molecule is present in order to achieve tight stereospecific binding of NAD, the chemistry being entirely associated with the nicotinamide moiety. The 4-position of the nicotinamide ring is pro-chiral, the nicotinamide ring having two stereochemically distinct "faces". The elegant experiments of Cornforth, Vennesland, Westheimer and others have shown that a given dehydrogenase adds hydride ion in a stereospecific fashion to the 4-position of the nicotinamide ring. The reactions employed to demonstrate this were as follows:

$$CH_3CD_2OH + NAD^+ \underset{\substack{\text{dehydrogenase} \\ \text{(ADH)}}}{\overset{\text{alcohol}}{\rightleftharpoons}} CH_3CDO + NADD + H^+$$

The NADD was isolated and incubated with unlabelled acetaldehyde and enzyme:

$$CH_3CHO + NADD \overset{\text{ADH}}{\rightleftharpoons} CH_3CHDOH + NAD^+$$

All the deuterium was transferred from the NADD to the acetaldehyde, demonstrating unequivocally that in the first reaction the deuterium had been added stereospecifically to NAD^+ and that it had been removed stereospecifically and completely in the second reaction. Stereospecific transfer to the aldehyde would result in optically active deuterioethanol and this was indeed found to be the case, the deuterioethanol having a specific rotation of $-0.28° \pm 0.03°$. The stereoisomers of monodeuterated NAD may be represented diagrammatically by:

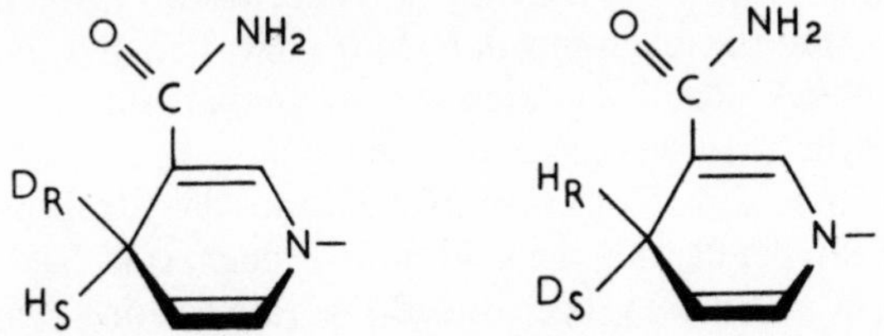

(5.5)

Nearly all dehydrogenases have been shown, by means of experiments of the type outlined above, to be stereospecific with respect to one or other of the A or B stereoisomers; lipoyl dehydrogenase is an exception to this general rule.

The catalytic apparatus that must be provided by the enzyme to enable hydride transfer to occur in a properly aligned system is in principle quite simple. In the case of substrate oxidation, a basic species is required to abstract a proton from the substrate (the most acidic proton on the reaction coordinate), and the transfer of a hydride ion from the carbon of the substrate to NAD^+ then results from the induced electron flow in the system. It is unlikely that the reaction involves much charge accumulation on the incipient hydride ion. Thus the chemical mechanism of substrate oxidation and reduction of a typical dehydrogenase can be represented by the scheme (5.6).

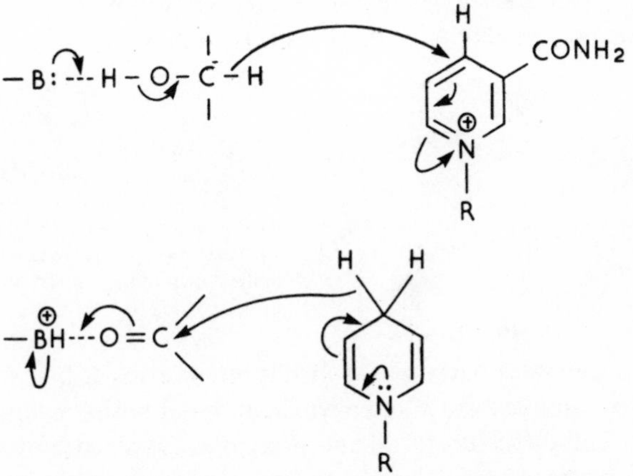

(5.6)

This scheme is somewhat similar to the chymotrypsin mechanism (chapter 9) where a base (Asp 194: His 57) removes a hydroxyl proton from serine 195, the serine oxygen then attacking the substrate carbonyl centre. The "equivalent" hydroxyl group is a component of the dehydrogenase substrate rather than part of the enzyme as in chymotrypsin. In both cases it seems likely that the reaction will be fully concerted. In other words it is unlikely that much negative charge will accumulate on the serine 195 or substrate (dehydrogenase) oxygen atoms.

The kinetic mechanism of horse liver alcohol dehydrogenase has been studied in great detail and the addition of coenzyme has been shown to occur first in an ordered mechanism. The rate-limiting step is enzyme–NADH dissociation but in the reverse direction the rate of reduction of acetaldehyde is rate-limited by the chemical hydride transfer step. The hydride transfer step is rate-limiting in yeast alcohol dehydrogenase since enzyme–product dissociation is fast. The kinetic isotope effect k_H/k_D is 3–5 when deuterated substrate or NADD is used in place of normal substrate or NADH. The basic species B in scheme (5.6) which is responsible for initiating hydride transfer is an ionizable water molecule bound to a zinc atom in the case of the horse liver enzyme, and a histidine residue in the yeast enzyme (see chapter 9).

The substrate binding/catalytic domains of dehydrogenases are variable but an important common feature, representing the NAD^+/NADH binding region, is present in all dehydrogenases (see chapter 3).

3. Flavin nucleotides

The flavin nucleotides have the structures shown below (5.7), and are closely related to riboflavin, an important B vitamin.

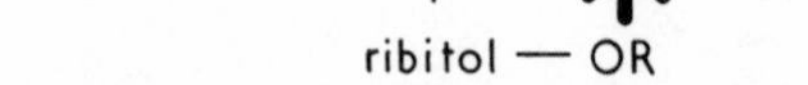

R = H riboflavin
R = Ⓟ flavin mononucleotide (FMN)
R = ⓅⓅAd flavin dinucleotide (FAD)
The nitrogen atoms (N-1, N-5) marked with arrows are those to which hydrogen is added when the flavin is reduced.

(5.7)

The flavin nucleotides participate in what often seems at first sight to be a bewilderingly complex range of enzyme-catalysed redox reactions. In this section we shall concentrate on the principles involved and omit much detail. It is perhaps useful briefly to describe some of the categories of reaction in which the flavoprotein enzymes participate. The categories, three of which are given below, are best described in terms of the *nature* and *reaction mode* of the second substrate, which is the electron acceptor when the first substrate is oxidized.

(a) *The electron acceptor is not oxygen.* NAD^+ or $NADP^+$ is the electron acceptor, for example lipoamide dehydrogenase (5.8):

$$\left[\begin{matrix} -SH \\ \\ -SH \end{matrix}\right. + NAD^+ \rightleftharpoons \left[\begin{matrix} -S \\ | \\ -S \end{matrix}\right. + NADH + H^+ \qquad (5.8)$$

(b) *The respiratory chain is the electron acceptor*, i.e. ubiquinone or cytochrome—the direct electron acceptor is again not oxygen.
(c) *Oxygen is the electron acceptor* and one atom of oxygen is incorporated into the product, for example the amino acid oxidases (5.9):

$$R{-}CH(NH_3^+)(CO_2^-) + O_2 + H_2O \rightleftharpoons R{-}C({=}O)(CO_2^-) + NH_4^+ + H_2O_2 \qquad (5.9)$$

The above classification is incomplete but does give an indication of the types of reaction in which the flavin nucleotides participate. The enzymes which catalyse these reactions are frequently referred to as *flavoproteins* because of the tight binding of the flavin nucleotides to proteins. The dissociation constants range from 10^{-8} M to 10^{-11} M, and so flavin nucleotides are usually thought of as prosthetic groups tightly bound to the enzyme under all normal circumstances. Enzymes of class (a) and (c) do not require metal atom participation whilst class (b) require iron.

The most important difference between NAD^+- and FAD-mediated redox processes is apparent in the mechanisms of the reductive processes. NAD^+ is reduced in a single two-electron process whilst FAD is reduced as a result of two one-electron (free radical) transfers. The reduction process can be depicted as:

$$AH_2 + FAD \rightleftharpoons AH^*FADH^* \rightleftharpoons A + FADH_2$$

The flavin moiety of the diradical intermediate has a structure of the type (5.10):

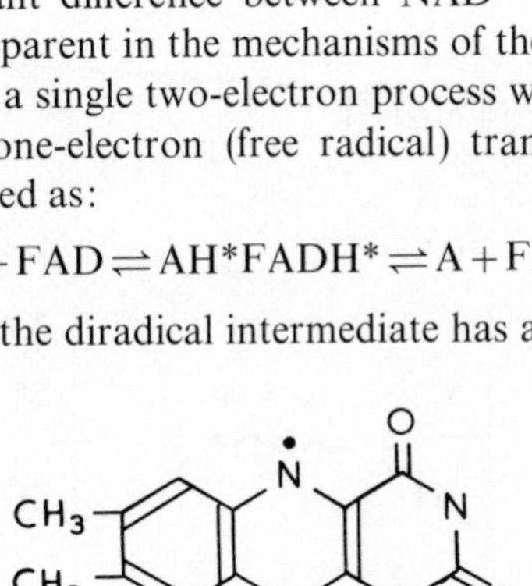

(5.10)

The second hydrogen radical then adds to the radical nitrogen to give $FADH_2$.

The redox potential (E'_0) of the unbound $FAD/FADH_2$ couple is -0.185 V, considerably less negative than the -0.32 V of the NADH/NAD^+ couple. This is to be expected since flavoproteins mediate electron transport between NAD and the cytochromes. FAD when bound to protein may have a variable redox potential which apparently depends upon neighbouring groups in the particular protein to which it is bound. Metal atoms and disulphide bonds adjacent to the FAD are important in this respect. The intermediate, variable value of the redox potential enables flavoproteins to engage in a wide variety of reactions which (as we have seen above) includes redox coupling to NAD at one extreme and to oxygen at the other; they are thus able to span a very large portion of the redox scale. Flavoprotein enzymes are generally membrane- or particle-bound and this adds another dimension to the complexity of a complete description of these enzymes.

The importance of the participation of enzyme groups in the flavoproteins is well exemplified by one of the proposed mechanisms of lipoamide dehydrogenase (5.11):

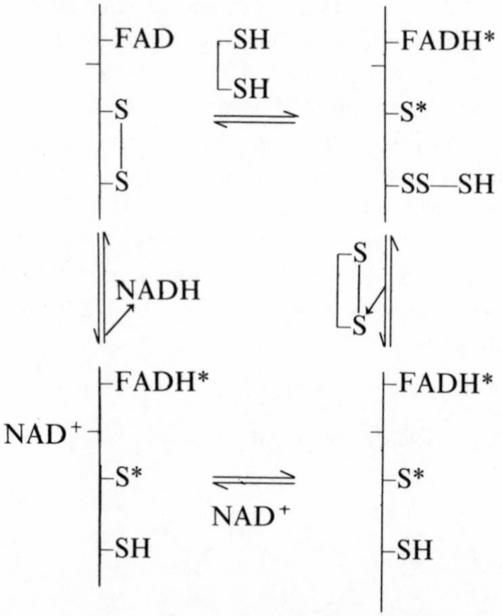

	─FAD ─S │ ─S	$\underset{}{\overset{\text{SH, SH}}{\rightleftharpoons}}$	─FADH* ─S* ─SS—SH
	⇅ NADH		⇅ S—S
NAD^+─	─FADH* ─S* ─SH	$\underset{NAD^+}{\rightleftharpoons}$	─FADH* ─S* ─SH

(5.11)

There are many ways in which a mechanism of this type may be written. In the form given here the electrons are shared between the disulphide bond adjacent to the FAD at the active site, and the FAD itself. The fully reduced form $FADH_2$ does not appear in this particular formulation.

The oxidation of reduced flavin by molecular oxygen occurs via a hydroperoxide intermediate, the hydroperoxide (—OOH) group being attached to the isoalloxazine ring at positions 4a or 9a (5.7). In hydroxylase reactions where water rather than peroxide is the product (one atom of oxygen is incorporated into the substrate), ring opening at N-5 or N-10 probably occurs. In oxidase-catalysed reactions, α-carbanion addition to positions 4a, 9a or 10a occurs, for example the lactate oxidase-catalysed reaction

$$\text{lactate} + O_2 \xrightarrow{\text{FMN}} \text{pyruvate} + H_2O_2.$$

Thus a variety of modes of direct chemical participation of the iso-alloxazine ring are seen in flavoprotein-catalysed reactions.

4. Coenzyme A

The structure (5.12) and biosynthesis of coenzyme A (CoA, CoA–SH) are rather complex.

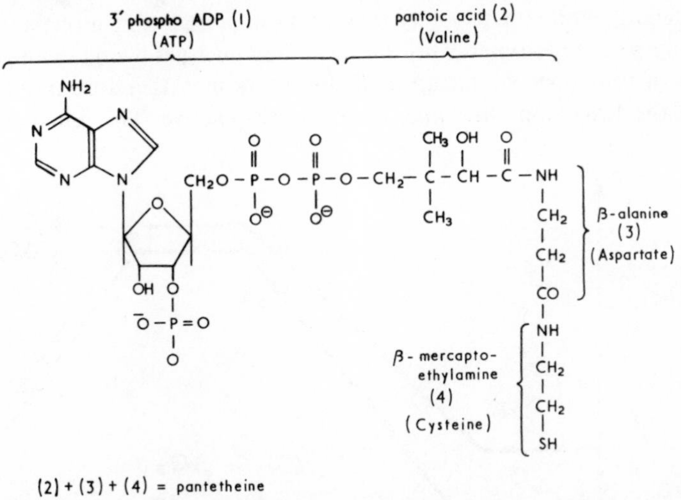

3′ phospho ADP (1) (ATP)
pantoic acid (2) (Valine)
$CH_2O-P(=O)(O^\ominus)-O-P(=O)(O^\ominus)-O-CH_2-C(CH_3)_2-CH(OH)-C(=O)-NH-CH_2-CH_2-CO-NH-CH_2-CH_2-SH$
β-alanine (3) (Aspartate)
β-mercapto-ethylamine (4) (Cysteine)

(2) + (3) + (4) = pantetheine

(5.12)

The sulphydryl group of the β-mercaptoethylamine moiety is the key functional group of the molecule and readily forms thiol ester derivatives of carboxylic acids under the influence of thiokinase enzymes. Coenzyme A serves as a carrier of acyl groups in a number of enzyme-catalysed reactions. It is pertinent to consider why acyl groups are carried in the form of their thiol esters rather than as oxygen esters.

Sulphur forms double bonds to carbon less readily than does oxygen because the carbon and sulphur nuclei are further apart and allow less π-overlap. The high nucleophilicity of the sulphur atom of thiourea as compared with the oxygen atom of urea is a good illustration of this. This means that resonance forms of the type that are important in the stabilization of esters (5.13) do not contribute to resonance stabilization of thiol esters.

$$\text{I} \quad R-C(=O)-\ddot{O}-R' \longleftrightarrow R-C(-O^{\ominus})=O^{\oplus}-R'$$

$$\text{II} \quad R-C(=O)-\ddot{S}-R' \not\longleftrightarrow R-C(-O^{\ominus})=S^{\oplus}-R' \qquad (5.13)$$

The lack of resonance stabilization of thiol esters implies a higher ground state energy than in oxygen esters, although the transition states for hydrolysis for both types of ester have similar energies owing to loss of this mode of resonance stabilization. Thiol esters may therefore more easily reach the transition state and so are more reactive. The schematic free

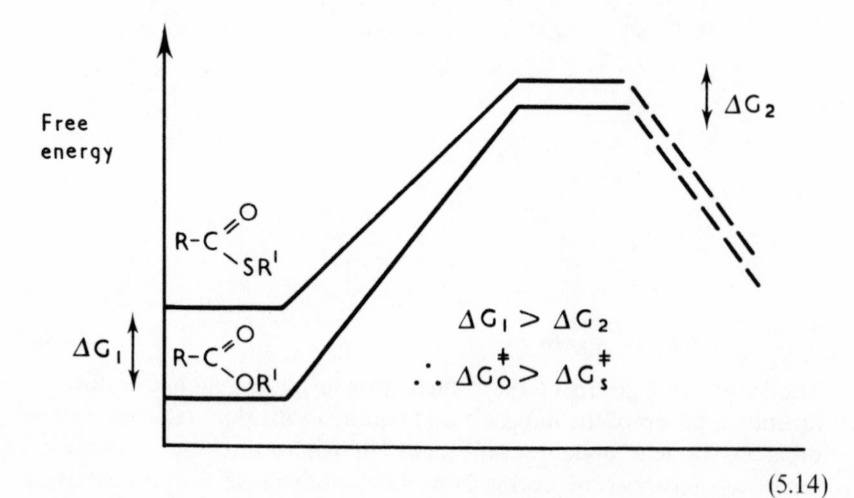

(5.14)

energy profiles (5.14) for the reaction of oxygen and thiol esters assist in the appreciation of the points made above.

Carbanion formation at the α-carbon atom of thiol esters is more favourable than for oxygen esters. Both the carbanions are resonance stabilized in essentially the same way (5.15)

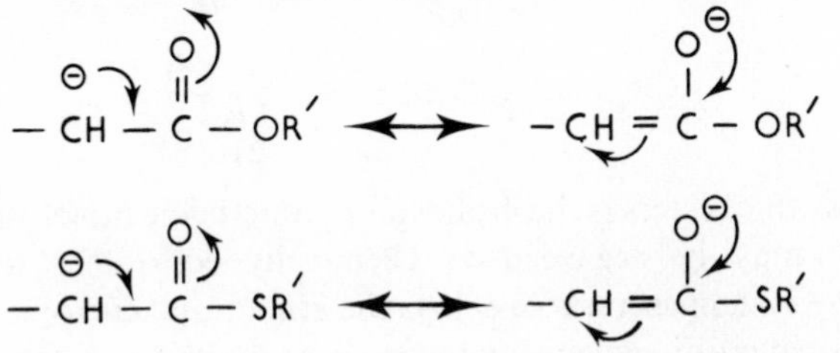

$$-\overset{\ominus}{C}H-\overset{O}{\overset{\|}{C}}-OR' \longleftrightarrow -CH=\overset{O^{\ominus}}{\overset{|}{C}}-OR'$$

$$-\overset{\ominus}{C}H-\overset{O}{\overset{\|}{C}}-SR' \longleftrightarrow -CH=\overset{O^{\ominus}}{\overset{|}{C}}-SR' \qquad (5.15)$$

However, the carbonyl group of the thiol ester has more double bond character than that of the oxygen ester (see (5.13)). Therefore the resonance stabilization of the type shown above will be more favourable for thiol than for oxygen esters.

The biosyntheses of *acyl-CoA* compounds are catalysed by thiokinases via an enzyme-bound activated acyl intermediate, usually an acyl-adenylate, and thus the available free energy of ATP is used to allow the thermodynamically unfavourable acyl-SCoA synthesis to take place.

The facile aminolysis of thiol esters is exploited in the acetylation of amino sugars by *acetyl-SCoA*.

$$\text{glucosamine} + \text{acetyl-SCoA} \rightleftharpoons N\text{-acetyl glucosamine} + \text{CoASH}$$

The electrophilicity of the thiol ester carbonyl group and the α-carbon acidity are important in the formation of *aceto-acetyl-SCoA* (5.16).

$$2CH_3-C\begin{matrix} \nearrow O \\ \searrow SCoA \end{matrix} \rightleftharpoons CH_3\cdot CO\cdot CH_2\cdot CO\cdot SCoA \qquad (5.16)$$

The mechanism of the thiolase which catalyses this reaction involves intermediate enzyme acylation followed by attack of the second molecule of acetyl-CoA on the thioacyl enzyme (5.17). The enzyme probably provides general base assistance for the (carbanion) attack on the acyl enzyme.

Fatty acid synthesis *de novo* is catalysed by a multi-enzyme complex wherein the acyl components are bound not to CoA–SH but to acyl carrier protein (ACP). The portion of ACP to which the acyl derivatives

$$\text{E—S}^{\ominus} + \text{CH}_3\text{C(=O)SCoA} \rightleftharpoons \text{E—S—C(=O)CH}_3 \ldots \rightleftharpoons \text{CH}_3\text{COCH}_2\text{CO·SCoA} + \text{E—S}^{\ominus}\ (\text{BH}^{\oplus}) \tag{5.17}$$

are bound as thiol esters is the 4-phospho-pantetheine moiety of CoA–SH. Thus ACP may be regarded as chemically equivalent to CoA–SH, although the system as a whole enjoys the kinetic and entropic advantages that arise when multi-enzyme complexes are involved in catalysing a series of sequential reactions.

The condensation reaction of fatty acid synthesis *de novo* also involves acyl enzyme formation. The acetyl-CoA component of the reaction is first transacylated to give acetyl-SACP which then reacts with an enzyme thiol group to yield acetyl-S-enzyme. The condensation is thus between malonyl-SACP and the acyl-enzyme. The product is acetoacetyl-SACP, and the enzyme is the leaving group. Malonyl-SACP loses carbon dioxide during the reaction and this loss is used to ensure that the reaction is very favourable in the direction of condensation. Malonic acid and malonate esters readily form carbanions in mildly basic conditions and so malonyl-SACP will more readily attack the acyl-enzyme than will acetyl-CoA. The condensation reaction may occur with loss of CO_2 (5.18) or with loss of bicarbonate (5.19).

$$^{\ominus}\text{O}_2\text{C—CH}_2\text{—C(=O)SACP} + \text{R—C(=O)—S—ENZ} \longrightarrow \text{RCOCH}_2\text{COSACP} + \text{CO}_2 \tag{5.18}$$

$$\text{ENZ—B:} + \text{H—OH} + \text{O}_2\text{C—CH}_2\text{—C(=O)SACP} + \text{R—C(=O)—SENZ} \longrightarrow \text{RCOCH}_2\text{CO SACP} + \text{HCO}_3^{\ominus} \tag{5.19}$$

(In the latter instance a general basic group will be provided by the enzyme to assist water insertion at the carbonyl carbon atom (5.19).)

Removal of hydrogen from fatty acyl-SACP by fatty acyl-SACP dehydrogenase is facilitated by acidity of the α-carbon protons. One would expect a concerted general base catalysed reaction involving little formal negative charge development on the α-carbon. The important acyltransferase role of CoA is demonstrated by the enzyme CoA transferase. The rather unusual mechanism shown in (5.20) has been proposed. The enzyme carboxyl group is here provided by the γ-carboxyl of a glutamate residue.

$$R_1COSCoA + E-CO_2^{\ominus} \rightleftharpoons \left(E-C(=O)-O-C(=O)R_1 \quad CoAS^{\ominus} \right) \rightleftharpoons E-C(=O)SCoA + R_1CO_2^{\ominus}$$

$$E-C(=O)SCoA + R_2CO_2^{\ominus} \rightleftharpoons E-CO_2^{\ominus} + R_2C(=O)SCoA \qquad (5.20)$$

The adenyl moiety of CoA–SH is far removed from the sulphydryl group and does not participate directly in the coenzymatic chemistry. The presence of this group must be explained on the basis of the utilization of the ubiquitous "adenine binding site" in order to enhance the localization of the coenzyme at the active site of the enzyme. Various derivatives of CoA–SH, lacking the adenyl moiety, will serve in many CoA–SH-requiring enzyme reactions albeit with less favourable kinetic parameters: *N*-acetyl-L-cysteamine is an extreme example of such a derivative.

5. Thiamine pyrophosphate (TPP)

Thiamine pyrophosphate (vitamin B_1) has the structure

$$\text{(4-amino-2-methylpyrimidin-5-yl)}-CH_2-\overset{\oplus}{N}\text{(thiazolium; 4-}CH_3\text{, S)}-CH_2-CH_2O-\overset{O}{\overset{\|}{P}}(O^{\ominus})-O-\overset{O}{\overset{\|}{P}}(O^{\ominus})-O^{\ominus} \qquad (5.21)$$

The thiazole ring has a relatively acidic proton at position 2, and this proton rapidly exchanges with deuterium when dissolved in deuterium oxide. This indicates that ylid (5.22) formation is relatively favourable.

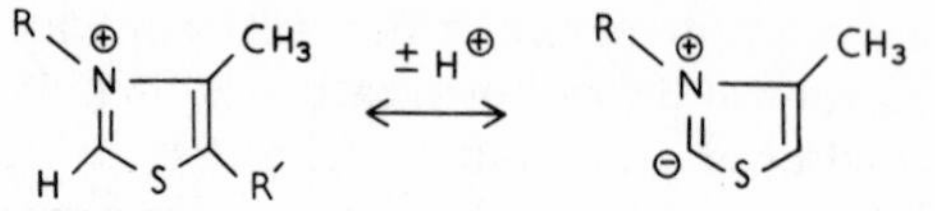

(5.22)

The carbanionic form of (5.22) will be active as a nucleophile, although the adjacent positively charged nitrogen will weaken the nucleophilicity compared with a simple non-ylid carbanion. Cyanide anion reacts as a carbanion and catalyses the benzoin condensation, as do *N*-ethyl thiazolium salts by the same general mechanism (5.23).

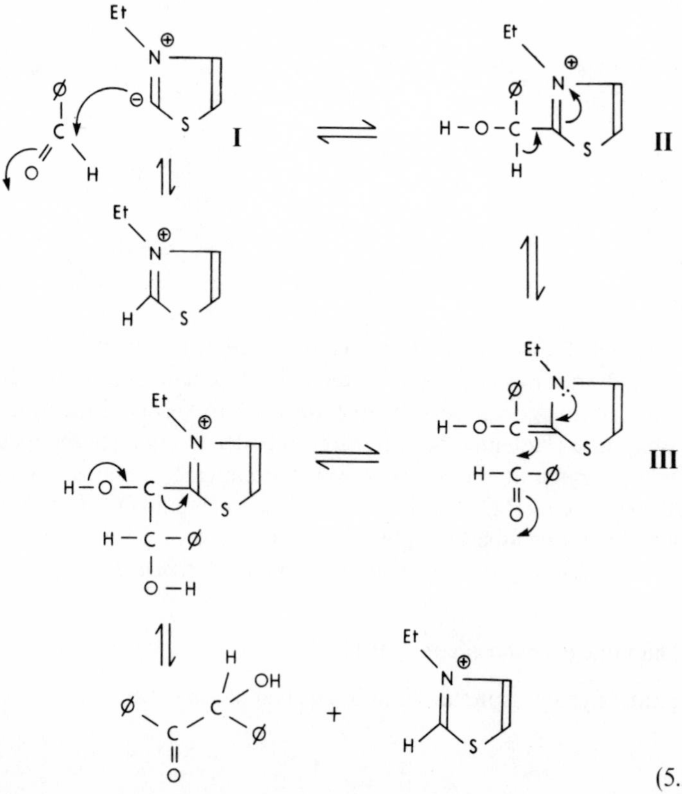

(5.23)

The properties required of a catalyst of this reaction are (a) sufficient nucleophilicity to add to the aldehyde carbonyl; (b) labilization of the proton attached to the carbon adjacent to the catalyst in the adduct, i.e. II and III above; and (c) sufficient leaving ability for the catalyst to be regenerated after the condensation.

In section 4, in the discussion of sulphur chemistry we mentioned that structures of the type (I)

$$=\overset{\ominus}{S}- \quad \text{(I)} \qquad =\overset{\oplus}{S}- \quad \text{(II)}$$

are relatively favourable in contrast to structures of the type (II). It thus seems likely that the thiazolium ylid will be further stabilized by contribution from the structure (5.24). In this structure the sulphur *d*-orbitals participate and the sulphur atom is non-nucleophilic.

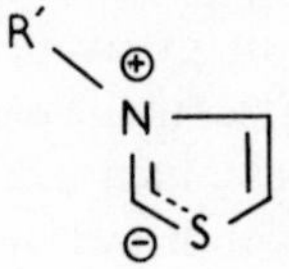

(5.24)

Model and enzymic studies strongly suggest that the mechanism (5.25) is substantially correct for the involvement of thiamine pyrophosphate in enzyme-catalysed reactions. Since pyruvate decarboxylase has been shown to catalyse H–D exchange at the 2-position of TPP, we include a general base able to assist proton removal at this position.

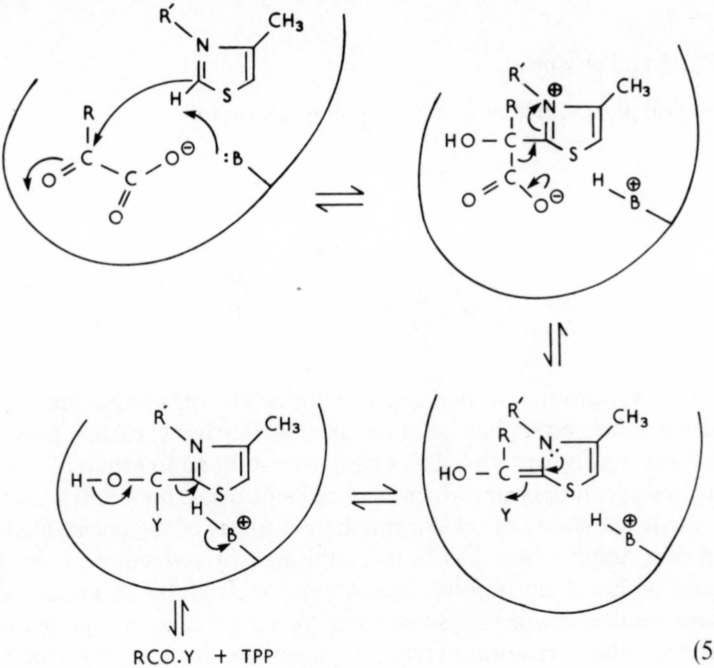

RCO.Y + TPP

(5.25)

Note that: (1) as in fatty acid synthesis, general base-catalysed attack of water on the carboxylate may initiate decarboxylation; (2) a resonance form of the adduct, prior to condensation with Y, may be drawn that has a negative charge on the carbon atom attached to position 2 of the thiazole ring. The resonance form will contribute to the structure of the adduct and render this carbon nucleophilic; (3) R may be a methyl group (as in pyruvate decarboxylase and dehydrogenase) or an hydroxymethyl group (as in transketolase) where instead of decarboxylation loss of glyceraldehyde-3-phosphate occurs; (4) Y may be a proton (as in pyruvate decarboxylase), lipoamide (as in pyruvate dehydrogenase) or ribose-5-phosphate (erythrose-4-phosphate) in the case of transketolase.

The scheme shown above has much in common with the mechanism given earlier for the benzoin condensation and it is apparent that the coenzymic function of thiamine pyrophosphate is particularly well understood in chemical terms.

We leave it to the reader to make a comparison between TPP-mediated reactions and the Schiff base intermediate mechanism of aldolases, where the amino group is provided by an ε-amino group of polypeptide lysine; the mechanisms have much in common.

6. Pyridoxal phosphate

Pyridoxal phosphate has a relatively simple structure (5.26).

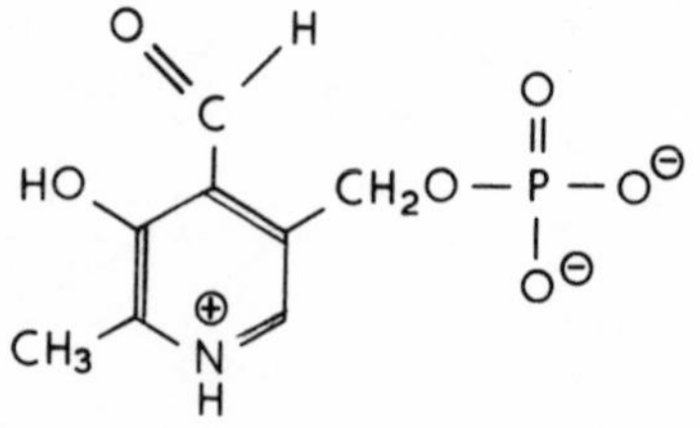

O H C HO $CH_2O-P(=O)(O^{\ominus})-O^{\ominus}$ CH_3 $\overset{\oplus}{N}$ H (5.26)

The aldehyde group is a good electrophilic centre, the pyridinium nitrogen enhancing the electrophilicity of the aldehyde carbonyl carbon atom. The coenzyme readily forms a Schiff's base with an ε-amino group of a specific lysine residue in enzymes which require its presence for catalytic activity. The existence of one of the enzyme-bound forms of the coenzyme in the form of a Schiff's base has been established by reduction of the holoenzyme with sodium borohydride. A completely stable secondary amine linkage results between a lysine residue of the protein and the pyridoxal. Enzymes which require pyridoxal phosphate for activity catalyse a wide range of reactions involved in amino acid metabolism, including racemization, transamination, decarboxylation and dehydration; indeed the coenzyme has a versatility comparable with that of ATP.

Extensive kinetic studies, in particular of transamination, have enabled the definition of a minimal kinetic mechanism

$$E_{ald} + AA_1 \rightleftharpoons X_1 \rightleftharpoons X_2 \rightleftharpoons E_{am} + KA_1$$
$$E_{am} + KA_2 \rightleftharpoons Y_1 \rightleftharpoons Y_2 \rightleftharpoons E_{ald} + AA_2$$

where E_{ald} refers to the enzyme-bound Schiff's base. This relates to, for example, the reaction

$$\underset{AA_1}{\text{glutamate}} + \underset{KA_2}{\text{pyruvate}} \rightleftharpoons \underset{KA_1}{\alpha\text{-ketoglutarate}} + \underset{AA_2}{\text{aspartate}}$$

The enzyme form E_{am} which arises as a result of the first half-reaction represents the enzyme-bound but freely dissociable amine form of the coenzyme, known as pyridoxamine. Steady-state, stopped-flow and temperature-jump kinetic studies have allowed the determination of the magnitude of ten of the twelve rate constants in the above scheme as well as setting lower limits for the remaining two. The bimolecular rate constants all have values of approximately $10^7\ M^{-1}\ sec^{-1}$ while those relating to the interconversion of the intermediates X_1, X_2, Y_1 and Y_2 have values between 10 and $10^2\ sec^{-1}$. Thus it is clear that the interconversion of these intermediates will be rate-limiting in the overall mechanism.

The study of non-enzymic model reactions has played a very important role in the development of our understanding of the mechanism of pyridoxal-dependent reactions. Early studies showed that in the presence of trivalent metal cations, pyridoxal will catalyse each of the reactions mentioned above at 100° in aqueous solution near neutral pH. In general these reactions are rather non-specific, several reactions occurring simultaneously, but some degree of specificity can be achieved by careful choice of pH and metal ion. More recent studies using the *minimal catalytically effective* model 3-hydroxy-4-pyridine aldehyde in the absence of metal ions have demonstrated that in carefully chosen conditions specific quantitative transamination is catalysed. The reactivity of the catalyst is very much higher when the pyridine nitrogen is in the cationic pyridinium form as compared with the neutral form, and the rate-limiting step of the reaction (aldimine $\rightleftharpoons$ ketimine) is accelerated in the presence of imidazole or acetate general base catalysts. Since the enzyme-catalysed reactions require no metal atom, these recent model studies (which demonstrate considerable specificity) approach closely the most likely mechanism for the enzyme-catalysed reactions. The absolute requirement for the hydroxy group at the 3-position of 4-pyridine aldehyde may be explained in terms of hydrogen bonding between the hydroxyl group and the aldimine/ketimine nitrogen. This interaction assists in the maintenance of overall coplanarity of the substrate-pyridoxal system so ensuring a high degree of delocalization and "system polarizability". It is important to remember that the initial reaction of an amino acid with enzyme bound coenzyme is a transimination reaction, this type of reaction being much faster in model reactions than equivalent Schiff's base formation from aldehyde and amine.

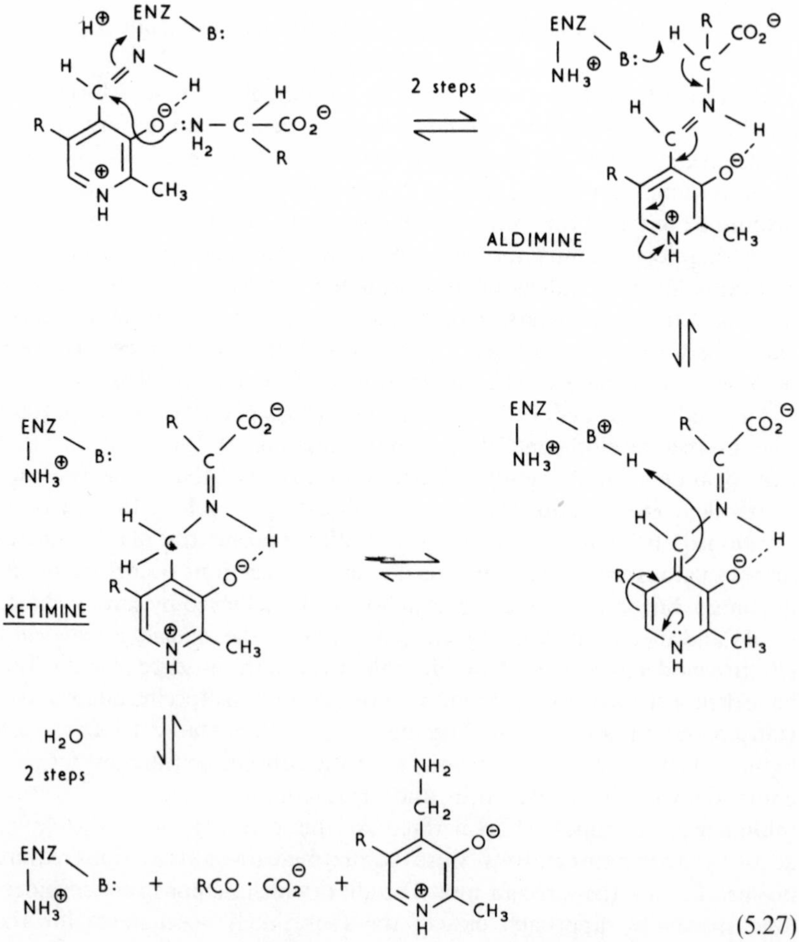

H⊕ ENZ B: H C N H R O⊖ :N — C — $CO_2^⊖$ H_2 H R N⊕ H CH_3

2 steps ⇌

ENZ $NH_3^⊕$ B: H R C $CO_2^⊖$ N H H C R O⊖ N⊕ H CH_3

ALDIMINE

⇌

ENZ $NH_3^⊕$ B: R $CO_2^⊖$ C N H C H H R O⊖ N⊕ H CH_3

KETIMINE

⇌

ENZ $NH_3^⊕$ B⊕ H R $CO_2^⊖$ C N H C H R O⊖ N H CH_3

⇌

H_2O

2 steps

ENZ $NH_3^⊕$ B: + $RCO \cdot CO_2^⊖$ + NH_2 CH_2 R O⊖ N⊕ H CH_3

(5.27)

Thanks to many model and enzyme studies, the mechanism of pyridoxal-dependent enzymes has been elucidated. A mechanism for the transaminases has been established in the greatest detail. The first "half" reaction (i.e. α-amino acid ⇌ keto acid) is shown above (5.27), the second "half" being essentially the reverse of the first. Tetrahedral intermediate species are omitted for clarity.

Spectroscopic evidence indicates the intermediacy of the quinonoid form between the aldimine and ketimine forms. An enzyme-bound basic species mediates proton transfer between the substrate α-carbon and the pyridoxal 4-carbon atom. The course of the reaction, always via aldimine and ketimine intermediates, is probably governed by the intra-complex orientation of the groups attached to the α-carbon of the substrate. The group most likely to be lost from the aldimine complex is that which is oriented in a plane perpendicular to the plane of the conjugated aldimine–pyridine system. These considerations can be represented diagrammatically (5.28).

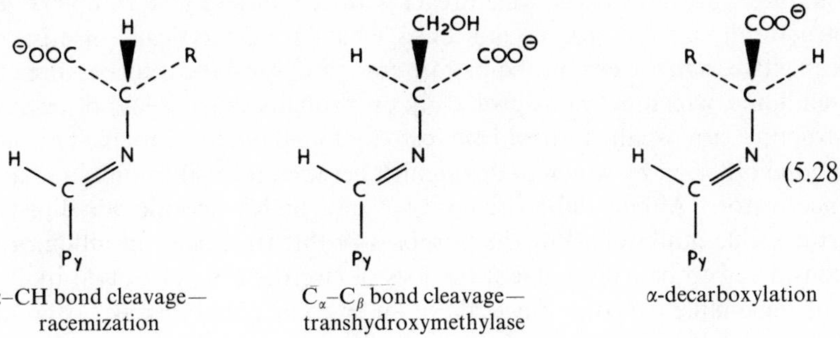

(5.28)

α-CH bond cleavage—racemization | C_α–C_β bond cleavage—transhydroxymethylase | α-decarboxylation

The situation is more complex than described here since, for example, serine can undergo enzyme-catalysed αβ dehydration as well as transhydroxymethylation. Thus detailed steric constraints and accurate catalytic group orientation around the substrate α-carbon atom are likely to be vital in defining the specificity of the pyridoxal-dependent enzymes. When in the protonated form, the positively charged nitrogen of the pyridine ring acts as an electron sink, conjugated to the substrate α-carbon, thus assisting labilization of groups attached to this atom as well as providing electron density from the lone pair in the quinonoid intermediate.

Although remarkable advances have been made in the detailed understanding of the pyridoxal-dependent enzyme reactions by means of model reactions, it should be remembered that the enzyme reactions proceed more rapidly than the model reactions by many orders of magnitude.

CHAPTER SIX

EFFECTS OF INHIBITORS AND pH

Inhibitors

Reversible and irreversible inhibition

Inhibitors are substances which react with enzymes to give complexes or structurally altered enzyme molecules which are catalytically inactive or less active than the enzyme–substrate complex. Substances such as urea or guanidine, which act non-specifically on proteins causing loss of tertiary structure, are usually termed *inactivators* or *denaturants*. Specificity is not a good criterion by which to distinguish between reversible inhibitors and inactivators. Affinity labels (chapter 9) are highly specific site-directed irreversible inhibitors. For the purposes of this treatment an inhibitor is considered to be irreversible if the loss of enzyme activity caused by it is not regainable over the timescale of an enzyme activity assay. Although this is an operational definition, borderline cases are rare in practice and the distinction is usually clear. The discussion which follows will be concerned with reversible inhibition.

Simple reversible inhibition

The general mechanism for simple reversible inhibition is given in figure 6.1. The inhibitor, I, forms complexes both with the free enzyme giving the complex EI, and with the enzyme–substrate complex giving ESI. Both inhibitor complexes are catalytically inactive. The inhibitor lowers the rate by reducing the amount of free enzyme available for reaction with S due to formation of EI, and by reducing the amount of productive ES complex due to formation of ESI.

An important restriction placed on the general mechanism of figure 6.1

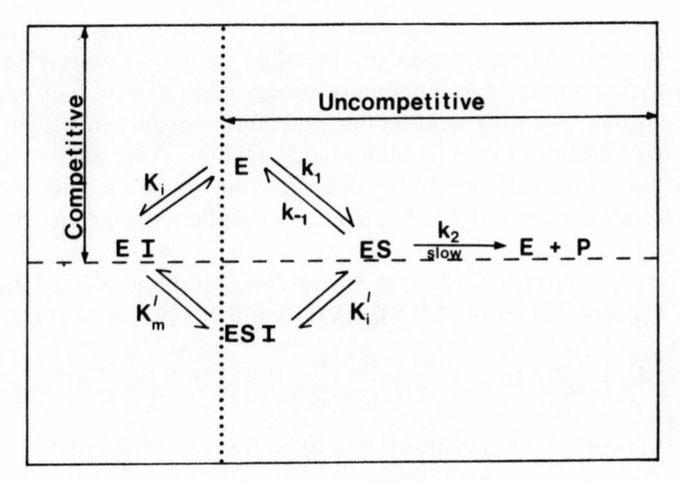

Figure 6.1 General mechanism for simple reversible inhibition.

is that the productive step

$$\mathrm{ES} \xrightarrow{k_2} \mathrm{E} + \mathrm{P}$$

is much slower than any of the other steps. This means that all steps prior to the productive step are at equilibrium characterized by the following dissociation constants:

$$K_i = \frac{(\mathrm{E})(\mathrm{I})}{(\mathrm{EI})} \qquad K_m = \frac{(\mathrm{E})(\mathrm{S})}{(\mathrm{ES})}$$

$$K'_i = \frac{(\mathrm{ES})(\mathrm{I})}{(\mathrm{ESI})} \qquad K'_m = \frac{(\mathrm{EI})(\mathrm{S})}{(\mathrm{ESI})}$$

Note that although K_m cannot usually be assumed to be an equilibrium constant (see chapter 4) it can be considered to be so in the context of this mechanism.

It can easily be shown that any three of the four constants are sufficient to define the system. For example, K'_m can be described in terms of K_m, K_i and K'_i:

$$\frac{K_m K'_i}{K_i} = \frac{\cancel{(\mathrm{E})}(\mathrm{S})}{\cancel{(\mathrm{ES})}} \cdot \frac{(\mathrm{EI})}{\cancel{(\mathrm{E})}\cancel{(\mathrm{I})}} \cdot \frac{\cancel{(\mathrm{ES})}\cancel{(\mathrm{I})}}{(\mathrm{ESI})} \equiv K'_m$$

If the equilibrium restriction is not placed on the general mechanism and the steady-state assumption is applied, the rate equation turns out to be highly complex, containing squared terms in both substrate and inhibitor concentrations. As most cases of reversible inhibition appear to obey simple hyperbolic kinetics, the equilibrium assumption is probably reasonable. A further assumption is that the total inhibitor concentration is equal to (I), the concentration of free (unbound) inhibitor. This will be approximately true if the total inhibitor concentration is much greater than the total enzyme concentration. (A similar assumption was made in chapter 4 regarding substrate concentration.)

Using the constants K_i, K'_i and K_m the concentrations of E, EI and ESI can be expressed in terms of (ES), (S), (I) and the appropriate constant(s):

$$(\mathrm{E}) = \frac{K_m}{(\mathrm{S})}\cdot(\mathrm{ES}) \quad (\mathrm{EI}) = \frac{K_m}{(\mathrm{S})}\cdot\frac{(\mathrm{I})}{K_i}\cdot(\mathrm{ES}) \quad (\mathrm{ESI}) = \frac{(\mathrm{I})}{K'_i}\cdot(\mathrm{ES})$$

These expressions are substituted into the conservation equation,

$$\mathrm{E}_0 = (\mathrm{E})+(\mathrm{EI})+(\mathrm{ESI})+(\mathrm{ES})$$

which is solved for (ES). The rate of the productive step is given by $d(\mathrm{P})/dt = k_2(\mathrm{ES})$ and the overall rate equation for the mechanism is

$$v_0 = \frac{d(\mathrm{P})}{dt} = \frac{V\mathrm{S}_0}{K_m(1+(\mathrm{I})/K_i)+\mathrm{S}_0(1+(\mathrm{I})/K'_i)} \tag{6.1}$$

As the terms containing the inhibitor concentration appear in the denominator of equation (6.1) it can be seen that increasing (I) will result in a decreased rate for a fixed substrate concentration. Dividing numerator and denominator of equation (6.1) by $1+\mathrm{I}/K'_i$ gives

$$v_o = \frac{\dfrac{V}{1+(\mathrm{I})/K'_i}\cdot\mathrm{S}_0}{K_m\left(\dfrac{1+(\mathrm{I})/K_i}{1+(\mathrm{I})/K'_i}\right)+\mathrm{S}_0} \tag{6.2}$$

which is of the same form as the Michaelis–Menten equation

$$v_0 = \frac{V^{app}\mathrm{S}_0}{K_m^{app}+\mathrm{S}_0} \tag{6.3}$$

The symbols V^{app} and K_m^{app} are used because their values depend on the value of (I). The *true* values of V and K_m are unchanged and V^{app} and K_m^{app} obtained from measurements of v_0 at different S_0 at a fixed (I) will be *apparent* constants. Reversible inhibition is classified according to the way the value of (I) affects V^{app}, K_m^{app} and the ratio V^{app}/K_m^{app}.

Competitive inhibition

A competitive inhibitor is one which acts by increasing K_m^{app} leaving V^{app} unaffected. This type of inhibition results from mutually exclusive binding of inhibitor and substrate to the free enzyme as illustrated by the following mechanism:

$$\mathrm{EI} \underset{+\mathrm{I}}{\overset{K_i}{\rightleftharpoons}} \mathrm{E} \underset{k_{-1}}{\overset{k_1(\mathrm{S})}{\rightleftharpoons}} \mathrm{ES} \xrightarrow{k_2} \mathrm{E}+\mathrm{P} \tag{6.4}$$

Mechanism (6.4) is a special case of the mechanism of figure 6.1 in which the ESI complex does not exist. This is equivalent to saying that $K_i' = \infty$. Thus the $(\mathrm{I})/K_i'$ terms of equation (6.1) equal zero and can be dropped leaving

$$v_0 = \frac{V\mathrm{S}_0}{K_m(1+(\mathrm{I})/K_i)+\mathrm{S}_0} \tag{6.5}$$

which is the rate equation for competitive inhibition. Equation (6.5) can also be derived directly from mechanism (6.4).

Comparison of equations (6.5) and (6.3) show that $V^{app} = V$, $K_m^{app} = K_m(1+(\mathrm{I})/K_i)$ and $V^{app}/K_m^{app} = V/K_m(1+(\mathrm{I})/K_i)$. As equation (6.5) is of the same form as the Michaelis–Menten equation, the variation of v_0 with S_0 at fixed (I) will show normal hyperbolic behaviour (figure 6.2a). A feature of competitive inhibition is that it is *substrate-antagonized*, i.e. the degree of inhibition for a fixed value of (I) decreases with increasing S_0. This may not be obvious from figure 6.2(a) but becomes apparent if one considers the *fractional inhibition*, *i*, which is defined as

$$i = 1 - \frac{v_0^i}{v_0}$$

where v_0^i and v_0 are the velocities in the presence and absence of inhibitor respectively. For competitive inhibition

$$i = \frac{(\mathrm{I})/K_i}{1+(\mathrm{I})/K_i+\mathrm{S}_o/K_m}$$

At fixed (I), *i* increases with decreasing S_0, approaching $i = (\mathrm{I})K_i/(1+(\mathrm{I})/K_i)$ as S_0 approaches zero; as S_0 increases, *i* decreases, and approaches zero as S_0 approaches infinity (figure 6.2b). Percent inhibition is given by 100*i*. Sometimes one sees competitive inhibitors characterized by an "I_{50} value", which is the concentration of inhibitor required to reduce the uninhibited

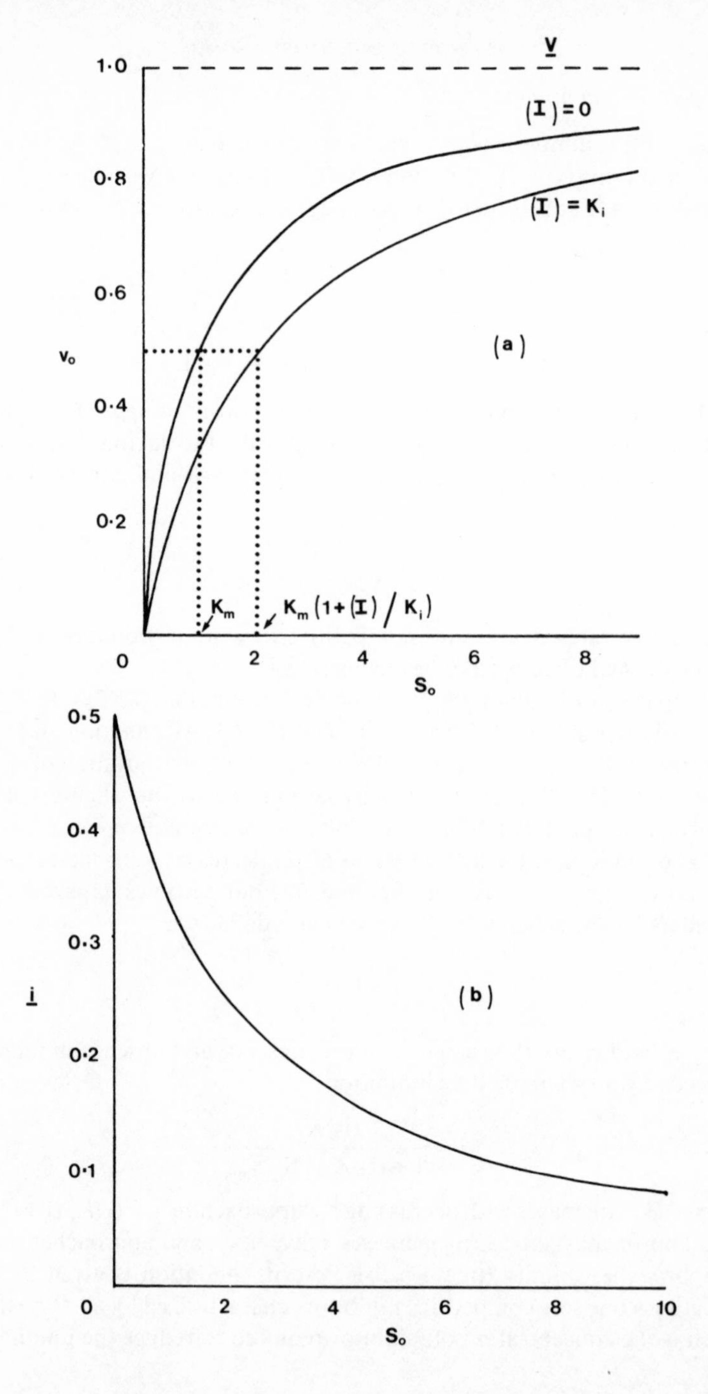
V
1·0
(I) = 0
0·8
(I) = K_i
0·6
(a)
v_o
0·4
0·2
K_m
K_m (1 + (I) / K_i)
0
2
4
6
8
S_o
0·5
0·4
0·3
i
(b)
0·2
0·1
0
2
4
6
8
10
S_o

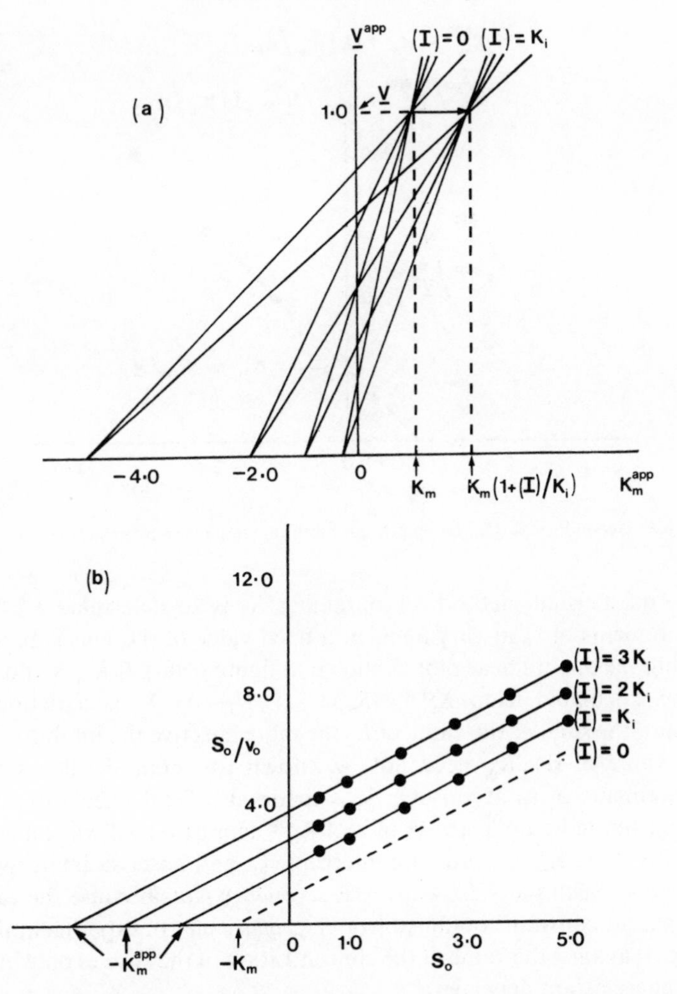

Figure 6.3 Diagnostic plots for competitive inhibition calculated assuming $K_m = K_i = V = 1$. (a) Direct linear plot. (b) Half-reciprocal plot.

rate by 50% ($i = 0.5$). The variation of i with S_0 shows that I_{50} values have no significance unless the substrate concentration is given; even then they cannot be used to calculate K_i unless K_m is known.

Figure 6.2 Competitive inhibition. Curves calculated assuming $K_m = K_i = V = 1$. (a) Variation of v_0 with S_0 in absence and presence of inhibitor. (b) Variation of fractional inhibition, i with S_0; $(I) = K_i$.

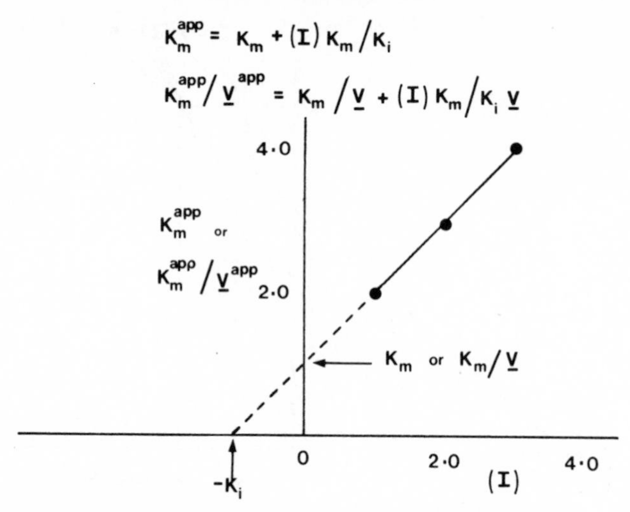

Figure 6.4 Secondary plot for competitive inhibition. Data taken from figure 6.3b.

The most usual method for obtaining K_i is to determine K_m^{app} from measurements of v_0 at varying S_0 at a fixed value of (I). This is best done by using the direct linear plot as shown in figure 6.3(a). If K_m is known, K_i can be calculated from $K_m^{app} = K_m(1 + (I)/K_i)$. As K_i is a dissociation constant, the smaller the value of K_i the more effective the inhibitor at any given value of (I). K_m need not be known to obtain K_i if a series of measurements of v_0 at varying S_0 is made at several fixed values of (I) as displayed in figure 6.3(b). A plot of K_m^{app} against (I) will yield a straight line with slope K_m/K_i, ordinate intercept K_m and abscissa intercept $-K_i$ (figure 6.4). Such plots are known as *secondary plots* because the value of an apparent constant (obtained from a *primary plot* of experimental data) is plotted against the value of the concentration of the species on which the apparent constant depends.

Graphical diagnosis of competitive inhibition is straightforward and can be made by using the direct linear plot or the half-reciprocal plot. In the presence of a competitive inhibitor the common intersection point of a direct linear plot is shifted to the right along a line intersecting V on the V^{app} axis and parallel with the K_m^{app}-axis (figure 6.3(a)). In the half-reciprocal plot the slope is $1/V^{app}$ (see chapter 4), and as V^{app} is unchanged by a competitive inhibitor, a pattern of parallel lines is obtained which is characteristic of competitive inhibition (figure 6.3(b)).

In competitive inhibition, substrate and inhibitor cannot bind simul-

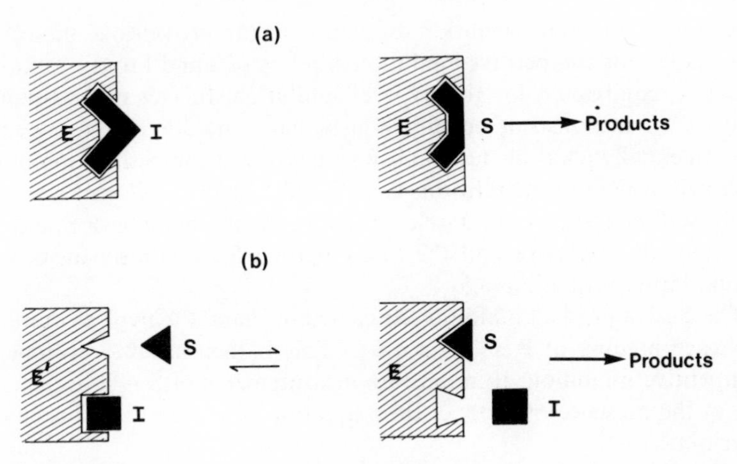

Figure 6.5 Schematic illustration of competitive inhibition. (a) Classical mechanism in which S and I bind mutually exclusively at the same site. (b) Alternative mechanism which assumes two conformers of the enzyme E and E′. E can bind S but not I; E′ can bind I but not S.

taneously to the enzyme. The simplest explanation for this is that inhibitor and substrate bind at the same site, the binding of one preventing the binding of the other. Indeed most competitive inhibitors are structurally similar to the substrate, and some authors define a competitive inhibitor as one which interacts with the same groups on the enzyme as the substrate and thus *competes* with the substrate for the active site. This is schematically illustrated in figure 6.5(a). One of the earliest reported examples of competitive inhibition is the action of malonate on succinate dehydrogenase. This enzyme catalyses the oxidation of succinate to fumarate. Malonate, like succinate, is a dicarboxylate and presumably binds to the active site preventing the binding of succinate. However, malonate cannot undergo dehydrogenation and forms a dead-end complex.

$$^{-}OOC{-}CH_2{-}CH_2{-}COO^{-} \xrightarrow[\text{dehydrogenase}]{\text{succinate}} {^{-}OOC}(H)C{=}C(H)COO^{-}$$

succinate → fumarate

$$^{-}OOC{-}CH_2{-}COO^{-}$$

malonate

Although the representation of figure 6.5(a) provides a molecular explanation for competitive inhibition, binding of S and I to the same site is not a requirement for this type of inhibition. In fact it is extremely laborious to demonstrate unambiguously that S and I do bind to the same site. An equally plausible mechanism giving rise to competitive inhibition is shown in figure 6.5(b). In this case S and I bind to different sites and might well be structurally unrelated. However, the binding of S and I is still mutually exclusive and the rate equation for this mechanism is of similar form to equation (6.5).

The case of product inhibition discussed in chapter 4 involves mutually exclusive binding of P and S with E. This is then another example of competitive inhibition. If initial rate measurements of S → P are carried out in the presence of product then equation (4.41) applies. This can be rewritten as

$$v_0 = \frac{V^{S'} S_0}{K_m^{S'}(1 + (P)/K_m^{P'}) + S_0}$$

which is identical in form to equation (6.5).

Uncompetitive inhibition

In the presence of an uncompetitive inhibitor both K_m^{app} and V^{app} are decreased by the same factor and thus V/K_m is unaffected. Such inhibition can arise from the following mechanism in which the inhibitor combines only with the enzyme substrate complex giving a dead-end (catalytically inactive) ESI complex:

$$\mathrm{E} \underset{k_{-1}}{\overset{k_1(\mathrm{S})}{\rightleftharpoons}} \mathrm{ES} \xrightarrow{k_2} \mathrm{E} + \mathrm{P}; \quad \mathrm{ES} \underset{K_i'}{\overset{+\mathrm{I}}{\rightleftharpoons}} \mathrm{ESI} \tag{6.6}$$

Reference to figure 6.1 shows that mechanism (6.6) is a special case of the general mechanism in which the EI complex does not exist, i.e. $K_i = \infty$. Therefore the $(I)/K_i$ term of equation (6.1) equals zero and can be dropped leaving

$$v_0 = \frac{V S_0}{K_m + S_0(1 + (I)/K_i')} \tag{6.7}$$

which is the rate equation for mechanism (6.6).

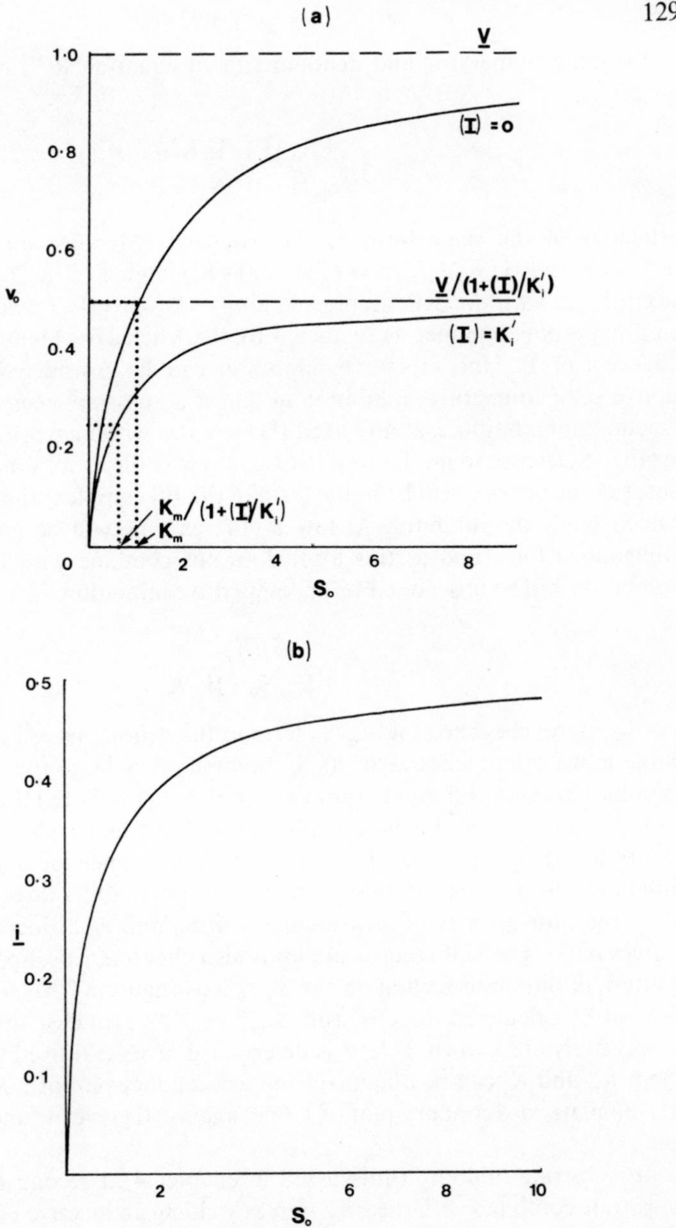

Figure 6.6 Uncompetitive inhibition. Curves calculated assuming $K_m = K_i' = V = 1$. (a) Variation of v_0 with S_0 in absence and presence of inhibitor. (b) Variation of fractional inhibition, i with S_0; $(I) = K_i'$.

Dividing numerator and denominator of equation (6.7) by $1+(\mathrm{I})/K_i'$ gives

$$v_0 = \frac{V\mathrm{S}_0/(1+(\mathrm{I})/K_i')}{K_m/(1+(\mathrm{I})/K_i')+\mathrm{S}_0} \tag{6.8}$$

which is of the same form as the Michaelis–Menten equation. Thus $V^{app} = V/(1+(\mathrm{I})/K_i')$, $K_m^{app} = K_m/(1+(\mathrm{I})/K_i')$ and $V^{app}/K_m^{app} = V/K_m$ in accordance with the definition given above. At fixed (I), v_0 varies with S_0 in a hyperbolic manner as predicted by the Michaelis–Menten equation (figure 6.6(a)). Uncompetitive inhibition can be considered to be the converse of competitive inhibition in that it is *substrate-enhanced*, i.e. the fractional inhibition, i, at any fixed (I) *increases* with increasing S_0 (figure 6.6(b)). Reference to mechanism (6.6) shows why this is so. When $\mathrm{S}_0 \gg K_m$ most of the enzyme will be in the form of the ES complex; this is the form which binds the inhibitor. At low S_0 the enzyme will be mostly in the unliganded form and as this form does not combine with I, very little inhibition will be observed. For uncompetitive inhibition

$$i = \frac{(\mathrm{I})/K_i'}{1+K_m/\mathrm{S}_0+(\mathrm{I})/K_i'}$$

As S_0 approaches zero the K_m/S_0 term in the denominator becomes very large and i approaches zero; as S_0 becomes very large the K_m/S_0 term becomes insignificant and i approaches $((\mathrm{I})/K_i')/(1+(\mathrm{I})/K_i')$ (figure 6.6(b)).

K_m^{app} and V^{app} can be obtained from the direct linear plot as shown in figure 6.7(a). The plot has a characteristic appearance for uncompetitive inhibition in that the common intersection point shifts directly towards the origin for a series of measurements of S_0 and v_0 at increasing fixed values of (I). The half-reciprocal plot is also characteristic, appearing as a pattern of lines intersecting on the S_0/v_0 axis (figure 6.7(b)). The value of K_i' can be calculated directly from K_m^{app} or V^{app} provided that K_m or V respectively are known. If K_m^{app} is determined at several fixed values of (I) both K_m and K_i' can be obtained from a secondary plot of $1/K_m^{app}$ against (I). Similarly a secondary plot of $1/V^{app}$ against (I) gives V and K_i' (figure 6.8).

In substrate inhibition (discussed in chapter 4) a second molecule of substrate combines with the ES complex yielding an inactive ES_2 complex. Thus substrate inhibition can be considered to be a case of uncompetitive inhibition. That this is so is seen by rearrangement of the rate equation (4.47) for substrate inhibition,

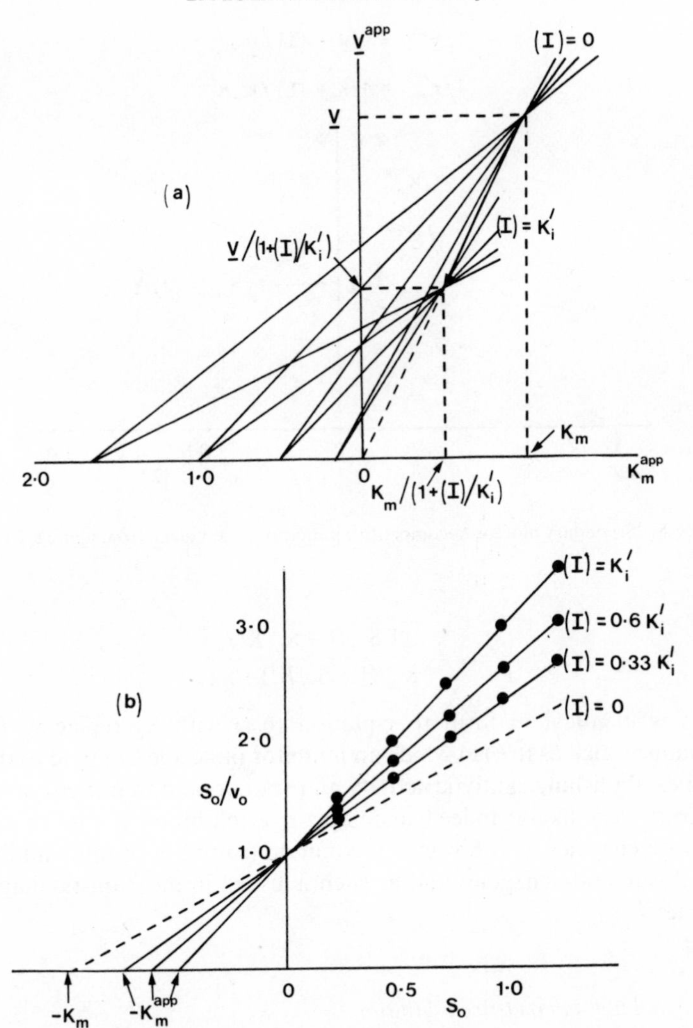

Figure 6.7 Diagnostic plots for uncompetitive inhibition calculated assuming $K_m = K_i' = V = 1$. (a) Direct linear plot. (b) Half-reciprocal plot.

$$v_0 = \frac{V}{1 + K_m/S_0 + S_0/K'} \tag{4.47}$$

Multiplying top and bottom by S_0 and dividing top and bottom by $(1 + S_0/K')$ gives

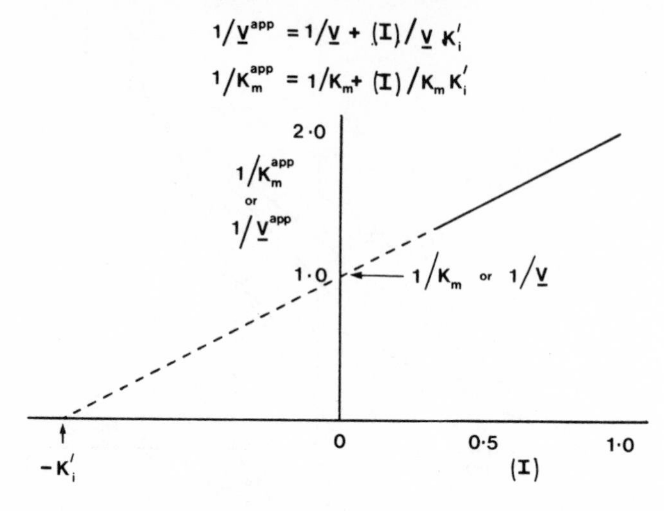

Figure 6.8 Secondary plot for uncompetitive inhibition. Data taken from figure 6.7b.

$$v_0 = \frac{VS_0/(1+S_0/K')}{K_m/(1+S_0/K')+S_0}$$

which is identical in form to equation (6.8) with S_0 replacing (I). A mechanism such as (6.6) in which an inhibitor molecule can bind to the ES complex abolishing catalytic activity and yet not bind to free enzyme may not seem very likely; indeed uncompetitive inhibition is rare in single-substrate enzymes. It is however a common feature of product inhibition of multisubstrate enzymes and as such is useful in mechanistic diagnosis (chapter 7).

Mixed and non-competitive inhibition

Competitive and uncompetitive inhibition represent the two pure forms of inhibition ("pure" because only one type of enzyme–inhibitor complex is involved). If both EI and ESI are formed, the general mechanism of figure 6.1 applies and the rate equation is equation (6.1). This type of inhibition is known as *mixed inhibition* because it combines the features of the two pure forms. Mixed inhibition is characterized by dependence of both V^{app} and V^{app}/K_m^{app} on (I). Reference to equation (6.2) shows that $V^{app} = V/(1+(I)/K_i')$, $K_m^{app} = K_m(1+(I)/K_i)/(1+(I)/K_i')$ and

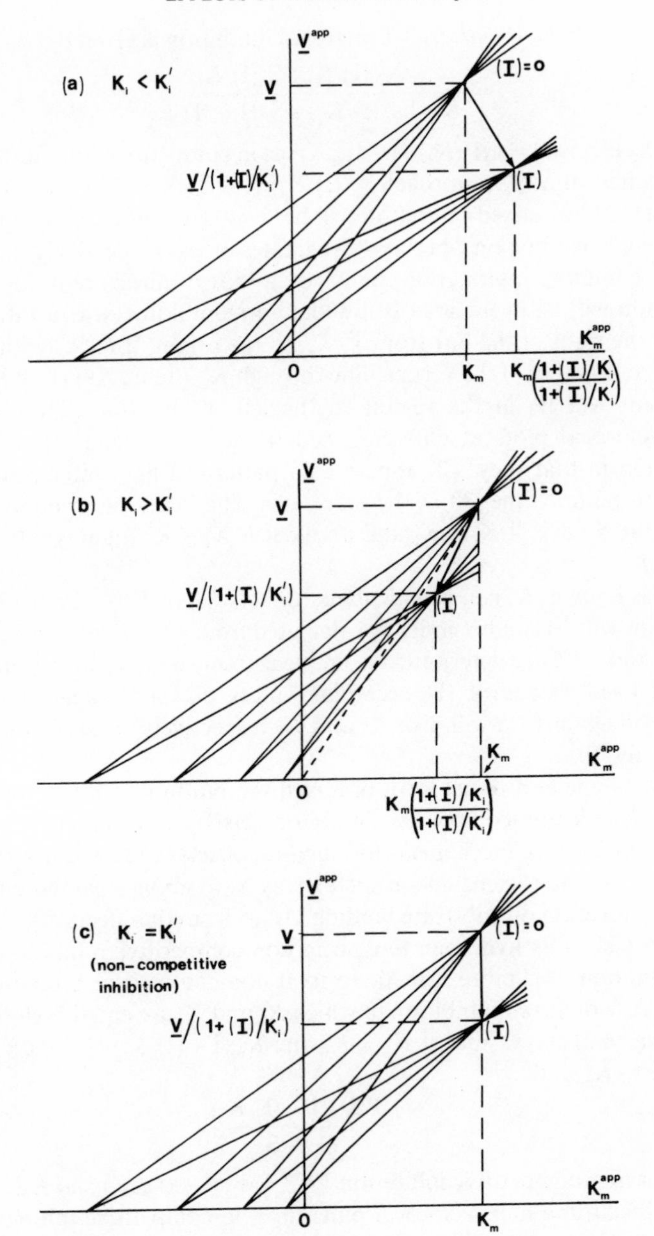

Figure 6.9 Direct linear plots for mixed inhibition.

$V^{app}/K_m^{app} = V/K_m(1+(I)/K_i)$. Fractional inhibition is given by

$$i = \frac{K_m(I)/K_i + S_0(I)/K_i'}{K_m(1+(I)/K_i)+S_0(1+(I)/K_i')} \tag{6.9}$$

When S_0 is zero, $i = ((I)/K_i)/(1+(I)/K_i)$ as in competitive inhibition; as S_0 approaches infinity, i approaches $((I)/K_i')/(1+(I)/K_i')$ as in uncompetitive inhibition. Thus mixed inhibition will be substrate-enhanced or substrate-inhibited depending on whether K_i is greater or less respectively than K_i'.

The common intersection point of a direct linear plot for mixed inhibition will lie in the area below the horizontal line drawn through V and to the right of the line from V, K_m to the origin. If $K_i < K_i'$ the point will lie to the right of the vertical line through K_m (figure 6.9(a)); if $K_i > K_i'$ the point will lie in the region to the left of this line (figure 6.9(b)). Half-reciprocal plots at different fixed (I) are also diagnostic of mixed inhibition in that they will appear as a pattern of lines intersecting at a common point to the left of the S_0/v_0 axis. The intersection point will be below the S_0 axis if $K_i < K_i'$ and above it if $K_i > K_i'$ (figures 6.10(a) and 6.10(b)).

If V is known, K_i' can be calculated directly from V^{app}; if both K_m and V are known, K_i can be similarly calculated from V^{app}/K_m^{app}. Alternatively, if K_m^{app} and V^{app} are determined from measurements of v_0 at varying S_0 at several fixed values of (I), secondary plots of $1/V^{app}$ against (I) and K_m^{app}/V^{app} against (I) will give K_i' and K_i respectively, as shown in figure 6.11(a) and (b).

There is one additional form of mixed inhibition which is termed *non-competitive*. A non-competitive inhibitor acts by lowering V^{app}, leaving K_m^{app} unaffected. A mechanism for such an effect would require that the inhibitor bind to the enzyme in such a way as to abolish catalytic activity but have no effect on substrate binding. Apart from the effects of very small inhibitors such as hydrogen ions, true non-competitive inhibition is not very common. It is more realistic to treat non-competitive inhibition as a special case of mixed inhibition in which K_i and K_i' are equal. Reference to equation (6.2) shows that if $K_i = K_i'$, the $K_m(1+(I)/K_i)/(1+(I)/K_i')$ term reduces to K_m and

$$v_0 = \frac{VS_0/(1+(I)/K_i)}{K_m+S_0}$$

Thus for non-competitive inhibition $V^{app} = V/(1+(I)/K_i')$ and $K_m^{app} = K_m$. Direct linear and half-reciprocal plots for non-competitive inhibition are shown in figures 6.9(c) and 6.10(c). A peculiar feature of non-competitive inhibition is that the fractional inhibition does not vary with S_0. If

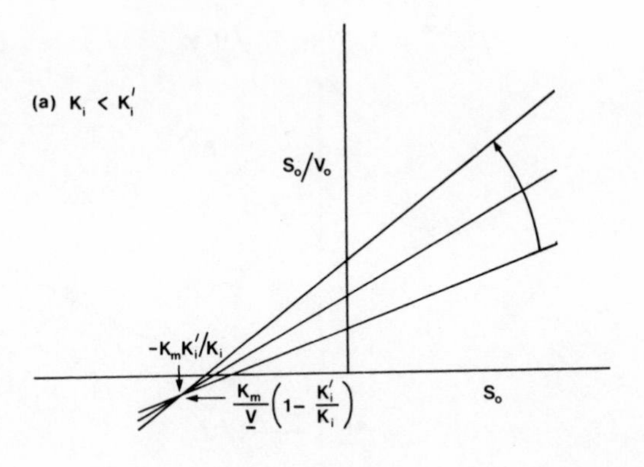

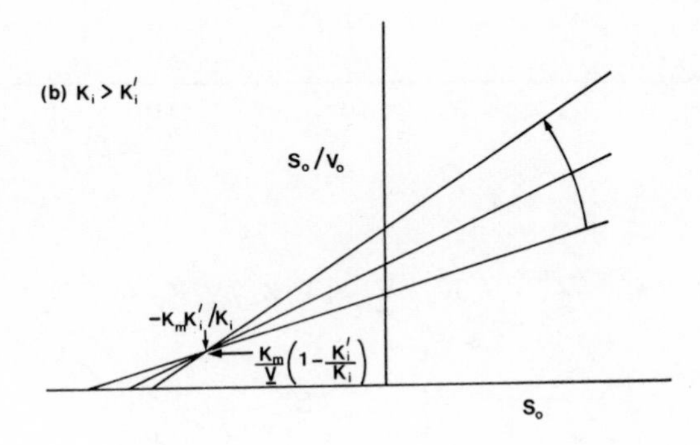

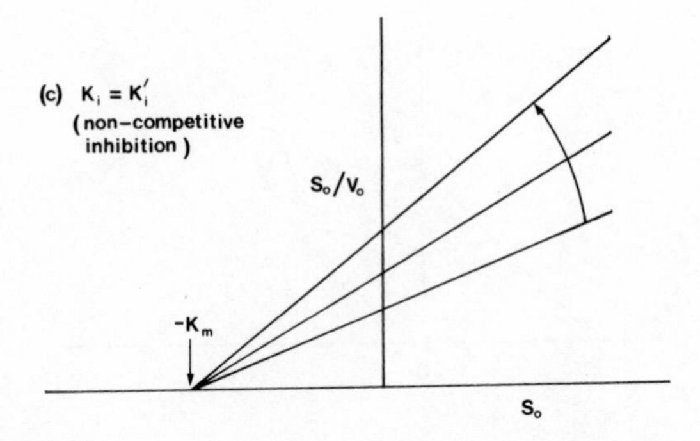

Figure 6.10 Half-reciprocal plots for mixed inhibition. Arrow indicates increasing values of fixed (I).

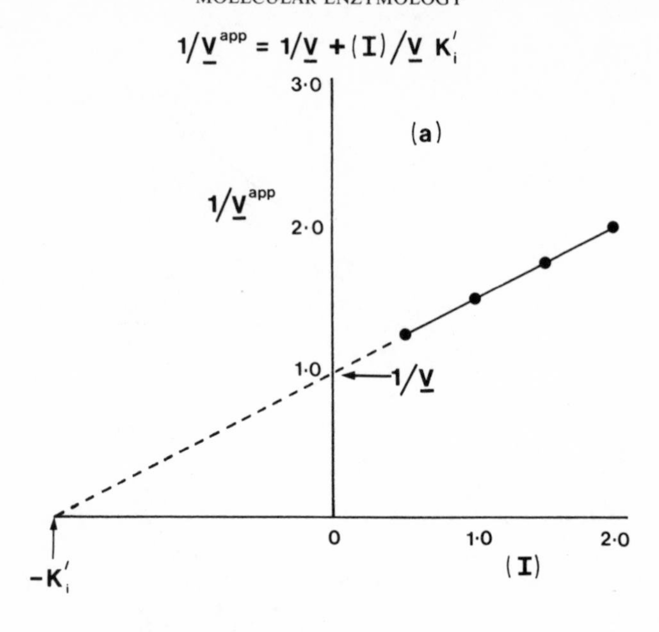

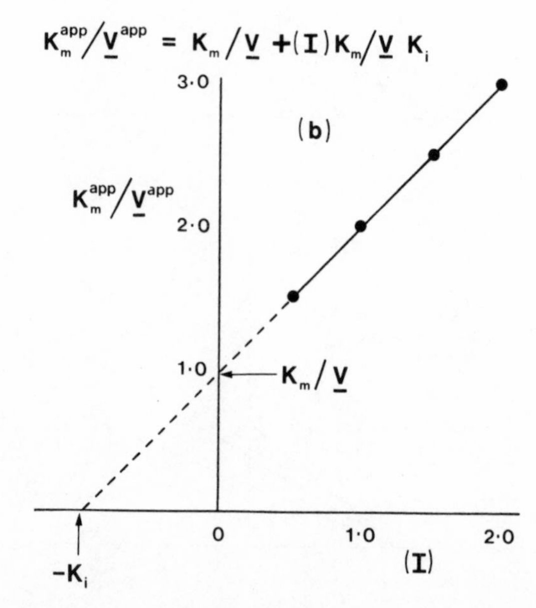

Figure 6.11 Secondary plots for mixed inhibition. Calculated assuming $K_i = K_m = V = 1$, $K_i' = 2$. (a) Plot of $1/V^{app}$ against (I). (b) Plot of K_m^{app}/V^{app} against (I).

$K_i = K_i'$, equation (6.9) reduces to

$$i = \frac{(\mathrm{I})/K_i}{1 + (\mathrm{I})/K_i}$$

As S_0 does not appear in the equation for fractional inhibition, i is independent of substrate concentration in the presence of a non-competitive inhibitor.

Other plotting methods

For the cases of inhibition described above it is possible to obtain K_i and K_i' directly from experimental data without using secondary plots. Inversion of equation (6.1) for mixed inhibition gives

$$\frac{1}{v_0} = \frac{K_m}{VS_0} + \frac{1}{V} + \left(\frac{K_m/K_i}{VS_0} + \frac{1/K_i'}{V}\right)(\mathrm{I}) \tag{6.10}$$

Written in this way it is seen that a plot of $1/v_0$ against (I) will be a straight line. If measurements of v_0 are made at various values of (I) at two fixed substrate concentrations, S_0^1 and S_0^2, we can write

for S_0^1
$$\frac{1}{v_0^1} = \frac{K_m}{VS_0^1} + \frac{1}{V} + \left(\frac{K_m/K_i}{VS_0^1} + \frac{1/K_i'}{V}\right)(\mathrm{I}) \tag{6.10a}$$

for S_0^2
$$\frac{1}{v_0^2} = \frac{K_m}{VS_0^2} + \frac{1}{V} + \left(\frac{K_m/K_i}{VS_0^2} + \frac{1/K_i'}{V}\right)(\mathrm{I}) \tag{6.10b}$$

To find the value of the (I) coordinate of the intersection point of the two lines described by equations (6.10a) and (6.10b) the two are equated and the $1/V$ terms cancel immediately:

$$\frac{K_m}{VS_0^1} + \left(\frac{K_m/K_i}{VS_0^1} + \frac{1/K_i'}{V}\right)(\mathrm{I}) = \frac{K_m}{VS_0^2} + \left(\frac{K_m/K_i}{VS_0^2} + \frac{1/K_i'}{V}\right)(\mathrm{I})$$

Gathering the (I) terms gives

$$(\mathrm{I})\left(\frac{K_m}{VK_i}\left(\frac{1}{S_0^2} - \frac{1}{S_0^1}\right)\right) = \frac{K_m}{V}\left(\frac{1}{S_0^1} - \frac{1}{S_0^2}\right)$$

and therefore $(\mathrm{I}) = -K_i$. Thus a plot of $1/v_o$ against (I) at different fixed S_0 gives an intersection in the second (if $K_i < K_i'$) or third (if $K_i > K_i'$) quadrant which provides the value of K_i for mixed inhibition. If the inhibition is competitive ($K_i' = \infty$) the terms containing K_i' disappear and the same result is obtained (but the intersection point is always in

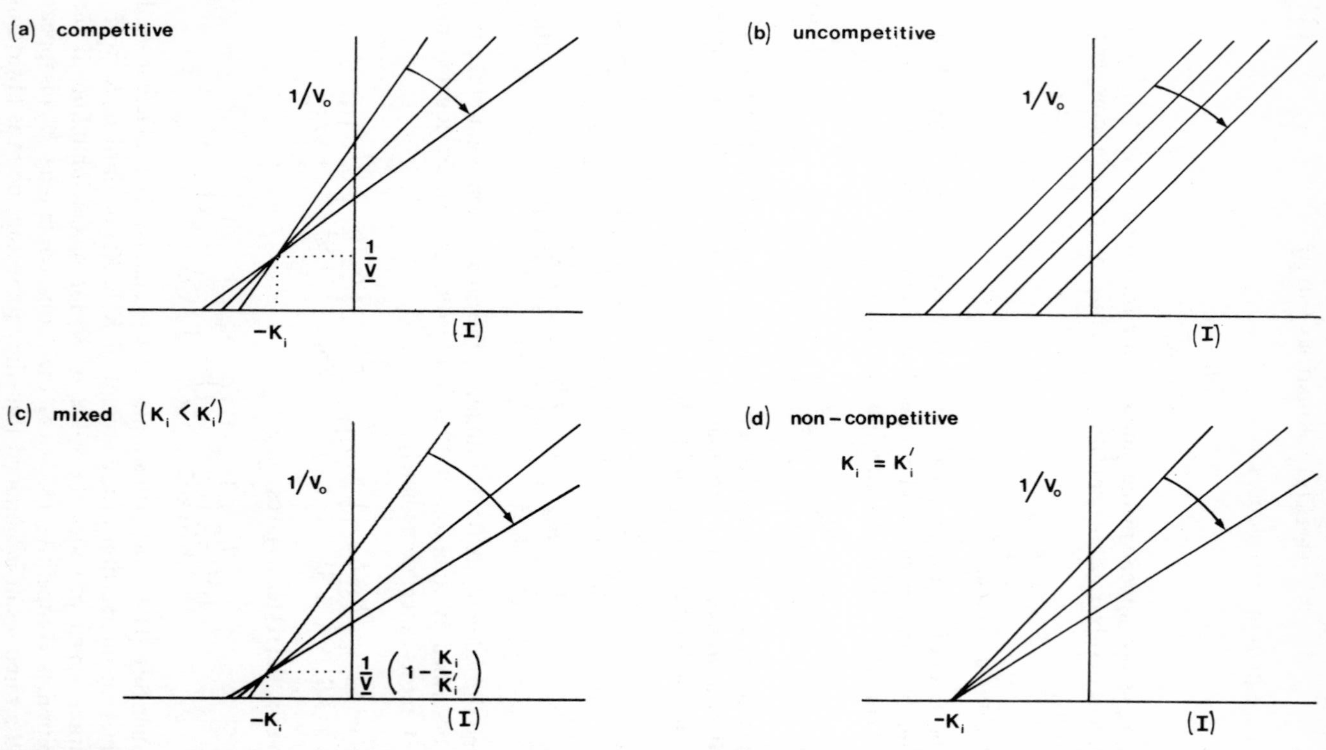

Figure 6.12 Dixon plots of $1/v_0$ against (I) for different types of simple inhibition. Arrow indicates increasing values of fixed S_0.

the second quadrant). In uncompetitive inhibition $K_i = \infty$, the (I) terms in equations (6.10a) and (6.10b) become equal, and parallel lines are obtained. Non-competitive inhibition gives an intersection point on the (I) axis. This method, which was suggested by Dixon, is illustrated in figure 6.12.

A similar approach, proposed by Cornish-Bowden, uses plots of S_0/v_0 against (I) to obtain K_i'. Multiplying equation (6.10) by S_0 gives

$$\frac{S_0}{v_0} = \frac{K_m}{V} + \frac{S_0}{V} + \left(\frac{K_m/K_i}{V} + \frac{S_0/K_i'}{V}\right)(\text{I}) \tag{6.11}$$

Applying a derivation entirely analogous to that described for the Dixon plot shows that at the intersection point of S_0/v_0 versus (I) plots, $(\text{I}) = -K_i'$. In competitive inhibition $K_i' = \infty$, the (I) terms of equation (6.11) are equal, and the lines are parallel. Non-competitive inhibition gives lines intersecting on the (I) axis, as in the Dixon plot (figure 6.13).

Competitive substrates

Some enzymes are rather non-specific regarding parts of the substrate molecule. For example, chymotrypsin has a preference for non-polar side chains but will catalyse the hydrolysis of a wide variety of amides and esters. Substrate analogues are frequently employed as competitive inhibitors and correlation of the structure of such analogues with their effectiveness as an inhibitor can give valuable information concerning the structural features involved in substrate binding. Sometimes the resemblance of the analogues to the true (i.e. physiological) substrate may be sufficient for them to act as substrates themselves.

Let us consider the case of an enzyme acting in the presence of a mixture of two alternative substrates, S and S′:

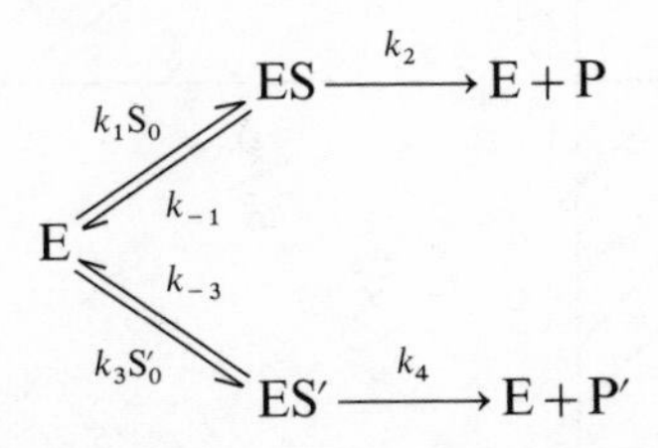

In this mechanism, binding of S and S′ is mutually exclusive and each substrate behaves as a competitive inhibitor of the other. Studies of mixed substrates are usually carried out in a manner similar to those involving

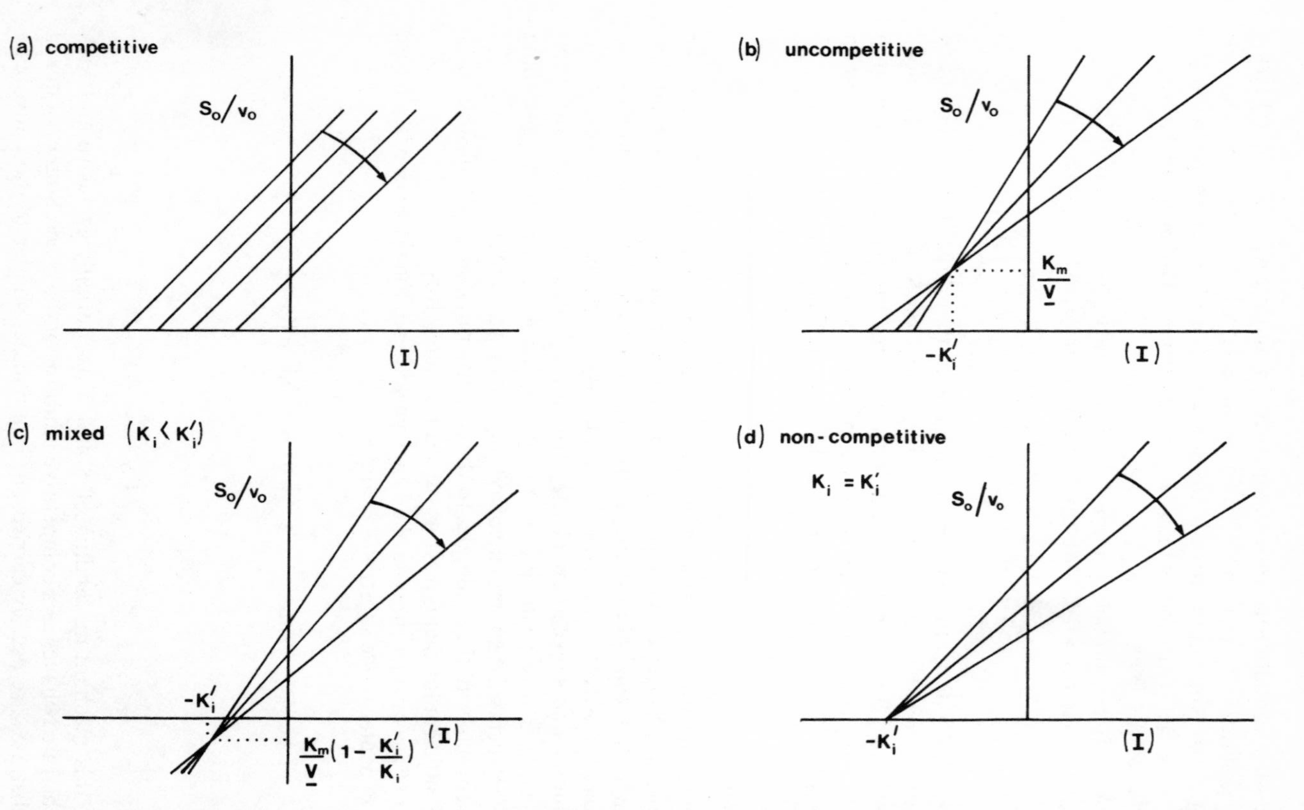

Figure 6.13 Cornish-Bowden plots of S_0/v_0 against (I) for different types of simple inhibition. Arrow indicates increasing values of fixed S_0.

inhibitors. That is, the velocity is measured at several concentrations of one substrate, say S, the concentration of S′ being held constant. The rate equation for this mechanism will depend on the assay method used. The assay may be sensitive to both P and P′. For example if the enzyme is a non-specific phosphatase and the reaction is followed by measuring the production of inorganic phosphate, P and P′ are identical. In this case $v_0 = d(\mathrm{P})/dt + d(\mathrm{P}')dt = k_2(\mathrm{ES}) + k_4(\mathrm{ES}')$ and the rate equation is

$$v_0 = \frac{V\mathrm{S}_0/K_\mathrm{m} + V'\mathrm{S}'_0/K'_\mathrm{m}}{1 + \mathrm{S}_0/K_\mathrm{m} + \mathrm{S}'_0/K'_\mathrm{m}} \tag{6.12}$$

Where $V = k_2\mathrm{E}_0$, $V' = k_4\mathrm{E}_0$, $K_\mathrm{m} = (k_{-1} + k_2)/k_1$ and $K'_\mathrm{m} = (k_{-3} + k_4)/k_3$. Note that equation (6.12) cannot be written in the form of the Michaelis–Menten equation and thus the variation of the rate with S_0, S'_0 being held constant, will not follow normal hyperbolic kinetics and half-reciprocal plots of S_0/v_0 against S_0 will be curved.

Now consider the situation where the assay is sensitive only to P. In the case of phosphatase, S might be *p*-nitrophenyl phosphate, S′ α-glycerophosphate and the assay a spectrophotometric one which follows the appearance of *p*-nitrophenolate at 400 nm. The rate equation can be derived directly from equation (6.12) which can be rewritten as

$$v_0 = \frac{V\mathrm{S}_0}{K_\mathrm{m}\left(1 + \dfrac{\mathrm{S}_0}{K_\mathrm{m}} + \dfrac{\mathrm{S}'_0}{K'_\mathrm{m}}\right)} + \frac{V'\mathrm{S}'_0}{K'_\mathrm{m}\left(1 + \dfrac{\mathrm{S}_0}{K_\mathrm{m}} + \dfrac{\mathrm{S}'_0}{K'_\mathrm{m}}\right)}$$

or

$$v_0 = \frac{V\mathrm{S}_0}{K_\mathrm{m}\left(1 + \dfrac{\mathrm{S}'_0}{K'_\mathrm{m}}\right) + \mathrm{S}_0} + \frac{V'\mathrm{S}'_0}{K'_\mathrm{m}\left(1 + \dfrac{\mathrm{S}_0}{K_\mathrm{m}}\right) + \mathrm{S}'_0} \tag{6.12a}$$

Each of the two main terms of equation (6.12a) is of the same form as equation (6.5) for competitive inhibition, the first term representing the rate of production of P and the second the rate of production of P′. Indeed the same results would be obtained if there were two enzymes each one specific for one substrate and competitively inhibited by the other.

The rate of P production is then

$$v_0 = \frac{V\mathrm{S}_0}{K_\mathrm{m}(1 + \mathrm{S}'_0/K'_\mathrm{m}) + \mathrm{S}_0}$$

and the kinetic pattern obtained at several fixed values of S'_0 will be identical to that for competitive inhibition. However, the "K_i" obtained will *not* be a true dissociation constant but will be the Michaelis constant for S′. This is true even if $k_4 \ll k_2$. This approach can be used to obtain the K_m of a substrate for which no convenient assay is available by carrying out kinetic measurements in a mixed system with a conveniently assayable chromogenic competitive substrate.

Non-productive binding

It was pointed out in chapter 4 that binding of substrate to enzyme is likely to be multidentate. With the exception of the case of substrate inhibition discussed in chapter 4, we have considered that the substrate is always bound to the enzyme in such a way that the groups on the enzyme are correctly aligned with the complementary groups of the substrate so that the resulting ES complex will be able to break down to product. Such productive binding can be called *eutopic*, from the Greek "good place". Let us consider the case where the substrate might bind to the active site in a misoriented fashion so that the resulting ES complex is catalytically inactive. The effect of such *dystopic* ("bad place") or *non-productive* binding can be expressed in the following mechanism

$$\begin{array}{l} \mathrm{E} \underset{k_{-1}}{\overset{k_1(\mathrm{S})}{\rightleftharpoons}} \mathrm{ES} \xrightarrow{k_2} \mathrm{E+P} \\ \mathrm{E} + \mathrm{S} \overset{K'}{\rightleftharpoons} \mathrm{SE} \end{array} \tag{6.13}$$

where $K' = (\mathrm{E})(\mathrm{S})/(\mathrm{SE})$ is the dissociation constant of the dystopic complex. Mechanism (6.13) is entirely analogous to mechanism (6.4) for competitive inhibition but in this case it is the substrate itself which acts as the inhibitor. The rate equation can thus be written immediately,

$$v_0 = \frac{V\mathrm{S}_0}{K_m(1+\mathrm{S}_0/K')+\mathrm{S}_0} \tag{6.14}$$

which can be compared to equation (6.5) for competitive inhibition. Because inhibitor and substrate are identical, equation (6.14) has to be rearranged to cast it into the form of the Michaelis–Menten equation:

$$v_0 = \frac{V\mathrm{S}_0}{K_m+\mathrm{S}_0(1+K_m/K')}$$

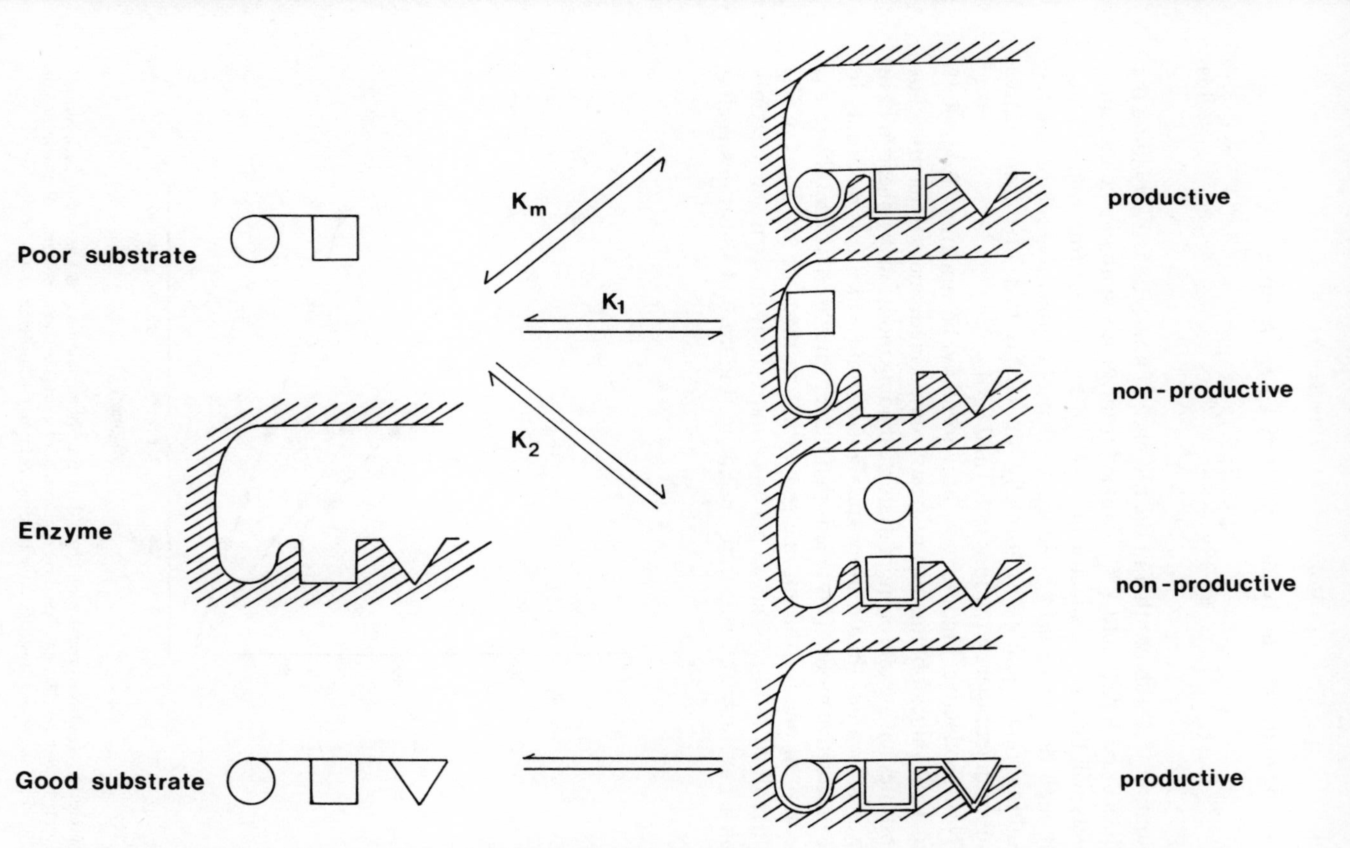

Figure 6.14 Schematic illustration of non-productive binding; the observed Michaelis constant for the "poor" substrate will be $K_m^{app} = K_m/(1 + K_m/K_1 + K_m/K_2)$. Adapted from Jencks (1966).

Dividing numerator and denominator by $1+K_m/K'$ gives

$$v_0 = \frac{VS_0/(1+K_m/K')}{K_m/(1+K_m/K')+S_0} \tag{6.15}$$

Equation (6.15) shows that in the presence of non-productive binding the values of both K_m^{app} and V^{app} obtained from steady-state kinetic measurements will be less, by a factor of $1+K_m/K'$, than their true values. Thus, although the mechanism is similar to that of competitive inhibition, the resulting rate equation is formally similar to that of uncompetitive inhibition (compare equations (6.15) and 6.8)).

Non-productive binding will not normally be apparent from K_m^{app} and V^{app} values obtained from steady-state kinetic measurements. Thus substrates having one or more dystopic binding modes may appear to have lower K_m values than a more specific substrate which can bind only in the productive mode. This means that judgements as to which substrate is the natural one for the enzyme should not be made on the basis of K_m values alone. Non-productive binding is more likely to occur with enzymes which have a broad substrate specificity. Figure 6.14 illustrates such a

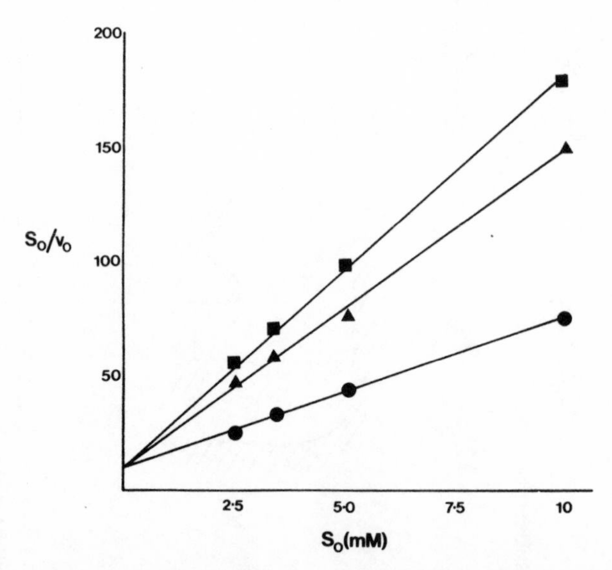

Figure 6.15 Half-reciprocal plot of oxidation of phenylalanine (●), *m*-iodo- (▲), and *p*-iodophenylalanine (■), catalysed by L-amino acid oxidase. The common ordinate intercept (K_m^{app}/V^{app}) suggests non-productive binding. Adapted from Zeller *et al.* (1975).

situation for a "good" substrate which binds only eutopically and a "poor" one which has two dystopic binding modes. An example of this is the L-amino acid oxidase of snake venom. If phenylalanine is taken as the "good" substrate it can be seen (figure 6.15) that substitution of a bulky substituent in the benzene ring decreases both V and K_m leaving the ratio K_m/V unaffected. It is worth pointing out that contamination of substrate by a competitive inhibitor (i.e. a D/L substrate mixture) has the same effect on the kinetics as dystopic complex formation.

Transition-state analogues

About 30 years ago Pauling asserted that the structure of the active site of an enzyme is complementary not to that of the substrate but to the transition state of the substrate in the reaction

$$S \xrightarrow{E} P$$

He further proposed that the enzyme binds the transition state much more tightly than it does the substrate.

In order to see why this should be so we must compare the uncatalysed reaction with the enzyme-catalysed one.

$$\begin{array}{ccccc} E+S & \overset{K_u^{\ddagger}}{\rightleftharpoons} & E+S^{\ddagger} & \longrightarrow & E+P \\ K_S \updownarrow & & K_{S^{\ddagger}} \updownarrow & & \\ ES & \underset{}{\overset{K_c^{\ddagger}}{\rightleftharpoons}} & ES^{\ddagger} & \longrightarrow & E+P \end{array} \tag{6.16}$$

The top line of mechanism (6.16) represents the uncatalysed reaction

$$S \xrightarrow{k_u} P$$

with a unimolecular rate constant k_u. The E terms are superfluous to this reaction but are included to relate it to the enzyme-catalysed reaction

$$ES \xrightarrow{k_c} E+P$$

with unimolecular rate constant k_c on the bottom line. According to transition state theory (chapter 1) an equilibrium exists between the ground state of the reactant S and its transition state $S^{\ddagger}$ governed by the equilibrium constant $K_u^{\ddagger} = (S^{\ddagger})/(S)$ for the uncatalysed reaction and $K_c^{\ddagger} = (ES^{\ddagger})/(ES)$ for the catalysed one. The corresponding rate constants are given by

$$k_u = K_u^{\ddagger}\, kT/h$$

$$k_c = K_c^{\ddagger}\, kT/h$$

and so

$$k_u/k_c = K_u^{\ddagger}/K_c^{\ddagger} \qquad (6.17)$$

where k is the Boltzmann constant, T the absolute temperature and h Planck's constant, and the ratio of the rates of the uncatalysed to the catalysed reaction is given by equation (6.17). K_S and $K_{S^{\ddagger}}$ are the dissociation constants of the substrate and its transition state, respectively, from their complexes with the enzyme, ES and $ES^{\ddagger}$. Mechanism (6.16) constitutes a thermodynamic "box" and, as with the mechanism of figure 6.1 any one of the four constants is defined by the other three. Thus

$$K_{S^{\ddagger}} = K_S K_u^{\ddagger}/K_c^{\ddagger}$$

and from equation (6.17),

$$K_{S^{\ddagger}} = K_S k_u/k_c \qquad (6.18)$$

Equation (6.18) predicts that the dissociation constant of the $ES^{\ddagger}$ complex will be smaller (by a factor k_u/k_c) than the dissociation constant of the ES complex; in other words the transition state binds much more tightly to the enzyme than does the substrate. This argument assumes no involvement of induced fit in binding. For most one-substrate enzymes the ratio of the catalysed to the uncatalysed rate ("the rate ratio") can be very high, 10^{10} or greater, and this should then be the ratio of the dissociation constants of the ES complex to that of the $ES^{\ddagger}$ complex (the "binding ratio"). Indeed the rate of the uncatalysed reaction may be immeasurably slow so that only a minimal estimate of the binding ratio can be made.

A *transition-state analogue* is a stable compound whose structure resembles the transition state of the substrate portion of the $ES^{\ddagger}$ complex. Because the transition state is by definition the least stable species along the reaction coordinate, its structure cannot be determined directly but can only be inferred from a knowledge of the mechanism of the reaction. A number of inhibitors purporting to resemble the transition state have been synthesized and found to have binding affinities some 2 to 4 orders of magnitude greater than that of the substrate. The discrepancy between these values and the predicted binding ratio of $\sim 10^{10}$ probably results from small structural differences between the analogue and the true transition state.

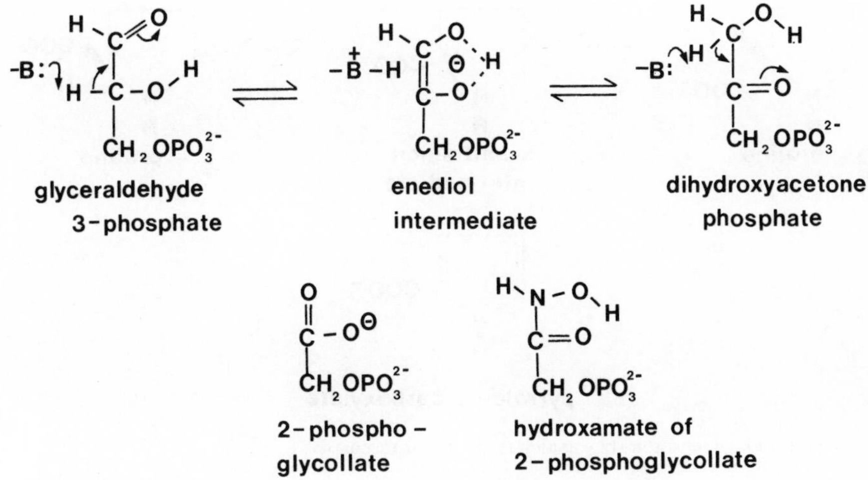

glyceraldehyde 3-phosphate ⇌ enediol intermediate ⇌ dihydroxyacetone phosphate

2-phosphoglycollate; hydroxamate of 2-phosphoglycollate

Figure 6.16 Transition-state analogues for triose phosphate isomerase.

As an example we may consider the isomerization of glyceraldehyde-3-phosphate to dihydroxyacetone phosphate catalysed by triose phosphate isomerase. The reaction is thought to go via an enediol, the proton transfer between carbons 1 and 2 of the triose being mediated by a basic group on the enzyme. The reaction, shown in figure 6.16, includes a postulated structure for the transition state. Also shown are the structures of 2-phosphoglycollate and its hydroxamate which have been proposed as transition-state analogues and found to have enzyme dissociation constants some 600 and 30 times less, respectively, than the dissociation constant of the substrate. Another example is the racemization of D- or L-proline catalysed by proline racemase. At some stage during the reaction the α-carbon assumes a planar configuration and so the substrate binding site must be such that interaction with a molecule having a planar α-carbon is favoured over interaction with molecules with a tetrahedral α-carbon. The planar compound pyrrole-2-carboxylate is a powerful inhibitor of proline racemase ($K_i \sim 300$ times smaller than K_m) and can be considered to be a transition-state analogue (figure 6.17).

Transition-state analogues are tightly-bound highly-specific competitive inhibitors for which a number of applications exist. As the complexes formed with the enzyme are stable, they may be used to trap the enzyme in a state which mimics its conformation during the catalytic event and allows structural investigations by physical methods. Their specificity and

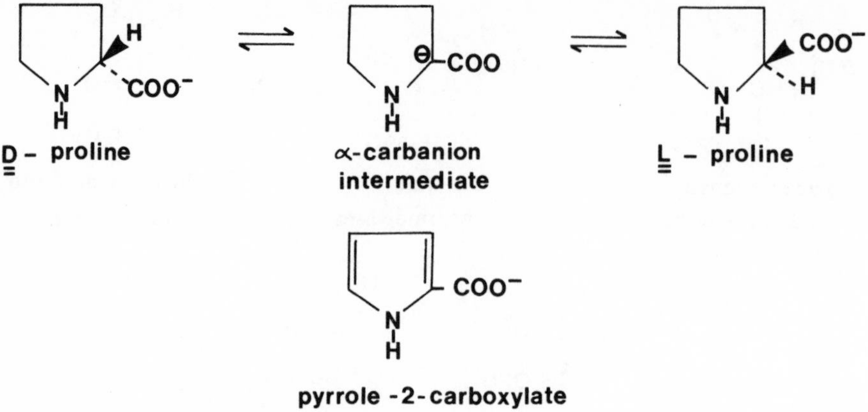

D - proline ⇌ α-carbanion intermediate ⇌ L - proline

pyrrole -2-carboxylate

Figure 6.17 Transition-state analogue for proline racemase.

potency as enzyme inhibitors can be used to advantage in metabolic and pharmacological studies and may result in the development of useful chemotherapeutic agents.

Partial inhibition

In the general inhibition mechanism of figure 6.1 the ESI complex was assumed to be catalytically inert. If we remove this restriction we obtain

$$\begin{array}{ccccc} E & \underset{+S}{\overset{K_m}{\rightleftharpoons}} & ES & \xrightarrow{k} & E+P \\ K_i \Updownarrow +I & & K_i' \Updownarrow +I & & \\ EI & \underset{}{\overset{+S}{\rightleftharpoons}} & ESI & \xrightarrow{k'} & EI+P \end{array} \tag{6.19}$$

Mechanism (6.19) is similar to that of figure 6.1 but has been further generalized to allow the ESI complex to break down to product. The productive steps with rate constants k and k' are assumed to be slow relative to the reversible steps; thus the latter can be taken to be at equilibrium described by the indicated dissociation constants as before.

The rate equation of mechanism (6.19) is

$$v_0 = \frac{E_0 S_0 (k + k'(I))/(1 + (I)/K_i')}{K_m\left(\dfrac{1+(I)/K_i}{1+(I)/K_i'}\right) + S_0} \tag{6.20}$$

which is of the same form as the Michaelis–Menten equation (6.3) with

$$K^{app} = \frac{K_m(1+(I)/K_i)}{(1+(I)/K_i')}$$

and $V^{app} = E_0(k+k'(I))/(1+(I)/K_i')$. Thus K_m^{app}, V^{app} and V^{app}/K_m^{app} all depend on (I) and a graphical pattern typical of mixed inhibition will be obtained in half-reciprocal plots of S_0/v_0 upon S_0. However, there is an important difference between equation (6.20) and equation (6.2) for mixed inhibition described at the beginning of this chapter. Both K_m^{app}/V^{app} and $1/V^{app}$ obtained from equation (6.2) are linear functions of (I). This applies also to the special cases of competitive and uncompetitive inhibition discussed above and for this reason these inhibition types are sometimes referred to as *linear inhibition.* But from equation (6.20)

$$1/V^{app} = (1+(I)/K_i')/E_0(k+k'(I))$$

and

$$K_m/V^{app} = K_m(1+(I)/K_i)/E_0(k+k'(I))$$

These equations describe a *hyperbolic* rather than a linear relationship of $1/V^{app}$ and K_m^{app}/V^{app} to (I) and thus inhibitions described by equation (6.20) and those derived from it are known as *hyperbolic inhibition.* This means that secondary plots of $1/V^{app}$ and K_m^{app}/V^{app} against (I) will be curved. Another distinction between mechanism (6.19) and that of figure 6.1 results from the ability of the ESI complex to break down to product. As a consequence there will always be some active complex present and inhibition never approaches 100% no matter how high (I) is raised. For this reason such inhibition is also known as *partial inhibition.* Equation (6.20) thus represents partial or hyperbolic mixed inhibition.

Depending on the ratios K_i/K_i' and k/k', various special cases of partial inhibition can be formulated which have their counterparts in simple linear inhibition. For example if $k = k'$ (i.e. the inhibitor does not affect the rate of the productive step) then equation (6.20) reduces to

$$v_0 = \frac{kE_0S_0}{K_m\left(\dfrac{1+(I)/K_i}{1+(I)/K_i'}\right)+S_0} \qquad (6.21)$$

From equation (6.21), $V^{app} = kE_0$ $(=V)$ and is unaffected by (I) and $K_m^{app} = K_m(1+(I)/K_i)/(1+(I)/K_i')$. If $K_i' > K_i$, then half-reciprocal plots will yield a pattern of parallel lines as for competitive inhibition.* Equation (6.21) thus describes *partial competitive inhibition.* Equations for partial non-competitive inhibition ($K_i/K_i' = 1 < k/k'$) and partial uncompetitive inhibition ($K_i/K_i' = k/k'$) can be set up which will also give typical patterns in half-reciprocal plots. But in all cases of partial inhibition the variation of K_m^{app}/V^{app} or $1/V^{app}$ with (I) will be hyperbolic and both Dixon and Cornish-Bowden plots will be curved.

Effects of pH

The sensitivity of enzyme activity to hydrogen concentration has long been known. Enzymes are usually irreversibly inactivated at pH below 4 and above 10. This loss of activity is caused by denaturation due to disruption of the tertiary structure of the protein. But for most enzymes there exists a pH region in which the variation of rate with pH is reversible. It is this phenomenon with which we shall be mainly concerned.

Enzymes, being proteins, are polyelectrolytes containing many acidic and basic groups on the amino-acid side chains. When the number of positive and negative charges on these groups are equal the protein will have no net charge and will have zero mobility in an electric field. The pH at which this occurs is called the isoelectric point. As pointed out by Haldane in 1930, the pH region of the isoelectric point often does not correspond to the pH range over which enzymes show maximal activity. It is therefore reasonable to assume that enzyme activity depends on proton ionizations of particular groups on the molecule.

pH *profiles*

Over the range in which enzyme activity is reversibly affected by pH, a plot of activity vs. pH (known as a pH *profile*) is often *bell-shaped* (figure 6.18). The low-pH part of the curve is conventionally called the rising and the high-pH part the falling arm or limb; the pH at which activity is highest is termed the *optimum* pH or pH_{opt}. Let us assume that there are two ionizing groups both involved in conferring activity on the enzyme. One group is required in the form of its conjugate acid (protonated), A–H, and the other

* Note that if $K_i > K_i'$, K_m^{app} will be *smaller* than K_m and, as V^{app} is unchanged, I will act as an *activator* rather than an inhibitor. In fact, mechanism (6.19) can be treated as a general mechanism for activators as well as inhibitors, and cases can be conceived (e.g. $K_i/K_i' > k/k' > 1$) in which I behaves both as activator (at high S_0) and as inhibitor (at low S_0).

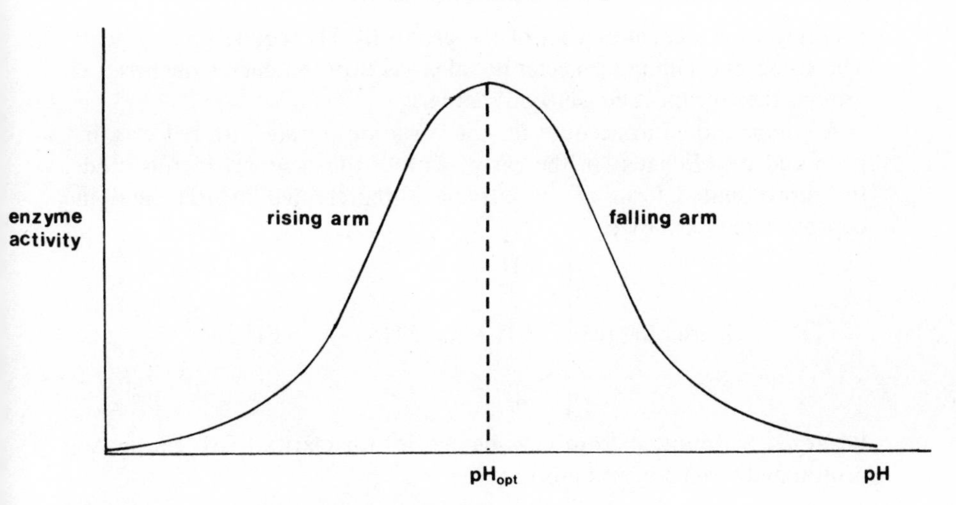

Figure 6.18 A typical enzyme activity–pH profile.

is required as its conjugate base (deprotonated), B. Let us consider the effect of pH on the ionization state of these groups (figure 6.19). The protonic equilibria are governed by the acid dissociation constants K_a and K_b. B–H is assumed to be a stronger acid than A–H and thus $K_b > K_a$. The rising arm on the low-pH side of the pH profile represents the increase in activity as the group required as a base (B) is generated by deprotonation. The falling arm on the high-pH side represents the loss in activity

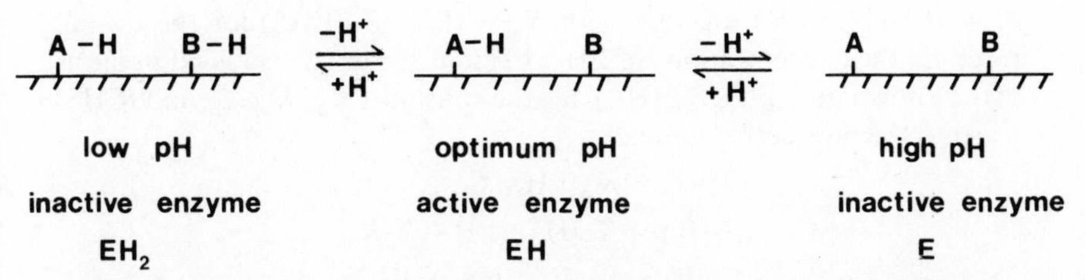

Figure 6.19 Effect of pH on ionizing groups on an enzyme. The diprotonated, monoprotonated, and unprotonated enzyme forms are EH_2, EH and E respectively. In this model the alternative monoprotonated form

A B–H

is assumed not to exist.

resulting from deprotonation of the group (A–H) required as an acid.* The rising and falling arms can be taken as titration curves of these two groups, the "titrant" being enzyme activity.

A kinetic model to account for the variation of rate with pH was first proposed by Michaelis in the early years of this century. In this model the diprotonated form of the enzyme is represented by EH_2 and the deprotonated form by E.

$$\text{increasing pH}\ \Big\downarrow \quad \begin{array}{c} EH_2 \\ K_b \ \upharpoonleft\!\downharpoonright \\ EH + S \rightleftharpoons EHS \xrightarrow{k_2} EH + P \\ K_a \ \upharpoonleft\!\downharpoonright \\ E \end{array}$$

(Charges are omitted from enzyme species for clarity.) EH, the monoprotonated form, corresponds to

```
A–H   B
 |____|
```

and is assumed to be the only species able to interact with substrate to form the enzyme–substrate complex. Comparison of this mechanism with mechanism (6.6) for competitive inhibition shows that they are very similar. In this case protonation of EH and binding of S are mutually exclusive and the proton thus acts as a competitive inhibitor. Similarly, protonation of E, which will not interact with S, is required to form EH, which will, and the proton thus acts as a compulsory activator with respect to E. Alternatively OH^- may be looked upon as the species which converts EH to the active form E.

Using the relevant dissociation constants $K_b = (EH)(H^+)/(EH_2)$ and $K_a = (E)(H^+)/(EH)$ together with $K_m = (EH)(S)/(EHS)$ for the overall reaction, the concentrations of EH_2, EH and E can be expressed in terms of the known quantities S_0, (H^+), E_0, the constants K_m, K_a, K_b, and (EHS) which is the productive complex:

$$(EH) = K_m(EHS)/S_0$$
$$(EH_2) = K_m(H^+)(EHS)/S_0K_b$$
$$(E) = K_mK_a(EHS)/S_0(H^+)$$

* The proposal that only two ionizing groups are involved in enzyme activity may seem unduly naive. However, it follows from the fact that A–H and B represent the ionizing groups whose pK values lie nearest to pH_{opt}. There may well be other groups on whose state of protonation enzyme activity is dependent but these will not be detected by this approach (unless their pK's are very close to pK_a and pK_b), because the enzyme will already be in an inactive form when the pH at which they protonate or deprotonate is reached.

The rate of the overall reaction is $d(\mathrm{P})/dt = k_2(\mathrm{EHS})$ and the conservation equation is

$$\mathrm{E}_0 = (\mathrm{E})+(\mathrm{EH})+(\mathrm{EH}_2)+(\mathrm{EHS})$$

Substituting the expressions above into the conservation equation, solving for (EHS) and substituting into the rate expression gives:

$$v_0 = \frac{d(\mathrm{P})}{dt} = \frac{V\mathrm{S}_0}{K_\mathrm{m}\left(1+\dfrac{K_\mathrm{a}}{(\mathrm{H}^+)}+\dfrac{(\mathrm{H}^+)}{K_\mathrm{b}}\right)+\mathrm{S}_0} \tag{6.22}$$

where $V = k_2\mathrm{E}_0$. Equation (6.22) has the form of the Michaelis–Menten equation where $K_\mathrm{m}^\mathrm{app} = K_\mathrm{m}(1+K_\mathrm{a}/(\mathrm{H}^+)+(\mathrm{H}^+)/K_\mathrm{b})$. If K_b and K_a differ sufficiently there will be a pH range, on the rising arm of the pH profile, over which BH can be deprotonated without affecting the protonation state of group A. In this range $(\mathrm{H}^+)/K_\mathrm{b} \gg K_\mathrm{a}/(\mathrm{H}^+)$ and

$$v_0 = \frac{V\mathrm{S}_0}{K_\mathrm{m}\left(1+\dfrac{(\mathrm{H}^+)}{K_\mathrm{b}}\right)+\mathrm{S}_0} \tag{6.23}$$

Equation (6.23) is of the same form as equation (6.5) for competitive inhibition and thus can be analysed by using Dixon plots of $1/v_0$ against (H^+) at different fixed S_0 or plots of S_0/v_0 on S_0 at different (H^+) as described earlier in this chapter. Thus (H^+) acts as a competitive inhibitor with inhibitor constant K_b.

The same considerations apply to the falling arm of the pH-profile. In this case over the high pH range, $K_\mathrm{a}/(\mathrm{H}^+) \gg (\mathrm{H}^+)/K_\mathrm{b}$ and

$$v_0 = \frac{V\mathrm{S}_0}{K_\mathrm{m}\left(1+\dfrac{K_\mathrm{a}}{(\mathrm{H}^+)}\right)+\mathrm{S}_0} \tag{6.24}$$

By invoking the expression for the ion product of water, $(\mathrm{H}^+)(\mathrm{OH}^-) = 10^{-14}$, equation (6.24) can be written as

$$v_0 = \frac{V\mathrm{S}_0}{K_\mathrm{m}\left(1+\dfrac{(\mathrm{OH}^-)}{10^{-14}/K_\mathrm{a}}\right)+\mathrm{S}_0}$$

which is again of the same form as equation (6.5). Thus OH^- acts as a competitive inhibitor over this pH range with a "K_i" of $10^{-14}/K_\mathrm{a}$.

The model of Michaelis described above predicts that $K_\mathrm{m}^\mathrm{app}$ and hence

$d(\mathrm{P})/dt$, varies with pH. However, V is pH-independent. Thus the dependance of *rate* on pH will decrease as S_0 increases and the rate will be pH-independent when $S_0 \gg K_m^{app}$. Such a situation is rarely observed. This limitation can be overcome if both the free enzyme and the enzyme–substrate complex are assumed to be capable of protonation and deprotonation to inactive species. The following model, proposed by Waley in 1953, is the most widely used:

$$
\begin{array}{ccccc}
EH_2 & & EH_2S & & \\
K_b \updownarrow & & \updownarrow K_b' & & \\
EH + S & \underset{k_{-1}}{\overset{k_1}{\rightleftharpoons}} & EHS & \xrightarrow{k_2} & P + EH \\
K_a \updownarrow & & \updownarrow K_a' & & \\
E & & ES & &
\end{array}
$$

EH_2S and ES are non-productive complexes and their protonic equilibria with EHS are characterized by the constants $K_b' = (\mathrm{EHS})(\mathrm{H}^+)/(\mathrm{EH_2S})$ and $K_a' = (\mathrm{E})(\mathrm{H}^+)/(\mathrm{EHS})$. The direct combination of EH_2 and E with S is not shown explicitly. This does not mean that these reactions do not occur. But if the vertical steps in the mechanism are assumed to be very rapid compared with the horizontal steps, the concentrations of E, EH_2, ES and EH_2S will be governed by the constants K_a, K_b, K_a' and K_b'. This assumption is a reasonable one because proton transfers between electronegative groups are usually very fast compared with other bond-making and breaking processes. The assumption that EH_2S and ES are inactive is justified if, as is often the case, the pH-profile is symmetrical and the enzyme activity tends to zero at high and low pH.

The rate equation corresponding to the general mechanism is easily derived using the same approach as for the simpler Michaelis model.

$$v_0 = \frac{d(\mathrm{P})}{dt} = \frac{V S_0}{K_m\left(1 + \dfrac{K_a}{(\mathrm{H}^+)} + \dfrac{(\mathrm{H}^+)}{K_b}\right) + S_0\left(1 + \dfrac{K_a'}{(\mathrm{H}^+)} + \dfrac{(\mathrm{H}^+)}{K_b'}\right)} \tag{6.25}$$

where $K_m = (k_{-1} + k_2)/k_1$. Rewriting equation (6.25) as

$$v_0 = \frac{V S_0 \Big/ \left(1 + \dfrac{K_a'}{(\mathrm{H}^+)} + \dfrac{(\mathrm{H}^+)}{K_b'}\right)}{K_m\left(\dfrac{1 + \dfrac{K_a}{(\mathrm{H}^+)} + \dfrac{(H^+)}{K_b}}{1 + \dfrac{K_a'}{\mathrm{H}^+} + \dfrac{(\mathrm{H}^+)}{K_b'}}\right) + S_0} \tag{6.26}$$

which is of the same form as the Michaelis–Menten equation with

$$K_m^{app} = K_m \left(\frac{1 + \dfrac{K_a}{(H^+)} + \dfrac{(H^+)}{K_b}}{1 + \dfrac{K'_a}{(H^+)} + \dfrac{(H^+)}{K'_b}} \right) \tag{6.27}$$

$$V^{app} = V \bigg/ \left(1 + \frac{K'_a}{(H^+)} + \frac{(H^+)}{K'_b} \right) \tag{6.28}$$

and

$$\frac{V^{app}}{K_m^{app}} = \frac{V}{K_m} \bigg/ \left(1 + \frac{K_a}{(H^+)} + \frac{(H^+)}{K_b} \right) \tag{6.29}$$

we see that K_m^{app} and V^{app} will not, in general, vary in the same way with pH. Specifically, at $S_0 \gg K_m^{app}$ the shape of the pH-profile will be determined by K'_a and K'_b, the ionization constants of the enzyme–substrate complex; at $S_0 \ll K_m^{app}$, K_a and K_b, the ionization constants of the free enzyme, determine the shape. As K_m^{app} is itself pH-dependent, a substrate concentration which is saturating at one pH may well not be at a different pH. Thus a pH-profile obtained at a single substrate concentration is rather useless for providing information on enzyme ionizations. A proper pH study should involve determination of K_m^{app} and V^{app} at each pH.

If K_b/K_a and K'_b/K'_a are sufficiently large, then at low pH such that $(H^+)/K_b \gg K_a/(H^+)$ and $(H^+)/K'_b \gg K'_a/(H^+)$ equation (6.26) becomes

$$v_0 = \frac{V S_0/(1 + (H^+)/K'_b)}{K_m \left(\dfrac{1 + (H^+)/K_b}{1 + (H^+)/K'_b} \right) + S_0} \tag{6.30}$$

which is of the same form as equation (6.2) describing mixed inhibition. Thus in this pH range (H^+) behaves as a mixed inhibitor and the data can be analysed as described earlier in this chapter. An analogous treatment at high pH such that $(H^+)/K_b$ and $(H^+)/K'_b$ are much smaller than $(H^+)/K_a$ and $(H^+)/K'_a$ respectively, gives

$$v_0 = \frac{V S_0/(1 + K'_a/(H^+))}{K_m \left(\dfrac{1 + K_a/(H^+)}{1 + K'_a/(H^+)} \right) + S_0} \tag{6.31}$$

Substitution of $10^{-14}/(OH^-)$ for (H^+) in equation (6.31) shows that at high pH OH^- acts as a mixed inhibitor and the same methods of analysis can be applied.

For the special case where $K_a = K'_a$ and $K_b = K'_b$ (i.e. the presence of bound substrate does not affect the ionization constants), we obtain

$$v_0 = \frac{VS_0/(1+K_a/(H^+)+(H^+)/K_b)}{K_m+S_0}$$

In this case, at the low and high pH regions, (H^+) and (OH^-) respectively will act as non-competitive inhibitors.

Analysis of pH *profiles*

It is evident from equations (6.28) and (6.29) that the pH-variation of V^{app} is governed by the values of K'_a and K'_b, the ionization constants of the enzyme–substrate complexes, whereas the pH-dependence of V^{app}/K_m^{app} reflects the ionizations on the free enzyme, i.e. K_a and K_b. The discussion which follows is mainly concerned with the determination of K'_a and K'_b (and V) from V^{app}. However, as equations (6.28) and (6.29) are identical in form, any method applicable to analysis of V^{app} will be valid for V^{app}/K_m^{app} and the determination of K_a, K_b and V/K_m.

If K'_a and K'_b are sufficiently far apart, the pH profile of V^{app} will have a flat top, and V_{opt}, which is the value of V^{app} at pH_{opt}, will be virtually equal to the true V. If data are available over a wide pH range then pK'_a and pK'_b can be simply read off the V^{app}–pH-profile as the values of pH at which V_{opt} is half maximal. This follows from equation (6.28), because if $K'_b \gg K'_a$ then $K'_a/(H^+)$ is negligible compared to $(H^+)/K'_b$ in the rising arm of the pH-profile and

$$V^{app} = \frac{V}{1+(H^+)/K'_b} \tag{6.32}$$

If we define the pH at which $V^{app}/V_{opt} = \frac{1}{2}$ on the rising and falling side as pH_b and pH_a respectively, then

$$1+(H^+)_b/K'_b = 2$$

and

$$(H^+)_b = K'_b$$

or

$$pH_b = pK'_b$$

On the falling side $(H^+)/K'_b$ is negligible compared to $K'_a/(H^+)$ and

$$V^{app} = \frac{V}{1+K'_a/(H^+)} \tag{6.33}$$

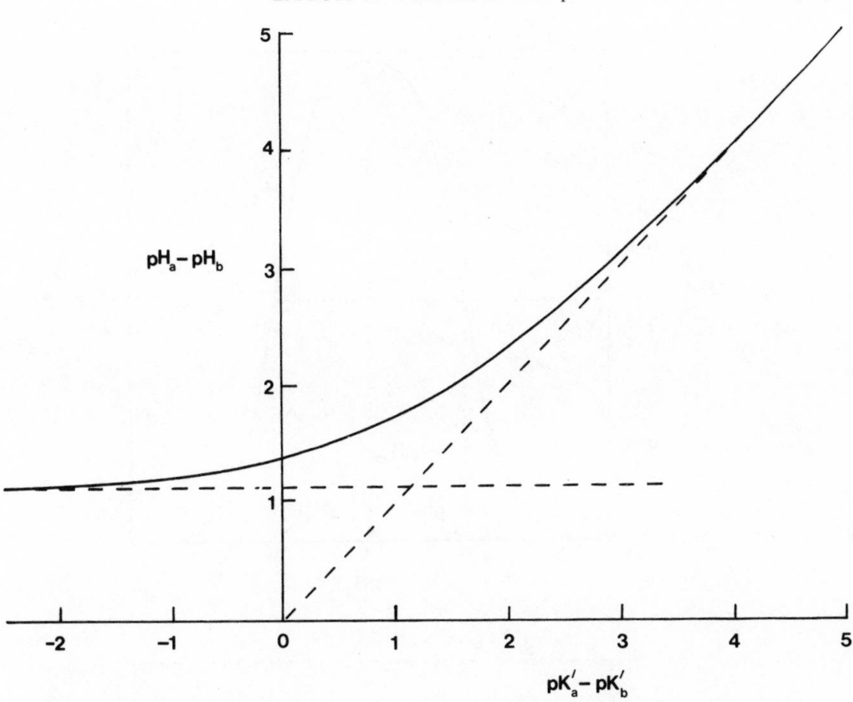

Figure 6.20 Relationship between the width of a pH at half height ($pH_a - pH_b$) and the difference between the molecular p*K* values ($pK'_a - pK'_b$); from Tipton and Dixon (1979).

At pH_a

$$1 + K'_a/(H^+)_a = 2$$

$$(H^+)_a = K'_a$$

or

$$pH_a = pK'_a$$

This procedure only works if the difference between pK'_a and pK'_b is sufficiently large, for only then will $V \simeq V_{opt}$. The difference $pK_a - pK_b$ will of course not be known *ab initio* but this can be determined from the relationship between the width of the pH-profile at half height ($pH_a - pH_b$) and the p*K* difference, $pK'_a - pK'_b$ shown in figure 6.20. Here it is seen that $pK'_a - pK'_b$ must be at least 3 and preferably > 3.5 to allow pK'_a and pK'_b to be estimated with good accuracy by this method. The application of the method is illustrated in figure 6.21(a). If pK'_a and pK'_b are not sufficiently well separated the pH profile will not have a flat top and V_{opt} will be less

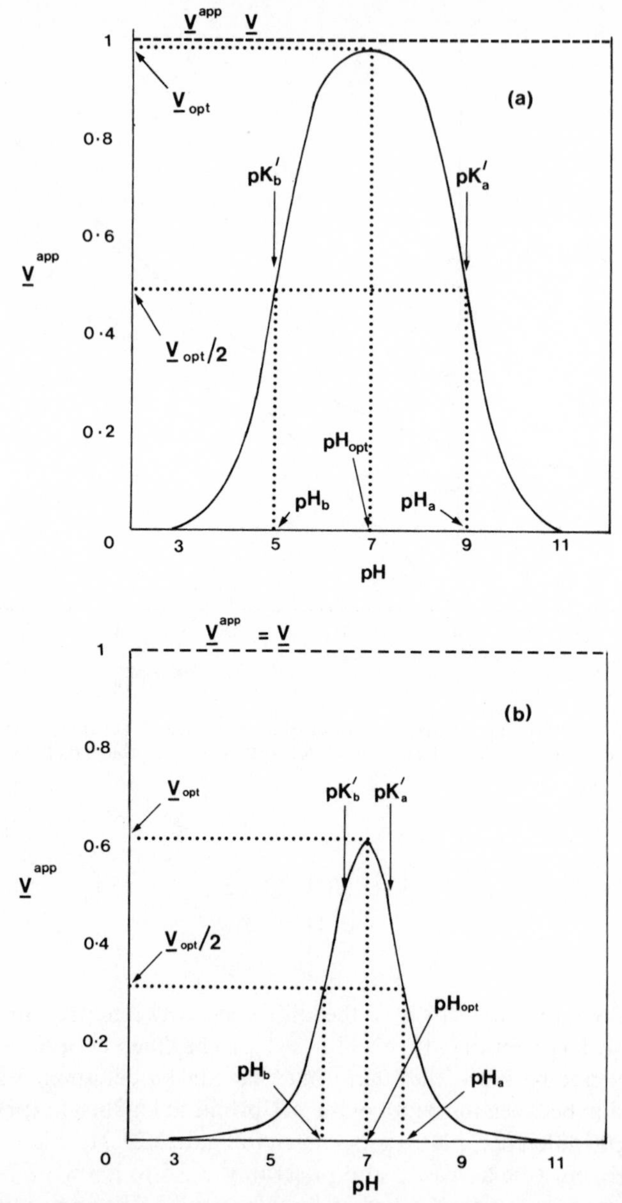

Figure 6.21 pH profiles of V^{app} ($V = 1$). (a) Calculated assuming $pK'_b = 5.0$, $pK'_a = 9.0$. (b) Calculated assuming $pK'_b = 6.5$, $pK'_a = 7.5$. pH_{opt} ($= (pK'_a + pK'_b)/2$) is 7.0 in both cases. Note the effect of pK difference on the shape of the curves.

than V. Consequently pH_a and pH_b will not provide good estimates of pK'_a and pK'_b (see figure 6.21(b)).

The most commonly used procedure for obtaining pK values is that of Dixon. This relies on plots of log V^{app} against pH. If logarithms are taken of both sides of equations (6.32) and (6.33) one obtains

$$\log V^{app} = \log V - \log(1 + (H^+)/K'_b)$$

and

$$\log V^{app} = \log V - \log(1 + K'_a/(H^+))$$

At sufficiently low and high pH $(H^+)/K'_b \gg 1$ and $K'_a/(H^+) \gg 1$ so that

$$\log V^{app} = \log V + pH - pK'_b \tag{6.34}$$

and

$$\log V^{app} = \log V - pH + pK'_a \tag{6.35}$$

and a plot of log V^{app} against pH will be a curve which is asymptotic to straight lines of slope +1 and −1. The pH values at which these lines cross a horizontal line drawn through or near V_{opt} are pK'_b and pK'_a. Dixon has devised a set of rules for constructing such logarithmic plots which are also applicable, under favourable circumstances, to the analysis of the more complex variation of K_m with pH (equation (6.27)) and which provide estimates of all four ionization constants. These are given in most specialized texts on enzyme kinetics and will not be repeated here. However, it should be noted that K'_a and K'_b must be reasonably well separated for the method to work successfully. More important and less widely realized is that for the relationships given by equations (6.34) and (6.35) to be valid, a wide pH range must be covered. Specifically, data must be obtained *at least* one pH unit below and above pK'_b and pK'_a respectively in order to be able to position the tangents to those portions of the curve that approach linearity.

A more direct, computational approach was proposed by Alberty and Massey who devised a method applicable to cases where pK'_a and pK'_b are not widely separated. This is based on the relationship between $(H^+)_{opt}$ and K'_a and K'_b. At pH_{opt} the slope of the tangent of the pH profile is zero. By differentiating equation (6.28) with respect to (H^+) and setting the derivative equal to zero one obtains

$$(H^+)^2_{opt} = K'_a K'_b \tag{6.36}$$

or

$$(H^+)_{opt} = \sqrt{K'_a K'_b} \tag{6.37}$$

Substituting equation (6.37) for (H^+) in equation (6.28) gives

$$V_{opt} = \frac{V}{1+2\sqrt{K'_a/K'_b}} \tag{6.38}$$

The relationship of $(H^+)_a$ and $(H^+)_b$ to K'_a and K'_b can be found from equations (6.28) and (6.38) realizing that at $(H^+)_a$ and $(H^+)_b$, $V^{app}/V_{opt} = \frac{1}{2}$:

$$\frac{V^{app}}{V_{opt}} = \tfrac{1}{2} = \frac{1+2\sqrt{K'_a/K'_b}}{1+K'_a/(H^+)+(H^+)/K'_b}$$

This gives a quadratic in (H^+) of the form $ax^2+bx+c=0$,

$$(H^+)^2 - \left(K'_b+4\sqrt{K'_aK'_b}\right)(H^+)+K'_aK'_b = 0$$

the roots of which are $(H^+)_a$ and $(H^+)_b$. The sum of the roots of a quadratic is given by $-b/a$ and so

$$(H^+)_a+(H^+)_b = K'_b+4\sqrt{K'_aK'_b}$$

Invoking equation (6.37) and rearranging gives

$$K'_b = (H^+)_a+(H^+)_b+4(H^+)^2_{opt}$$

As $(H^+)_a$, $(H^+)_b$ and $(H^+)_{opt}$ can all be estimated from the pH-profile, K'_b can be calculated and K'_a obtained from K'_b and equation (6.36). A convenient way of obtaining pK'_a and pK'_b based on this method has been devised by H. Dixon. If $pH_a - pH_{opt}$ (or $pH_{opt} - pH_b$) is defined as $\log q$ then pK'_a and pK'_b are given by $pH_{opt} \pm \log(q-4+1/q)$.

Alberty and Massey's method uses estimates of $(H^+)_{opt}$, $(H^+)_a$ (and $(H^+)_b$). $(H^+)_{opt}$ is fairly easy to obtain because if K'_a and K'_b are not well separated the pH profile will have a sharp top. However, good estimates of $(H^+)_a$ and $(H^+)_b$ require data which are good enough to allow a reasonable approximation to the true curve to be sketched. The following graphical method, based on that of Friedenwald and Maengwyn-Davies, does not use $(H^+)_a$ and $(H^+)_b$, and allows K'_b and V to be obtained directly from the data and $(H^+)_{opt}$. From equation (6.36) $K'_a = (H^+)^2_{opt}/K'_b$. Substitution into equation (6.28) and inversion gives

$$\frac{1}{V^{app}} = \frac{1}{V} + \frac{1}{VK'_b}\left(\frac{(H^+)^2_{opt}}{(H^+)} + (H^+)\right)$$

and so a plot of $1/V^{app}$ against $(H^+)+(H^+)^2_{opt}/(H^+)$ will be linear with ordinate intercept $1/V$ and abscissa intercept $-K'_b$ as shown in figure 6.22(b). K'_a can then be calculated from equation (6.36) as before.

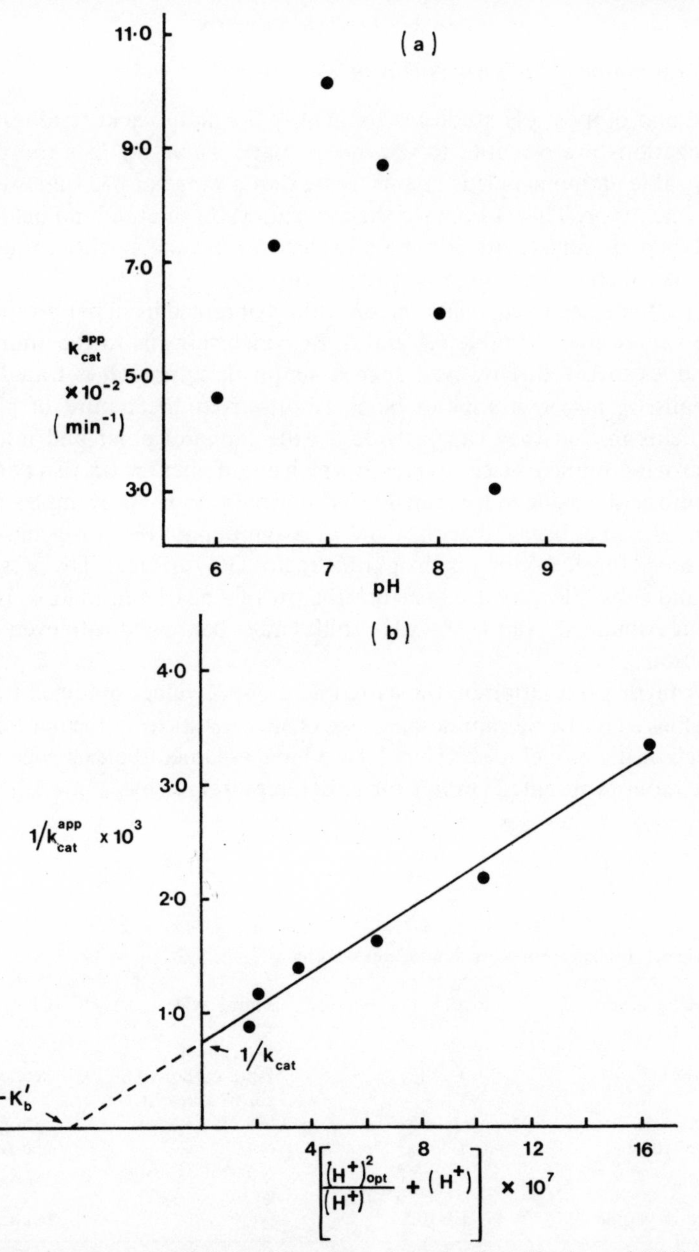

Figure 6.22 (a) pH profile of k_{cat}^{app} for ribonuclease; data taken from Herries *et al.* (1962). (b) Plot of $1/k_{cat}^{app}$ against $(H^+)+(H^+)^2_{opt}/(H^+)$ for the data of Herries *et al.* according to the method of Friedenwald and Maengwyn-Davies.

Interpretation of p*K's from* pH *profiles*

The aim of most pH studies is to identify the amino acid residues whose ionizations are essential to enzyme activity. Table 6.1 lists the pK's of ionizable amino acid side chains. Note that a range of pK values is given for each type. This is because the incorporation of an amino acid into a polypeptide subjects its side chain to electrostatic and neighbouring group effects which vary throughout the protein.

It is tempting to correlate the pK values obtained from pH profiles with the values given in table 6.1 and draw conclusions as to the amino acid residues involved. However, such a simplistic approach is fraught with pitfalls. A major stumbling block results from the nature of globular proteins in that they can provide a wide range of polar and non-polar microenvironments (see chapter 3) which might alter the pK of a particular amino acid residue to well outside its "normal" range. As examples we may cite an ε-amino group with a pK of 8 (glutamate dehydrogenase) or a carboxyl of pK 6.5 (in a hydrophobic region of lysozyme). The presence of bound substrate may itself perturb the pK of a nearby residue, so that pK values obtained from V^{app} – pH profiles must be treated with even greater caution.

A further limitation on the assignment of pK values obtained from pH profiles to particular amino-acid side chains results from the nature of the mechanistic model itself. Thus far we have assumed the existence of only one monoprotonated enzyme form, EH, represented by

```
A–H  B
 |    |
 ‾‾‾‾‾‾
```

Table 6.1 Ionizing groups on amino-acid side chains

Ionizing group	pK*	Amino acid	Charge type
Carboxyl	3–5–(7)	C-terminal asp(β-carboxyl) glu(γ-carboxyl)	neutral
Imidazole	(4)–5.5–7	his	cationic
Sulphydryl	8–9	cys	neutral
Amino	α7.6–8.5 ε(6)–9–10.5	N-terminal lys	cationic
Hydroxyl (phenolic)	10–10.5	tyr	neutral

* The pK range given is that normally found in proteins, the values given in parentheses being perturbed from their normal ones by unusual environments. Taken from W. W. Cleland (1977).

(figure 6.19). This is equivalent to saying that the alternative form

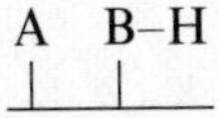

A B–H

does not exist. But in general there is no justification for this assumption. Considering the free enzyme as a dibasic acid we obtain the following scheme which takes both forms into account:

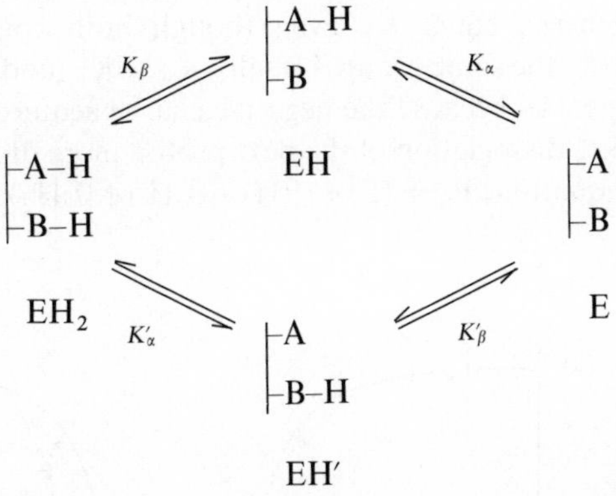

⊢A–H
⊢B
EH

K_β ⇌ K_α

⊢A–H
⊢B–H
EH_2

⊢A
⊢B
E

K'_α ⇌ K'_β

⊢A
⊢B–H
EH′

K_α, K_β, K'_α and K'_β are *microscopic* or *group* dissociation constants referring to proton dissociations from a particular group on the enzyme, the primed constants being those involving EH′, and α and β referring to proton dissociations from A–H and B–H respectively. These are the constants of interest, and what is required is the relationship between these and the experimentally determined constants K_a and K_b, which are termed *macroscopic* or *molecular* dissociation constants.*

Defining the microscopic constants:

$$K_\alpha = \frac{(E)(H^+)}{(EH)} \qquad (6.39)$$

$$K_\beta = \frac{(EH)(H^+)}{(EH_2)} \qquad (6.40)$$

$$K'_\alpha = \frac{(EH')(H^+)}{(EH_2)} \qquad (6.41)$$

$$K'_\beta = \frac{(E)(H^+)}{(EH')} \qquad (6.42)$$

* A similar set of ionizations can occur on the enzyme–substrate complex, and the following discussion applies equally to K'_a and K'_b.

The scheme is a thermodynamic box (see the discussion on transition-state analogues earlier) and

$$K_\alpha K_\beta = K'_\alpha K'_\beta \tag{6.43}$$

or

$$K_\beta / K'_\beta = K'_\alpha / K_\alpha \tag{6.44}$$

so any three of the four constants are sufficient to define the system. K_β will not, in general, equal K'_β even though both constants describe dissociations from the same group. Usually $K_\beta > K'_\beta$ (and hence $K'_\alpha > K_\alpha$ from equation (6.44)) because the negative charge acquired by the loss of the proton makes dissociation of the next proton more difficult. Using the conservation equation $E_0 = (E)+(EH)+(EH')+(EH_2)$ and equations

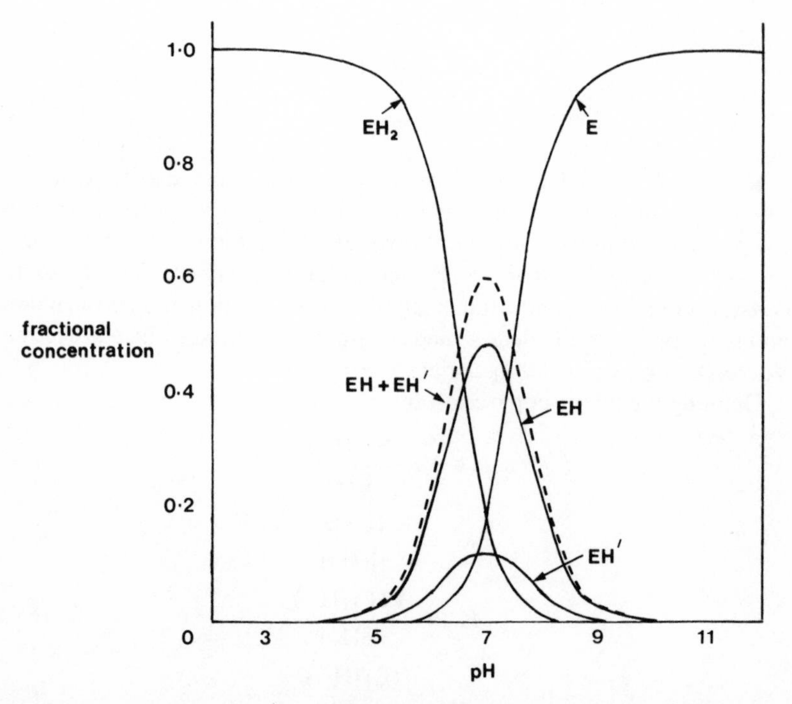

Figure 6.23 pH profiles of the enzyme forms EH_2, EH,ÆH′ and E (solid curves) calculated assuming $pK_\beta = 6.6$, $pK'_\beta = 6.8$, $pK_\alpha = 7.4$, $pK'_\alpha = 7.2$; ($pK_b = 6.5$, $pK_a = 7.5$). The dashed curve represents the sum EH + EH′. Concentrations are expressed as fractions of E_0 $(=(EH_2)+(EH')+(EH)+(E))$.

(6.39)–(6.41), expressions can be derived for the concentrations of each of the four species:

$$(\mathrm{E}) = \frac{\mathrm{E}_0 K_\alpha K_\beta/(\mathrm{H}^+)^2}{1 + \dfrac{K_\beta + K'_\alpha}{(\mathrm{H}^+)} + \dfrac{K_\beta K_\alpha}{(\mathrm{H}^+)^2}}$$

$$(\mathrm{EH}') = \frac{\mathrm{E}_0 K'_\alpha/(\mathrm{H}^+)}{1 + \dfrac{K_\beta + K'_\alpha}{(\mathrm{H}^+)} + \dfrac{K_\beta K_\alpha}{(\mathrm{H}^+)^2}}$$

$$(\mathrm{EH}_2) = \frac{\mathrm{E}_0}{1 + \dfrac{K_\beta + K'_\alpha}{(\mathrm{H}^+)} + \dfrac{K_\beta K_\alpha}{(\mathrm{H}^+)^2}}$$

$$(\mathrm{EH}) = \frac{\mathrm{E}_0 K_\beta/(\mathrm{H}^+)}{1 + \dfrac{K_\beta + K'_\alpha}{(\mathrm{H}^+)} + \dfrac{K_\beta K_\alpha}{(\mathrm{H}^+)^2}}$$

The variation of these with pH is illustrated in figure 6.23.

The salient point regarding figure 6.23 is that although (EH) and (EH′) vary with pH, their *ratio* is pH-independent. This is easily shown by dividing equation (6.40) by equation (6.41) whence it is seen that (EH)/(EH′) is a constant:

$$\frac{(\mathrm{EH})}{(\mathrm{EH}')} = \frac{K_\beta}{K'_\alpha}$$

This means that any pH-dependent effect ascribable to EH can also be ascribed to EH′. Thus the group required in its acid form might well be the one with the lower pK. The experimental method does not allow a distinction to be made between EH and EH′ as the active species.

Let us now consider the meanings of K_a and K_b. In general K_b is the dissociation constant of the first proton to dissociate (from EH_2), whether it comes from –AH (giving EH′) or –BH (giving EH). Thus

$$K_\mathrm{b} = \frac{((\mathrm{EH}) + (\mathrm{EH}'))\,(\mathrm{H}^+)}{(\mathrm{EH}_2)}$$

and from equations (6.40) and (6.41)

$$K_\mathrm{b} = K_\beta + K'_\alpha \qquad (6.45)$$

Similarly, K_a is the dissociation constant of the second proton, whether it comes from –AH (on EH) or –BH (on EH′) so that

$$K_a = \frac{(E)(H^+)}{(EH)+(EH')}$$

and from equations (6.39) and (6.42)

$$K_a = 1\bigg/\left(\frac{1}{K_\alpha} + \frac{1}{K'_\beta}\right) \qquad (6.46)$$

(Note that the shape of the pH profile is determined by K_a and K_b, not by the microscopic constants. Thus the dashed line of figure 6.23, representing (EH)+(EH′), is identical to the pH profile of figure 6.21(b).) If $K_\beta \gg K'_\alpha$, then $K_\alpha \ll K'_\beta$ (equation (6.43)) and (EH) will predominate over (EH′) and from equations (6.45) and (6.46), $K_b \simeq K_\beta$ and $K_a \simeq K_\alpha$.

Effect of substrate on pH *profiles*

As mentioned earlier, a pH–activity profile at one substrate concentration has little value. However, the way in which the pH profile changes with substrate concentration can provide clues to the enzyme mechanism, remembering that the pH profile at $S_0 \ll K_m^{app}$ is governed by ionizations on the free enzyme while that obtained at $S_0 \gg K_m^{app}$ reflects ionizations in the enzyme–substrate complex. If the ionizing groups responsible for enzyme activity are at the active site and interact with the substrate then one can make predictions regarding the relative acidity of these groups on the free enzyme compared to the enzyme–substrate complex. Thus a group interacting in its acid form with the substrate (say via a hydrogen bond) might be expected to be a weaker acid in the enzyme–substrate complex than on the free enzyme. The converse would apply to the acidity of a group acting as a base. In other words the enzyme–substrate complex is more easily protonated and deprotonated than the free enzyme and one might predict that $pK_b > pK'_b$ and $pK_a < pK'_a$. Evidence for the role of one of the two histidines at the active site of ribonuclease acting as a binding site for water was obtained by this type of analysis.

Another possible effect of the substrate arises from its own ionization. Most metabolic intermediates are themselves ionizable substances and may have pK values within the pH range over which enzyme activity is reversibly affected. Enzymes are usually specific for a particular ionic form of the substrate and so pH changes which affect enzyme activity may also affect the amount of "active" substrate. It is not difficult to incorporate pH effects resulting from substrate ionization into the rate equation, but it is more usual to determine the substrate pK separately by titration.

Effects of buffers, ions and solvent

Any pH study necessarily involves variation of the ratio of the concentration of conjugate base to conjugate acid of the buffer being used to control the pH. It is therefore well to check that the buffer ions themselves do not specifically interact with the enzyme. This applies equally to the counter ions as these sometimes have specific effects on enzyme activity. Thus chloride is an activator of salivary amylase and sodium ion inhibits pyruvate kinase. This can best be checked by using at least two different buffers over the pH range at overlapping pH.

Another factor that can affect pK values is ionic strength. The ionizable groups on amino-acid side chains given in table 6.1 are classified as cationic or neutral according to the charge on the conjugate acid. Thus $—COOH$ ($\rightleftharpoons —COO^- + H^+$) is a neutral acid and $—NH_3^+$ ($\rightleftharpoons —NH_2 + H^+$) is a cationic acid. The pK values of neutral acids are sensitive to variations in ionic strength whereas those of cationic acids are little affected. Therefore it is always wise to use buffers of constant ionic strength when carrying out pH studies. Organic solvents have similar effects on the pK of neutral and cationic acids. Studies of these effects, superimposed on the effects of the change in buffer pK, have provided valuable information on the charge types of ionizing residues at the active site of enzymes. The application of this method to ribonuclease is discussed in chapter 9.

Other effects

Irreversible inactivation may seem to be a trivial effect but it is often overlooked in pH studies. What has to be established is the *stability* of the enzyme over the pH range studied. This can be done by pre-incubating the enzyme at various pH values for the time required for an activity assay. The pre-incubation mixture is then adjusted to a pH at which the enzyme is known to be stable (often pH_{opt}) and assayed. The results of a classical study of this type are given in figure 6.24. Here it is seen that the decrease in activity on the falling limb of curve A must be due at least in part to enzyme instability.

Variation of rate with pH does not always imply a mechanism of the type discussed above. Bell-shaped curves can also result from a change in

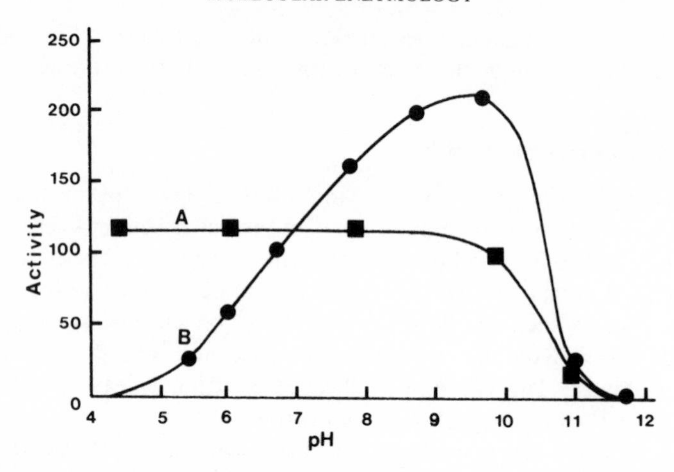

Figure 6.24 Influence of pH on the activity of tyramine oxidase. Curve A (■): activity of enzyme at pH 7.3 after being subjected for 5 min to various pH. Curve B (●): activity of enzyme at different pH. Taken from Hare (1928).

the rate-limiting step with pH. An example of such a reaction is semicarbazide formation in which the reactant is required in its unprotonated form but the resulting intermediate will not react further unless protonated. Although such a mechanism yields a bell-shaped pH profile, the pK values obtained from them will not necessarily be the pK's of the reactant and intermediate but may be partly determined by rate constants.

Consideration of all the caveats given above may lead one to wonder whether it is worth obtaining pH profiles. Nevertheless, such studies have often provided valuable clues regarding the nature of ionizing groups involved in enzyme catalysis. Provided that conclusions are backed up by confirmatory evidence (e.g. from chemical modification or spectrometric studies) pH studies can form a useful part of an enzyme kinetic investigation.

CHAPTER SEVEN

COMPLEX KINETICS AND COOPERATIVITY

Complex kinetics

Bisubstrate reactions

All the kinetic mechanisms we have discussed to this point refer to the uni-reactant transformation $S \rightarrow P$. However, most enzymes catalyse reactions between two or more substrates yielding two or more products, and the rate equations describing the mechanisms of such reactions are more complex than the ones which we have considered so far. In this treatment we shall restrict the discussion to reactions involving two substrates and two products. A number of excellent treatments of this subject are available (see bibliography) so we shall not attempt to provide an exhaustive summary, but rather aim to show the principles involved in obtaining mechanistic information.

The reaction which we are considering is

$$A + B \rightleftharpoons P + Q$$

Where such distinctions can usefully be made, P is the product structurally related to A, and Q the product related to B. Thus if the reaction is an ATP-dependent phosphorylation, A might be the substrate, P its phosphate ester and B and Q, ATP and ADP respectively. Protons may participate in the overall reaction but are not usually considered as reactants because of the complex effects of pH on enzyme activity (chapter 6). Water is frequently a reactant but is also not normally treated as such. The rate equations in this chapter will not be explicitly derived, their derivation following the principles used in chapters 4 and 6. A convenient method for deriving rate equations is given in chapter 10. Where the mechanism is one in which substrates must combine with enzyme in a particular order, A is the substrate which combines first. *Ab initio* it will not be known which of the two substrates is actually A but this can often be discovered from product inhibition patterns, binding studies or other independent experiments.

Kinetic studies of multisubstrate systems usually begin with measurements of initial rates in the absence of products. We shall consider a few common mechanisms to see what can be deduced from such studies.

Random rapid equilibrium mechanism

In this mechanism either substrate can bind to the enzyme giving binary complexes EA and EB which can then react with the other substrate to yield a ternary complex, EAB. The breakdown of this is the rate-limiting step of the overall reaction. As this step is slow, we can consider that all prior steps are in equilibrium characterized by the dissociation constants K_S^A, K_S^B, K_m^A, K_m^B.

$$\begin{array}{ccccccccc} & & \mathrm{EA} & & & & & \mathrm{EP} & \\ & {\scriptstyle K_S^A}\nearrow\!\!\swarrow & & \searrow\!\!\nwarrow{\scriptstyle K_m^B} & & & \nearrow\!\!\swarrow & & \searrow\!\!\nwarrow \\ \mathrm{E} & & & & \mathrm{EAB} \xrightarrow[\text{slow}]{k} \mathrm{EPQ} & & & & \mathrm{E} \\ & {\scriptstyle K_S^B}\searrow\!\!\nwarrow & & \nearrow\!\!\swarrow{\scriptstyle K_m^A} & & & \searrow\!\!\nwarrow & & \nearrow\!\!\swarrow \\ & & \mathrm{EB} & & & & & \mathrm{EQ} & \end{array} \quad (7.1)$$

The rate equation for this mechanism is

$$v_0 = \frac{V}{1 + \dfrac{K_m^A}{(A)} + \dfrac{K_m^B}{(B)} + \dfrac{K_S^A K_m^B}{(A)(B)}} \quad (7.2)$$

where $V = kE_0$. Note that the mechanism contains a thermodynamic box (see chapter 6) and therefore $K_S^A K_m^B = K_S^B K_m^A$. The steps to the right of the slow step are shown for completion but are superfluous to the rate equation in the absence of products. The mechanism is formally similar to the general inhibition mechanism but in this case the ternary complex is the catalytically active one. Creatine kinase is an enzyme which follows this mechanism.

If the equilibrium assumption is not applied and the reaction intermediates are assumed to be in the steady state then, as with the general inhibition mechanism, the rate equation becomes very complex and contains terms in $(A)^2$ and $(B)^2$. An enzyme following the random steady-state mechanism may exhibit unusual kinetic behaviour. This is discussed more fully in the second part of this chapter.

Compulsory order mechanism

This mechanism requires that A and B interact with the enzyme in that order; products are not released until a ternary (EAB) complex is formed. The products are also released in a specified order (here Q is the first released).

$$\mathrm{E} \underset{k_{-1}}{\overset{k_1(\mathrm{A})}{\rightleftharpoons}} \mathrm{EA} \underset{k_{-2}}{\overset{k_2(\mathrm{B})}{\rightleftharpoons}} \mathrm{EAB}\ (\mathrm{EPQ}) \xrightarrow[\ \ \mathrm{Q}\ \ ]{k_3} \mathrm{EP} \xrightarrow[\ \ \mathrm{P}\ \ ]{k_4} \mathrm{E} \tag{7.3}$$

A shorthand notation particularly suitable for ordered mechanisms is that proposed by Cleland in which substrate and product addition and release steps are represented by vertical arrows on a horizontal line representing the enzyme. The resulting enzyme species are given below the line. For mechanism (7.3) we have

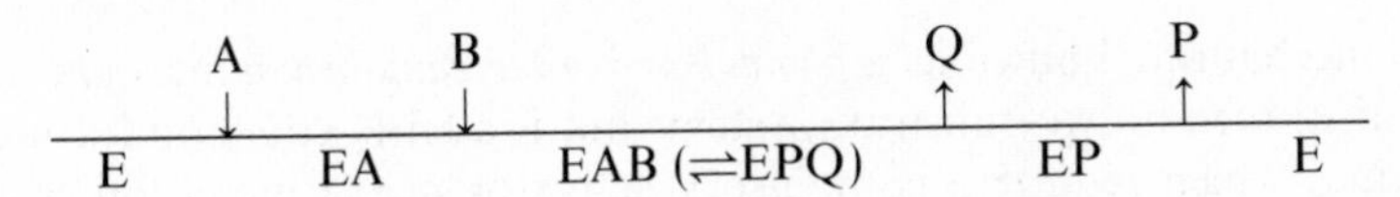

As both P and Q are released in irreversible steps (remember that (Q) and (P) are effectively zero under conditions of initial rate measurements) either $d(\mathrm{Q})/dt$ or $d(\mathrm{P})/dt$ can be taken as the overall rate. The rate equation is then

$$v_0 = \frac{V}{1 + \dfrac{K_\mathrm{m}^\mathrm{A}}{(\mathrm{A})} + \dfrac{K_\mathrm{m}^\mathrm{B}}{(\mathrm{B})} + \dfrac{K_\mathrm{S}^\mathrm{A} K_\mathrm{m}^\mathrm{B}}{(\mathrm{A})(\mathrm{B})}}$$

i.e. identical to equation (7.2). However, the meanings of K_m^A, K_m^B and k_cat ($=V/\mathrm{E}_0$) in terms of rate constants are not the same as in the random order equilibrium mechanism. K_m^A and K_m^B are *not* dissociation constants and the equality $K_\mathrm{S}^\mathrm{A} K_\mathrm{m}^\mathrm{B} = K_\mathrm{S}^\mathrm{B} K_\mathrm{m}^\mathrm{A}$ does not hold. Note that no EAB⇌EPQ step is explicitly incorporated into the mechanism. This is not to say that such a step does not exist, nor that it has no importance. It is simply that steady-state kinetics cannot provide information on unimolecular conversions of central complexes. If such a step were included in mechanism (7.3), the same rate equation would be obtained, albeit with the meaning of some of the kinetic constants altered. This is precisely the situation which obtains for the reversible one-substrate case for mechanisms with one and two central complexes, as discussed in chapter 4.

Steady-state kinetics are assumed for mechanism (7.3) but this need not be the case. If $K_S^A \gg K_m^A$ or $k_{-1} \gg k_{cat}$ then E, A and EA will be in thermodynamic equilibrium described by K_S^A. The subsequent steps are still treated as being in the steady state.

$$\begin{array}{c} \mathrm{A+E} \underset{}{\overset{K_S^A}{\rightleftharpoons}} \mathrm{EA} \underset{k_{-2}}{\overset{k_2(\mathrm{B})}{\rightleftharpoons}} \mathrm{EAB\ (EPQ)} \\ \mathrm{EAB} \xrightarrow{k_3} \mathrm{EP} + \mathrm{Q} \\ \mathrm{EP} \xrightarrow{k_4} \mathrm{E} + \mathrm{P} \end{array} \tag{7.4}$$

The rate equation for mechanism (7.4) is

$$v_0 = \frac{V}{1 + \dfrac{K_m^B}{(\mathrm{B})} + \dfrac{K_S^A K_m^B}{(\mathrm{A})(\mathrm{B})}} \tag{7.5}$$

This mechanism, known as a *compulsory order equilibrium* mechanism, is not often seen in two-substrate systems but is a fairly common feature of enzymes which require a compulsory activator (e.g. a metal ion) which must add to the enzyme prior to the addition of substrate B, and which cannot dissociate before product is released. As A is not a substrate it can remain bound to the enzyme between catalytic cycles, and the concentration of EA will be governed by (A) and K_S^A, enzyme concentration being negligible compared to (A) as is normal in steady-state kinetic studies. The Cleland notation makes this clear:

	A (K_S^A ↓)		B ↓		P ↑		B ↓		
E		EA		EAB (EPQ)		EA		EAB (EPQ)	etc.

Ping-pong (substituted enzyme) mechanism

In the classical ping-pong mechanism, binding of A and B is mutually exclusive and *no* ternary complexes are formed. Each substrate combines with the enzyme, and the corresponding product is released before the other substrate binds. As the overall reaction is between A and B it is obvious that A must transfer something to the enzyme for B to pick up.

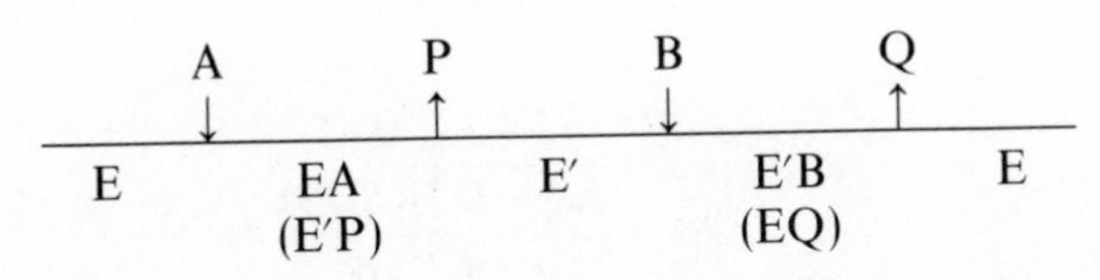

Thus the enzyme is altered by each half-reaction, the second half-reaction (B → Q) restoring the enzyme to the form suitable for the first half-reaction, A → P. Because of this the mechanism is called a *substituted enzyme* mechanism. The reciprocating nature of the two enzyme forms, E and E′, leads to the designation *ping-pong*. Steady-state kinetics cannot distinguish between E and E′ so the order of substrate addition is irrelevant in this case.

$$\mathrm{E} \underset{k_{-1}}{\overset{k_1(\mathrm{A})}{\rightleftharpoons}} \mathrm{EA} \xrightarrow{k_2} \mathrm{E}' + \mathrm{P}; \qquad \mathrm{E}' \underset{k_{-3}}{\overset{k_3(\mathrm{B})}{\rightleftharpoons}} \mathrm{EB} \xrightarrow{k_4} \mathrm{E} + \mathrm{Q} \tag{7.6}$$

The rate equation for mechanism (7.6) is

$$v_0 = \frac{V}{1 + \dfrac{K_{\mathrm{m}}^{\mathrm{A}}}{(\mathrm{A})} + \dfrac{K_{\mathrm{m}}^{\mathrm{B}}}{(\mathrm{B})}} \tag{7.7}$$

This sort of mechanism is often followed by enzymes that contain prosthetic groups acting as the group transfer agent. The pyridoxal phosphate-dependent transaminases are the best studied enzymes of this type.

Analysis of rate equations

In initial rate studies of multi-substrate enzymes the concentration of one substrate is varied holding the concentration of the other substrate (and cofactors if any) constant at different fixed levels. Usually the substrate concentrations are varied over a grid of values, e.g. five concentrations of A at each of five fixed concentrations of B. This allows for convenient display of results on linear plots. Under these conditions the kinetics of enzymes following equations (7.2), (7.5) and (7.7) will obey the Michaelis–Menten equation. Let us apply this method to equation (7.2). Regarding A as the variable substrate and B as the fixed substrate, equation (7.2) can be rearranged as

$$v_0 = \frac{V(\mathrm{A})/(1 + K_\mathrm{m}^\mathrm{B}/(\mathrm{B}))}{(\mathrm{A}) + \dfrac{K_\mathrm{m}^\mathrm{A} + K_\mathrm{S}^\mathrm{A} K_\mathrm{m}^\mathrm{B}/(\mathrm{B})}{1 + K_\mathrm{m}^\mathrm{B}/(\mathrm{B})}} \tag{7.8}$$

which is of the form of the Michaelis–Menten equation

$$v_0 = \frac{V^\mathrm{app} \mathrm{S}_0}{\mathrm{S}_0 + K_\mathrm{m}^\mathrm{app}}$$

where

$$V^\mathrm{app} = V/(1 + K_\mathrm{m}^\mathrm{B}/(\mathrm{B})) = V(\mathrm{B})/((\mathrm{B}) + K_\mathrm{m}^\mathrm{B}) \tag{7.9}$$

$$K_\mathrm{m}^\mathrm{app} = \frac{K_\mathrm{m}^\mathrm{A} + K_\mathrm{S}^\mathrm{A} K_\mathrm{m}^\mathrm{B}/(\mathrm{B})}{1 + K_\mathrm{m}^\mathrm{B}/(\mathrm{B})} = \frac{K_\mathrm{m}^\mathrm{A}(\mathrm{B}) + K_\mathrm{S}^\mathrm{A} K_\mathrm{m}^\mathrm{B}}{(\mathrm{B}) + K_\mathrm{m}^\mathrm{B}} \tag{7.10}$$

and

$$V^\mathrm{app}/K_\mathrm{m}^\mathrm{app} = V/(K_\mathrm{m}^\mathrm{A} + K_\mathrm{S}^\mathrm{A} K_\mathrm{m}^\mathrm{B}/(\mathrm{B})) = \frac{(V/K_\mathrm{m}^\mathrm{A})(\mathrm{B})}{(\mathrm{B}) + K_\mathrm{S}^\mathrm{A} K_\mathrm{m}^\mathrm{B}/K_\mathrm{m}^\mathrm{A}} \tag{7.11}$$

Note that, as with reversible inhibition and pH, $K_\mathrm{m}^\mathrm{app}$ is a more complicated function than V^app or $V^\mathrm{app}/K_\mathrm{m}^\mathrm{app}$. Note also that equations (7.9) and (7.11) are themselves in the form of the Michaelis–Menten equation and so using equation (7.11) $V/K_\mathrm{m}^\mathrm{A}$ and $K_\mathrm{S}^\mathrm{A} K_\mathrm{m}^\mathrm{B}/K_\mathrm{m}^\mathrm{A}$ can be obtained from a direct linear plot of $V^\mathrm{app}/K_\mathrm{m}^\mathrm{app}$ against (B). Similarly V and K_m^B can be obtained from a direct linear plot of V^app against (B); thus all four constants can be determined.

Alternatively, treating (B) as the variable substrate at different fixed levels of (A), rearrangement of equation (7.2) yields

$$v_0 = \frac{V(\mathrm{B})/(1 + K_\mathrm{m}^\mathrm{A}/(\mathrm{A}))}{(\mathrm{B}) + \dfrac{K_\mathrm{m}^\mathrm{B} + K_\mathrm{S}^\mathrm{A} K_\mathrm{m}^\mathrm{B}/(\mathrm{A})}{1 + K_\mathrm{m}^\mathrm{A}/(\mathrm{A})}} \tag{7.12}$$

where

$$V^\mathrm{app} = V/(1 + K_\mathrm{m}^\mathrm{A}/(\mathrm{A})) = V(\mathrm{A})/(\mathrm{A}) + K_\mathrm{m}^\mathrm{A} \tag{7.13}$$

$$K_\mathrm{m}^\mathrm{app} = \frac{K_\mathrm{m}^\mathrm{B} + K_\mathrm{S}^\mathrm{A} K_\mathrm{m}^\mathrm{B}/(\mathrm{A})}{1 + K_\mathrm{m}^\mathrm{A}/(\mathrm{A})}$$

and

$$V^\mathrm{app}/K_\mathrm{m}^\mathrm{app} = (V/K_\mathrm{m}^\mathrm{B})(\mathrm{A})/((\mathrm{A}) + K_\mathrm{S}^\mathrm{A}) \tag{7.14}$$

Similar analysis of the kinetic data according to equations (7.12), (7.13) and (7.14) will likewise yield values for all four constants.

Analogous treatment of equation (7.7) for the ping-pong mechanism shows that

$$v_0 = \frac{V(\mathrm{A})/(1+K_{\mathrm{m}}^{\mathrm{B}}/(\mathrm{B}))}{(\mathrm{A})+K_{\mathrm{m}}^{\mathrm{A}}/(1+K_{\mathrm{m}}^{\mathrm{B}}/(\mathrm{B}))} \tag{7.15}$$

and

$$v_0 = \frac{V(\mathrm{B})/(1+K_{\mathrm{m}}^{\mathrm{A}}/(\mathrm{A}))}{(\mathrm{B})+K_{\mathrm{m}}^{\mathrm{B}}/(1+K_{\mathrm{m}}^{\mathrm{A}}/(\mathrm{A}))} \tag{7.16}$$

From equation (7.15)

$$V^{\mathrm{app}} = V/(1+K_{\mathrm{m}}^{\mathrm{B}}/(\mathrm{B})) \tag{7.17}$$

$$V^{\mathrm{app}}/K_{\mathrm{m}}^{\mathrm{app}} = V/K_{\mathrm{m}}^{\mathrm{A}} \tag{7.18}$$

and from (7.16)

$$V^{\mathrm{app}} = V/(1+K_{\mathrm{m}}^{\mathrm{A}}/(\mathrm{A}) \tag{7.19}$$

$$V^{\mathrm{app}}/K_{\mathrm{m}}^{\mathrm{app}} = V/K_{\mathrm{m}}^{\mathrm{A}} \tag{7.20}$$

Thus $V/K_{\mathrm{m}}^{\mathrm{A}}$ and $V/K_{\mathrm{m}}^{\mathrm{B}}$ can be obtained from the primary data; analysis of V^{app} (equation (7.17) or (7.19)) gives V and thus values for $K_{\mathrm{m}}^{\mathrm{A}}$ and $K_{\mathrm{m}}^{\mathrm{B}}$.

The rate equation (7.5) for the compulsory order equilibrium mechanism (7.4) gives for variable (A) at fixed (B)

$$V^{\mathrm{app}} = V/(1+K_{\mathrm{m}}^{\mathrm{B}}/(\mathrm{B})) \tag{7.21}$$

$$K_{\mathrm{m}}^{\mathrm{app}} = \frac{K_{\mathrm{S}}^{\mathrm{A}}K_{\mathrm{m}}^{\mathrm{B}}/(\mathrm{B})}{1+K_{\mathrm{m}}^{\mathrm{B}}/(\mathrm{B})}$$

and

$$V^{\mathrm{app}}/K_{\mathrm{m}}^{\mathrm{app}} = V(\mathrm{B})/K_{\mathrm{S}}^{\mathrm{A}}K_{\mathrm{m}}^{\mathrm{B}} \tag{7.22}$$

With (B) as the variable substrate at fixed (A) we obtain

$$V^{\mathrm{app}} = V \tag{7.23}$$

$$K_{\mathrm{m}}^{\mathrm{app}} = K_{\mathrm{m}}^{\mathrm{B}}(1+K_{\mathrm{S}}^{\mathrm{A}}/(\mathrm{A}))$$

and

$$V^{\mathrm{app}}/K_{\mathrm{m}}^{\mathrm{app}} = V/K_{\mathrm{m}}^{\mathrm{B}}(1+K_{\mathrm{S}}^{\mathrm{A}}/(\mathrm{A})) \tag{7.24}$$

Significance of the kinetic constants

The meaning of the four kinetic constants V, $K_{\mathrm{m}}^{\mathrm{A}}$, $K_{\mathrm{m}}^{\mathrm{B}}$ and $K_{\mathrm{S}}^{\mathrm{A}}$ in terms of rate constants will, of course, depend on the mechanism from which the rate equation is derived. However, the kinetic constants can usefully be defined operationally. V is very straightforward. If both (A) and (B) become very large, i.e. approach saturating levels, the (A), (B) and (A)(B) terms in the denominator of equation (7.2) disappear and $v_0 = V$. So V is the *velocity when all substrates are at saturating levels*, i.e. the *maximum velocity*. Reference to equation (7.8) shows that when (B) is saturating, the (B) terms disappear and

$$v_0 = \frac{V(\mathrm{A})}{(\mathrm{A}) + K_\mathrm{m}^\mathrm{A}} \tag{7.25}$$

Similarly from equation (7.12) with (A) saturating

$$v_0 = \frac{V(\mathrm{B})}{(\mathrm{B}) + K_\mathrm{m}^\mathrm{B}}$$

Thus K_m^A and K_m^B are the *concentrations of* A *and* B *respectively giving half maximum velocity at saturating concentration of the other substrate* and are therefore the *Michaelis constants* for A and B.

The combined constant $K_\mathrm{S}^\mathrm{A}K_\mathrm{m}^\mathrm{B}$ contains the Michaelis constant for B and K_S^A. The significance of K_S^A can be understood by seeing what happens to equation (7.8) when (B) becomes very small. As (B) approaches zero, $V^\mathrm{app} \simeq V(\mathrm{B})/K_\mathrm{m}^\mathrm{B}$ from equation (7.9) and $K_\mathrm{m}^\mathrm{app} \simeq K_\mathrm{S}^\mathrm{A}$ from equation (7.10); thus equation (7.8) becomes

$$v_0 = \frac{(V(\mathrm{B})/K_\mathrm{m}^\mathrm{B})(\mathrm{A})}{(\mathrm{A}) + K_\mathrm{S}^\mathrm{A}}$$

So K_S^A is the limiting value of $K_\mathrm{m}^\mathrm{app}$ for (A) as (B) approaches zero. When (B) is very small the rate of reaction of (B) with (EA) in the compulsory order mechanism (7.3) will be very slow. In the random equilibrium mechanism (7.1) the extent of EAB formation will likewise be very small. Under such conditions E will be in virtual equilibrium with EA and for this reason, K_S^A is sometimes regarded as the dissociation constant of the EA complex. However, this will only be true where no steps intervene between the binding of A and the binding of B. If one or more unimolecular isomerizations of EA take place then K_S^A will reflect the total concentration of all EA binary complexes. (Thus for the sequence

$$\mathrm{E} + \mathrm{A} \underset{k_{-1}}{\overset{k_1}{\rightleftharpoons}} \mathrm{EA} \underset{k_{-2}}{\overset{k_2}{\rightleftharpoons}} \mathrm{E'A} \underset{k_{-3}}{\overset{k_3(\mathrm{B})}{\rightleftharpoons}} \mathrm{EAB} \rightleftharpoons$$

$K_\mathrm{S}^\mathrm{A} = (\mathrm{E})(\mathrm{A})/((\mathrm{EA}) + (\mathrm{E'A})) = k_{-1}k_{-2}/k_1(k_{-2} + k_2)$ and *not* k_{-1}/k_1.)

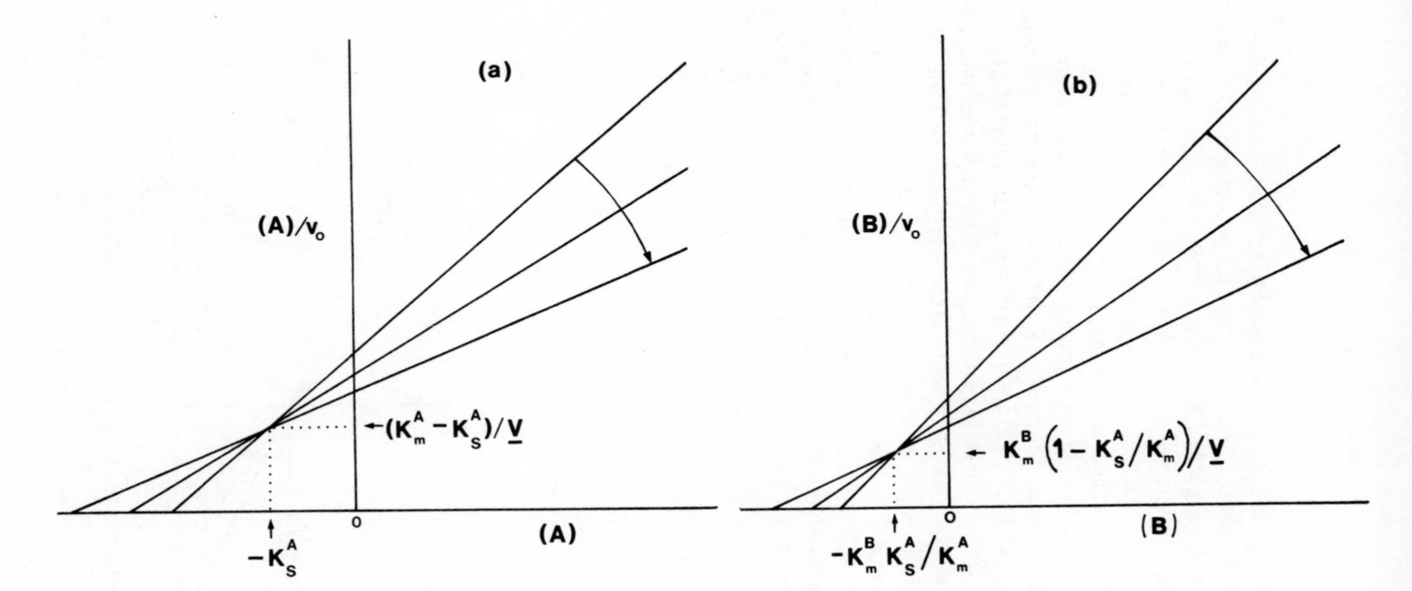

Figure 7.1 Half-reciprocal primary plots of (a) (A)/v_0 against (A), (b) (B)/v_0 against (B) for a compulsory order ternary complex mechanism. The random rapid equilibrium mechanism gives similar plots. In the example given here $K_m^A > K_s^A$. Arrow indicates increasing values of the fixed substrate concentration.

Plotting results

Multireactant kinetic data are usually plotted in one of the linear forms of the Michaelis–Menten equation as this gives a graphical display of the kinetic pattern. In a plot of S_0/v_0 against S_0 the slope has the value of $1/V^{app}$ and the y-intercept is K_m^{app}/V^{app}. The pattern resulting from the given rate equation can therefore be predicted from the equations for V^{app} and V^{app}/K_m^{app} given earlier in this chapter.

For enzymes following the general rate equation (7.2), V^{app} and V^{app}/K_m^{app} are different functions of the fixed substrate concentrations (equations (7.9), (7.11), (7.12) and (7.14)) so primary plots of $(A)/v_0$ against (A) and $(B)/v_0$ against (B) will give rise to a pattern of lines intersecting to the left of the S_0/v_0 axis. The intersection point will lie above the horizontal axis if $K_m^A > K_S^A$, below it if $K_m^A < K_S^A$, and on the axis if $K_m^A = K_S^A$ (figure 7.1). Secondary plots of $(B)/V^{app}$ and $(B)K_m^{app}/V^{app}$ against (B) or $(A)/V^{app}$ and $(A)K_m^{app}/V^{app}$ against (A) will also be linear. In the ping-pong mechanism (7.6), K_S^A is zero and the combined constant term is missing from the rate equation (7.7). As a result V^{app}/K_m^{app} is

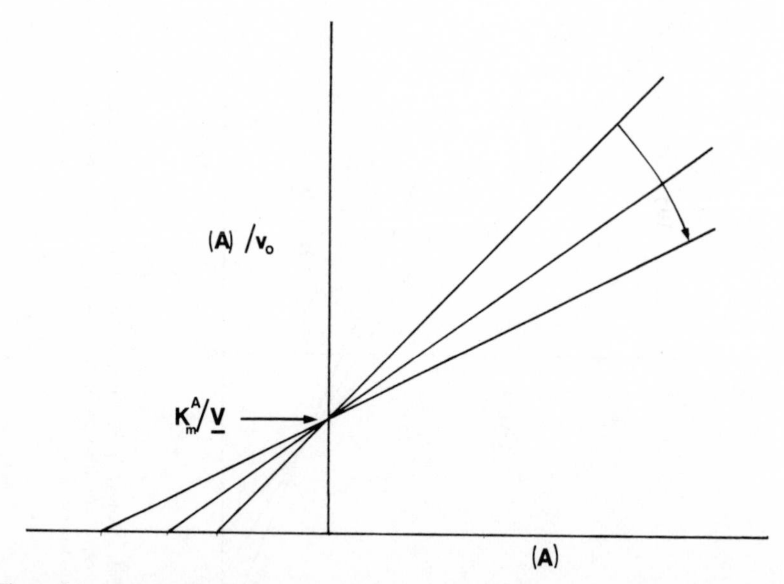

Figure 7.2 Half-reciprocal primary plot of $(A)/v_0$ against (A) for a ping pong mechanism. The plot of $(B)/v_0$ against (B) is similar, the value of the common ordinate intercept being K_m^B/V. Arrow indicates increasing fixed substrate concentration.

independent of the concentration of the fixed substrate (equations (7.18) and (7.20)) and the half-reciprocal plot for either substrate gives a pattern of lines intersecting on the vertical axis (figure 7.2). The rate equation (7.5) for the compulsory order equilibrium mechanism (7.4) lacks the K_m^A term. The pattern obtained will therefore depend on whether (A) or (B) is plotted as the variable substrate. If (A) is the variable substrate, a plot of (A)/v_0 against (A) will be a pattern of lines intersecting below the horizontal axis (figure 7.3(a)). However, unlike the other ternary complex mechanisms considered here, V^{app}/K_m^{app} is directly proportional to (B) (equation (7.22)) so a secondary plot of V^{app}/K_m^{app} against (B) will be a straight line through the origin with slope $V/K_S^A K_m^B$. Another distinguishing feature of this mechanism is that a plot of (B)/v_0 against (B) will give a set of parallel lines (figure 7.3(b)) because V^{app} is independent of (A) (equation (7.23)).

So the patterns of *primary plots* of (A)/v_0 against (A) at fixed (B) and (B)/v_0 against (B) at fixed (A) allow one to distinguish among mechanisms ((7.1) or (7.3)), (7.4) and (7.6). In the case of the compulsory order equilibrium mechanism (7.4) the order of addition can also be discerned from primary plots. However, primary plots will not enable a distinction to be made between the random equilibrium and the compulsory order mechanisms ((7.1) and (7.3)) nor do they reveal the order of substrate addition in the compulsory order mechanism. Also many other plausible mechanisms give identical rate equations to those discussed above, and therefore identical patterns of primary plots. These possibilities can often be distinguished by product inhibition studies.

Product inhibition

We have thus far assumed that all product release steps are irreversible. But a feature of multi-product reactions is that by studying the reaction in the presence of *one* product no significant reversal of the reaction will occur provided that initial rates are measured. Products inhibit by forming enzyme complexes which would be productive if the reaction were being assayed in the reverse direction. In the S → P reaction discussed in chapter 4 it was shown how the mutually exclusive binding of substrate and product at the active site gives rise to competitive inhibition by product. In multi-substrate multi-product reactions product will also inhibit by virtue of additional terms containing (P) or Q in the denominator of the rate equation but the inhibitory effects need not be competitive. The nature of the product terms often differs from one mechanism to another

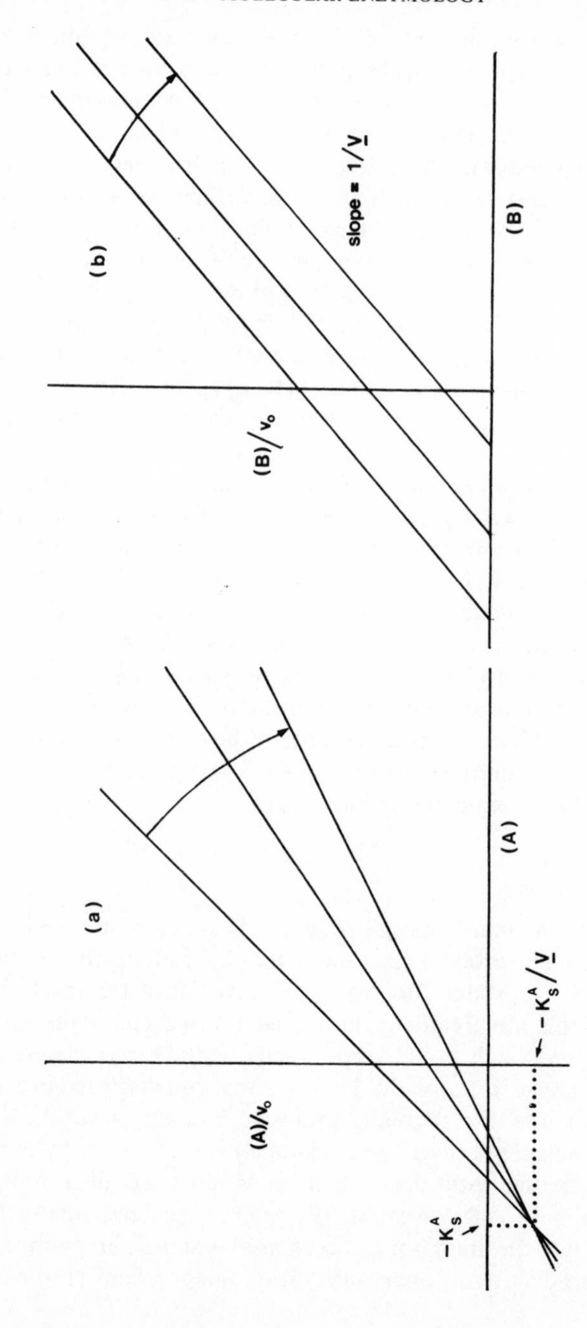

Figure 7.3 Half-reciprocal primary plots for a compulsory order equilibrium mechanism: (a) (A)/v_0 against (A); (b) (B)/v_0 against (B). For this mechanism the pattern of primary plots is not the same for A and B. Arrow indicates increasing fixed substrate concentration.

and so the inhibition patterns obtained with respect to substrates A and B can serve to eliminate mechanisms from consideration. In some cases, product inhibition patterns can be used to decide the binding order of substrates.

Cleland has devised a set of rules for predicting product inhibition patterns given the reaction mechanism. These are very convenient to use but it is always more judicious to derive the rate equations corresponding to the suspected mechanisms and work out the inhibition patterns for oneself. To show how this is done we shall obtain the expected inhibition patterns for the ping-pong mechanism in the presence of product P.

In the presence of P we must allow for reversal of the second step of mechanism (7.6) by incorporating the k_{-2}(P) step.

$$\mathrm{E} \underset{k_{-1}}{\overset{k_1(\mathrm{A})}{\rightleftharpoons}} \mathrm{EA} \underset{k_{-2}(\mathrm{P})}{\overset{k_2}{\rightleftharpoons}} \mathrm{E'} \underset{k_{-3}}{\overset{k_3(\mathrm{B})}{\rightleftharpoons}} \mathrm{EB} \xrightarrow{k_4} \mathrm{E} + \mathrm{Q}$$

The rate equation now becomes

$$v_0 = \frac{d(\mathrm{Q})}{dt} = \frac{V}{1 + \dfrac{K_\mathrm{m}^\mathrm{A}}{(\mathrm{A})} + \dfrac{K_\mathrm{m}^\mathrm{B}}{(\mathrm{B})}\left(1 + \dfrac{(\mathrm{P})}{K_\mathrm{i}}\right) + \dfrac{K_\mathrm{S}^\mathrm{A} K_\mathrm{m}^\mathrm{B}(\mathrm{P})}{K_\mathrm{i}(\mathrm{A})(\mathrm{B})}}$$

where K_i is the inhibition constant for P. Note that the K_S^A term, lacking in equation (7.7), is not zero in the presence of P. Regarding A as the varying substrate,

$${}_0 = \frac{V(\mathrm{A})/(1 + (K_\mathrm{m}^\mathrm{B}/(\mathrm{B}))(1 + (\mathrm{P})/K_\mathrm{i})}{(\mathrm{A}) + \dfrac{K_\mathrm{m}^\mathrm{A} + K_\mathrm{S}^\mathrm{A} K_\mathrm{m}^\mathrm{B}(\mathrm{P})/K_\mathrm{i}(\mathrm{B})}{1 + (K_\mathrm{m}^\mathrm{B}/(\mathrm{B}))(1 + (\mathrm{P})/K_\mathrm{i})}} \qquad (7.26)$$

and comparing this with equation (7.15) we see that mixed inhibition will be observed as both V^{app} and $K_\mathrm{m}^{\mathrm{app}}$ are affected by (P). The type of inhibition expressed may be dependent on the concentration of the fixed substrate. If (B) in equation (7.26) becomes very large, all (P) terms become negligible and the rate equation becomes identical to (7.25). Thus no inhibition will be observed at saturating (B). With (B) as the varying substrate at fixed (A)

$$v_0 = \frac{V(\mathrm{B})/(1 + K_\mathrm{m}^\mathrm{A}/(\mathrm{A}))}{(\mathrm{B}) + \dfrac{K_\mathrm{m}^\mathrm{B}(1 + (\mathrm{P})/K_\mathrm{i}) + K_\mathrm{S}^\mathrm{A} K_\mathrm{m}^\mathrm{B}(\mathrm{P})/K_\mathrm{i}(\mathrm{A})}{1 + K_\mathrm{m}^\mathrm{A}/(\mathrm{A})}}$$

Comparing with equation (7.16) we see that (P) affects K_m^{app} only so we have competitive inhibition. When (A) is saturating all (A) terms disappear, but not all (P) terms

$$v_0 = \frac{V(\mathrm{B})}{(\mathrm{B}) + K_m^{\mathrm{B}}(1 + (\mathrm{P})/K_i)}$$

so the inhibition is still competitive.

Table 7.1 gives the product inhibition patterns predicted for a few mechanisms. It can be seen that they allow a distinction to be made between the compulsory order and random equilibrium mechanisms and in the former case, also reveal the order of addition of substrates. The patterns in table 7.1 take no account of the possibility of *abortive complexes.* In ternary complex mechanisms these are enzyme complexes involving combination of product with substrate (usually the structurally non-related pair, i.e. EAQ or EBP). An example might be a phosphate ester–ATP–enzyme complex in a kinase-catalysed reaction. In a ping-pong mechanism abortive complexes could be formed by combination of product (or substrate) with the "wrong" form of the enzyme. Abortive complexes are dead-end complexes; i.e. they do not lie on the productive kinetic pathway. They can be expected to occur in many bisubstrate mechanisms and may alter the product inhibition patterns given in table 7.1. For example, in the random equilibrium mechanism presence of the abortive complex EPB in a product inhibition study using P would change the pattern to that found for Q in the ping-pong mechanism.

Thus although product inhibition studies can yield information not available from substrate kinetics, they do not always provide unambiguous answers. The Theorell–Chance mechanism, first proposed for horse liver alcohol dehydrogenase, is a compulsory order mechanism in which the interconversion and breakdown of the ternary complexes are so rapid as to make them kinetically undetectable.

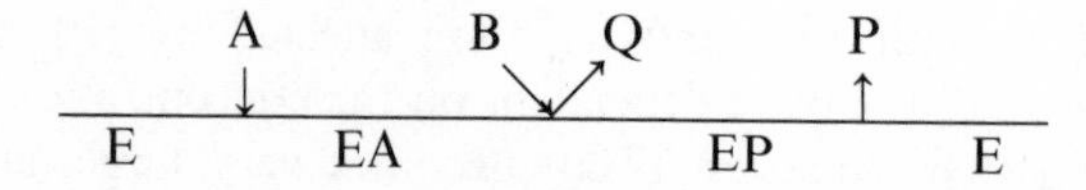

This mechanism gives a product inhibition pattern for P that is the converse of that predicted for Q with respect to A and B. Thus for this mechanism it is not possible to distinguish A from B by kinetic studies.

These caveats serve to show that steady-state kinetics will not always allow one to define the kinetic mechanism unambiguously. Such studies should always be backed up by other types of investigation. Whenever

Table 7.1 Product inhibition patterns and some Haldane relationships for three bisubstrate mechanisms $A + B \rightleftharpoons P + Q$

Mechanism	Varied substrate	(Fixed substrate)	Product inhibition P	Product inhibition Q	Haldane relationships
Random equilibrium (8.1)	A	(B unsat.)	Comp.	Comp.	$K_{eq} = \frac{V^f K_S^P K_m^Q}{V^r K_S^A K_m^B} = \frac{V^f K_S^Q K_m^P}{V^r K_S^A K_m^B}$
	A	(B sat.)	None	None	
	B	(A unsat.)	Comp.	Comp.	$= \frac{V^f K_S^P K_m^Q}{V^r K_S^B K_m^A} = \frac{V^f K_S^Q K_m^P}{V^r K_S^B K_m^A}$
	B	(A sat.)	None	None	
Compulsory order (8.3)	A	(B unsat.)	Comp.	Mixed	$K_{eq} = \frac{V^f K_S^P K_m^Q}{V^r K_S^A K_m^Q}$
	A	(B sat.)	Comp.	Uncomp.	
	B	(A unsat.)	Mixed	Mixed	
	B	(A sat.)	None	Mixed	
Ping-pong (substituted enzyme) (8.6)	A	(B unsat.)	Mixed	Comp.	$K_{eq} = \left(\frac{V^f}{V^r}\right)^2 \frac{K_m^P K_m^Q}{K_m^A K_m^B}$
	A	(B sat.)	None	Comp.	
	B	(A unsat.)	Comp.	Mixed	
	B	(A sat.)	Comp.	None	

$K_S^P, K_S^Q, K_m^P, K_m^Q$ are the dissociation and Michaelis constants respectively for P and Q. V^f is the maximum velocity in the forward direction, i.e. $A + B \rightarrow P + Q$; V^r is the maximum velocity in the reverse direction. Comp. = competitive, Uncomp. = uncompetitive, sat. = saturating, unsat. = unsaturating.

possible the kinetic constants should be obtained for both the forward and reverse reactions. If K_{eq} for the overall reaction is determined independently, the Haldane relationship (see chapter 4) can be used to check the kinetic constants for internal consistency. For certain mechanisms several Haldane relationships are possible (table 7.1) some of which may be specific for a particular mechanism. Adherance to these can sometimes provide supportive evidence for conclusions from kinetic data.

Further evidence can be obtained from independent binding experiments. For an enzyme following a random equilibrium mechanism, A should bind independently to the enzyme as also should B. The dissociation constants should agree with K_S^A and K_S^B calculated from kinetic data. If a compulsory order mechanism is obeyed B would not be expected to bind readily to free enzyme but A should and its dissociation constant should agree with K_S^A. The nature of the ping-pong mechanism predicts that it should be possible to isolate the two stable enzyme forms E and E′ as they would be expected to differ chemically. This has indeed been demonstrated with a number of such enzymes. Using the appropriate enzyme form, each half-reaction A → P and B → Q can be shown to take place independently, at a rate comparable to that of the overall reaction, the maximum conversion being stoichiometrically equivalent to the amount of active enzyme sites present. Another method for obtaining mechanistic information involves measurement of equilibrium reaction rates.

Equilibrium isotope exchange

At chemical (thermodynamic) equilibrium the net rate is zero but as the equilibrium is dynamic, substrate molecules are continually converted to product and vice versa. Because there is no *net* change in any of the reactant concentrations the interconversion velocities are usually measured by using substrates (or products) labelled with radioactive isotopes. In this method the enzyme-catalysed reaction is allowed to proceed to equilibrium. A small amount ("small" enough not to affect the equilibrium position) of radioactively labelled substrate is added and the initial rate of incorporation of label into product is determined by withdrawing aliquots of the reaction mixture at fixed times. Obviously a method for separating the components of the labelled exchange pair must be available. Kinetic isotope effects are assumed to be absent.

For a reaction such as $A+B \rightleftharpoons P+Q$ the nature of the isotopic exchange will depend on the reaction mechanism and the position of the label. Of the four exchanges $A \rightleftharpoons P$, $A \rightleftharpoons Q$, $B \rightleftharpoons P$ and $B \rightleftharpoons Q$ generally only three are possible. This is best illustrated by an example. In the reaction

$$\text{lactate} + NAD^{+} \rightleftharpoons \text{pyruvate} + NADH + H^{+}$$

^{14}C-labelled lactate can exchange with pyruvate, ^{14}C-labelled NAD^{+} can exchange with NADH and NAD—^{3}H can exchange with lactate but no exchange can occur between NAD^{+} and pyruvate. The variation of exchange rate with concentration of substrate provides a useful tool for discrimination between mechanisms particularly between the compulsory order and random mechanisms.

In order to study the variation of exchange rates with concentration of one of the substrates the concentration of the corresponding product must also be varied so as to keep the ratio of the two components constant and maintain the reaction at equilibrium. The reaction catalysed by heart muscle lactate dehydrogenase will be used as an example. The value of K_{eq} was determined independently where

$$K_{eq} = \frac{(\text{pyruvate})(NADH)(H^{+})}{(\text{lactate})(NAD^{+})}$$

The reaction mixture was set up so that (NADH)/(NAD^{+}) and (lactate)/(pyruvate) satisfied the equilibrium condition. Using ^{14}C-lactate and ^{14}C-NAD^{+} the initial rates of lactate $\rightleftharpoons$ pyruvate and $NAD^{+} \rightleftharpoons NADH$ were measured at varying lactate concentrations, keeping (lactate)/(pyruvate) constant. It was found that the lactate $\rightleftharpoons$ pyruvate rate increased hyperbolically with lactate concentration, whereas the $NAD^{+} \rightleftharpoons NADH$ rate increased initially but became progressively inhibited as lactate increased as shown in figure 7.4. This behaviour is consistent with the compulsory order ternary complex mechanism

$$\begin{array}{ccccc} & \overset{+NAD^{+}}{\nearrow} & E \cdot NAD^{+} & \xrightleftharpoons{+\text{lactate}} & E \cdot NAD^{+} \cdot \text{lactate} \\ E & & & & \updownarrow \\ & \underset{+NADH}{\searrow} & E \cdot NADH & \xrightleftharpoons{+\text{pyruvate}} & E \cdot NADH \cdot \text{pyruvate} \end{array}$$

The lactate–pyruvate exchange was faster than the NAD^{+}–NADH exchange which suggests that the rate-limiting step is coenzyme release. In this mechanism NAD^{+} and NADH cannot dissociate from the ternary complexes so their exchange will be inhibited when the enzyme is completely sequestered into ternary complexes by saturating lactate (and pyruvate) concentration. Lactate and pyruvate can exchange via the

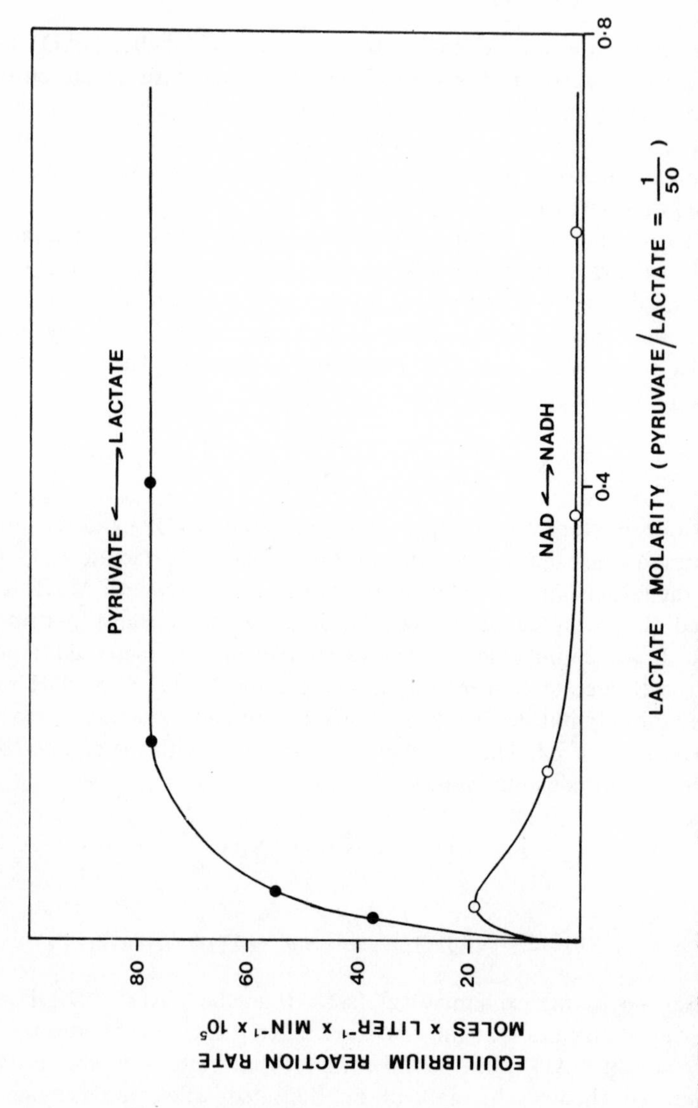

Figure 7.4 The effect of lactate and pyruvate concentrations on equilibrium reaction rates with bovine heart lactate dehydrogenase at pH 7.9. Redrawn from Silverstein and Boyer (1964).

ternary complexes; thus the exchange rate increases to a maximum as lactate approaches saturating levels. These results also establish the order of substrate addition and product release.

If the reaction proceeded by a random order equilibrium mechanism, both exchange rates would be equal because all substrate–product interconversions are governed by the same rate-limiting step. Neither exchange rate would show substrate inhibition. The ping-pong mechanism can be readily distinguished from ternary complex mechanisms by demonstrating exchange corresponding to each half reaction in the absence of the other substrate product couple.

None of the methods discussed above, used in isolation, can provide unequivocal evidence to pinpoint a particular mechanism. This is because the problem of kinetic ambiguity, discussed in chapter 4 for unireactant systems, exists to a much greater extent with multi-substrate reactions. By using a variety of approaches it should however be possible to bridge the gap between speculation and truth.

Cooperativity

Many of the enzymes of intermediary metabolism seem not to be involved in any special form of regulation of the flux of metabolites in a particular pathway. They do, of course, participate in regulation inasmuch as saturability limits the maximum rates. The response of rate towards changing substrate concentration is less than that of a second order reaction (with rate constant V_{max}/K_m) except at very low substrate concentration (relative to K_m) where the enzyme is not significantly saturated. It is easy to show by means of appropriate substitution in the Michaelis–Menten equation that an 81-fold change in the substrate concentration is required in order to alter the rate from $10 \rightarrow 90\%$ of V_{max}. This "control ratio" is markedly insensitive and will contribute to stability in circumstances in which substrate concentrations fluctuate. However, stability, although obviously vital to the organism as a whole, can reduce the ability of an organism to respond rapidly and sensitively to a stimulus. Thus stability towards random fluctuations is required but a sensitive concerted response to certain stimuli is highly beneficial, for example, in an avoidance response. Thus we come to ask the question: how may the response of rate be made more sensitive to a changing substrate concentration?

In order to answer this question we first consider the properties of the oxygen binding proteins, myoglobin and haemoglobin. The oxygen binding curves for these two haemo-proteins are shown in figure 7.5. The

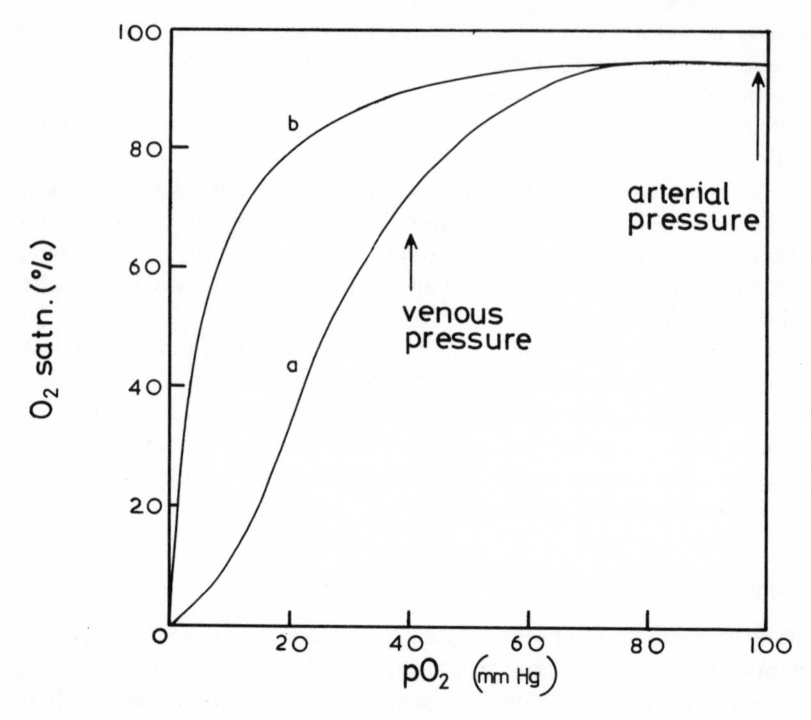

Figure 7.5 The oxygen binding curves for haemoglobin and myoglobin. The Adair constants (see 7.34) that characterize oxygen binding to haemoglobin are: (mm Hg) $K_1 = 25$, $K_2 = 8.5$, $K_3 = 10.4$, $K_4 = 0.12$, 25°, 0.05 M bis-tris buffer, pH 7.4, 0.1 M NaCl. The parameters that result from fitting to the MWC equation (see 7.37) are: $K_T = 20.4$ mm Hg; $K_R = 0.11$ mm Hg; $L = 1.9 \times 10^6$. The free energy difference between R- and T-forms varies between 16–36 kJ mole^{-1}, depending upon the conditions.

oxygen binding curve for myoglobin is plainly hyperbolic with the maximal loading/unloading sensitivity at an oxygen tension approaching zero. The curve for hæmoglobin is S-shaped or sigmoid and this confers two special functional properties upon the oxygen-transport system. Firstly, the protein is loaded and unloaded (the steep portion of the plot) at values of the oxygen tension well away from zero. Secondly, the slope of the plot in the region of half-saturation is greater than that of myoglobin.

Let us first rationalize these features in terms of the physiological functions of these proteins. Myoglobin acts as a buffer store of oxygen capacity in muscle. The oxygen stored by myoglobin is passed to cytochrome oxidase at the end-point of aerobic metabolism. It is well-known that muscle can contract vigorously for some time before it becomes anaerobic. When muscle becomes anaerobic during vigorous exercise myoglobin becomes completely depleted of oxygen which cannot be replaced at a rate sufficient to maintain aerobic metabolism. When the oxygen is depleted, the metabolic end product is lactic acid (which has been implicated in muscular pain generation). It is obvious that this is not a harmful process unless taken to excess. The situation in the brain is however quite different. There is no myoglobin in the brain and anaerobic metabolism is not possible—thus disaster in the form of unconsciousness followed later by brain damage occurs upon oxygen deprivation in the brain. Both muscle myoglobin and brain cytochrome oxidase are supplied with oxygen by the haemoglobin in the blood. Haemoglobin must yield its oxygen readily when the oxygen tension is reduced relative to that found in the alveoli of the lungs. Thus haemoglobin must have its greatest sensitivity to oxygen loading/unloading well away from zero oxygen tension and the degree of sensitivity must be as high as possible in the intermediate oxygen tension region. We can see from an examination of figure 7.5 that this is what is achieved by myoglobin and haemoglobin when their loading curves are considered in relation to the arterial and venous oxygen tensions, although the latter pressure is higher than might be expected.

The Hill equation

In 1910 A. V. Hill attempted to find an equation that related the loading or fractional saturation of haemoglobin to the oxygen tension. He did not know much about the nature of the protein except that it contained iron, and in particular he did not know whether it was homogeneous. He realized that sigmoid curves can be generated by functions of the form (7.27)

$$Y = \frac{x^n}{K + x^n} \tag{7.27}$$

In fact, such an equation can be derived for the case of the oxygenation of haemoglobin from a very simple conceptual model. Hill supposed that the oxygenation equilibrium could be described by (7.28)

$$\mathrm{Hb} + n\mathrm{O_2} \overset{K}{\rightleftharpoons} \mathrm{Hb(O_2)}_n \tag{7.28}$$

The dissociation constant for this process is given by (7.29):

$$K = \frac{(\mathrm{Hb})(\mathrm{O}_2)^n}{\mathrm{Hb}(\mathrm{O}_2)_n} \tag{7.29}$$

where n is the number of oxygen molecules bound by haemoglobin. These equations may be combined, together with the conservation equation for haemoglobin to give equation (7.30) which is known as the Hill equation.

$$Y = \frac{(\mathrm{O}_2)^n}{K + (\mathrm{O}_2)^n} \tag{7.30}$$

The Hill equation is usually expressed in terms of the fractional saturation Y, and h, the apparent, usually nonintegral, value of the index n in equation (7.30) which is known as the Hill constant. The Hill equation can be rearranged to give equation (7.31)

$$\frac{Y}{1-Y} = \frac{(\mathrm{O}_2)^h}{K} \tag{7.31}$$

which by taking logarithms allows a linear plot to be made of experimental data, according to equation (7.32)

$$\log\left(\frac{Y}{1-Y}\right) = h \log(\mathrm{O}_2) - \log K \tag{7.32}$$

A plot using equation (7.32) is shown in figure 7.6, and is known as a *Hill plot*. The data conform well to the equation in the region of 50% saturation $(Y/1-Y) = 1$ but deviation invariably occurs at the extremes. This deviation is not surprising since the model upon which the derivation of the Hill equation is based involves some sweeping assumptions. Firstly, it is assumed that the intermediate states of oxygenation are not populated, i.e. that states such as $\mathrm{Hb}(\mathrm{O}_2)$ do not exist and secondly that the dissociation of all oxygen molecules can be described by a single dissociation constant. We note that $h > 1$ since the equation (7.30) with $h(n) = 1$ is essentially the Michaelis–Menten equation, and gives a hyperbolic curve. The slope of the Hill plot (figure 7.6) for haemoglobin is ~ 2.8 which, since this value is non-integral, can be interpreted to mean that the model used to derive the Hill equation is approximate.

Since we now know that haemoglobin has four binding sites for oxygen, it is not unreasonable to suppose that h represents the minimum value of the number of binding sites available. The Hill equation may be used in the analysis of kinetic experiments by replacing Y by v/V_{max}, the fractional rate saturation.

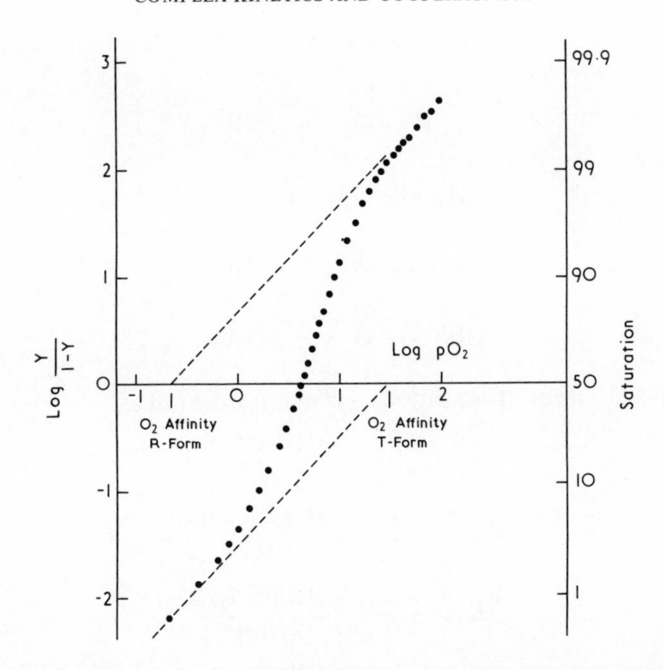

Figure 7.6 Hill plot of the oxygen binding data for haemoglobin, pH 7.4. Redrawn from Kilmartin, J. V., Imai, K. and Jones, R. T. (1975) *Erythrocyte Structure and Function*, A. R. Liss Inc., New York, p. 21.

Before proceeding to consider more advanced models of cooperativity, we note that the slope of a plot of Y against ligand for equation (7.30) has a slope of $(h/4)K$ when $Y = 0.5$. Thus the slope is directly proportional to h when $K = 1$ and the sensitivity increases as the degree of cooperativity increases.* Since the Hill equation is accurately fitted by data obtained from cooperative systems in the region of $Y = 0.5$, this conclusion has general if empirical validity.

The Adair equation

By 1925 it had been determined that haemoglobin is probably tetrameric, and Adair proposed a sequential mechanism for the oxygenation of

* When $K < 0.25$ the slope first decreases as h increases, but then increases, eventually tending to $h/4$ as h becomes large.

haemoglobin. In this mechanism shown in (7.33) the intermediate states may be populated.

$$\begin{aligned}
Hb+O_2 &\underset{k_1}{\overset{4k_1}{\rightleftharpoons}} HbO_2 \\
HbO_2+O_2 &\underset{2k_{-2}}{\overset{3k_2}{\rightleftharpoons}} Hb(O_2)_2 \\
Hb(O_2)_2+O_2 &\underset{3k_{-3}}{\overset{2k_3}{\rightleftharpoons}} Hb(O_2)_3 \\
Hb(O_2)+O_2 &\underset{4k_{-4}}{\overset{k_4}{\rightleftharpoons}} Hb(O_2)_4
\end{aligned} \qquad (7.33)$$

The four *intrinsic* dissociation constants are defined as:

$$\begin{aligned}
\tfrac{1}{4}K_1 &= \frac{1}{4}\frac{k_{-1}}{k_1} = \frac{(Hb)(O_2)}{(Hb(O_2))} \\
\tfrac{2}{3}K_2 &= \frac{2}{3}\frac{k_{-2}}{k_2} = \frac{(Hb(O_2))(O_2)}{(Hb(O_2)_2)} \\
\tfrac{3}{2}K_3 &= \frac{3}{2}\frac{k_{-3}}{k_3} = \frac{(Hb(O_2)_2)(O_2)}{(Hb(O_2)_3)} \\
4K_4 &= 4\frac{k_{-4}}{k_4} = \frac{(Hb(O_2)_3)(O_2)}{(Hb(O_2)_4)}
\end{aligned}$$

The statistical factors reflect the number of equivalent ways in which a given association or dissociation process may occur. The equilibrium constants K_1–K_4 will have the same numerical value when the site affinities are identical if the equilibrium constants are defined in this way. By means of the usual combination of the expressions for the equilibrium constants with appropriate conservation equations the concentrations of each of the species can be evaluated. These may be combined to give the Adair equation for four sites (7.34):

$$\begin{aligned}
Y &= \frac{\text{number of filled sites}}{\text{total number of sites}} \\
&= \frac{Hb(O_2)+2Hb(O_2)_2+3Hb(O_2)_3+4Hb(O_2)_4}{4(Hb+Hb(O_2)+Hb(O_2)_2+Hb(O_2)_3+Hb(O_2)_4)} \\
&= \frac{\dfrac{(O_2)}{K_1}+\dfrac{3(O_2)^2}{K_1K_2}+\dfrac{3(O_2)^3}{K_1K_2K_3}+\dfrac{(O_2)^4}{K_1K_2K_3K_4}}{1+\dfrac{4(O_2)}{K_1}+\dfrac{6(O_2)^2}{K_1K_2}+\dfrac{4(O_2)^3}{K_1K_2K_3}+\dfrac{(O_2)^4}{K_1K_2K_3K_4}}
\end{aligned} \qquad (7.34)$$

If the four intrinsic dissociation constants are equal, equation (7.34) simplifies to the simple hyperbolic expression (7.35)

$$Y = \frac{(O_2)}{K+(O_2)} \tag{7.35}$$

The Adair equation simplifies to the Hill equation (where $h = 4$) if $K_4 \ll K_1, K_2, K_3$.

$$Y = \frac{(O_2)^4}{K_1 K_2 K_3 K_4 + (O_2)^4} \tag{7.36}$$

Since $h = 2.8$ for the haemoglobin–oxygen binding data when plotted according to the Hill equation, it is apparent that K_4 is *not* much less than the other dissociation constants. We note that *positive* cooperativity is characterized by $K_1 > K_2 > K_3 > K_4$ and *negative* cooperativity by $K_1 < K_2 < K_3 < K_4$. In the former case the binding gets progressively tighter as the sites are filled whilst in the latter it becomes progressively weaker. The generality of the Adair equation is demonstrated by the possibility of mixed positive and negative cooperativity, i.e. $K_1 > K_2$ and $K_3 < K_4$. In fact the Adair equation represents a general equation for cooperativity; more advanced schemes based upon specific mechanistic proposals give rise to equations that may be expressed in the same general form as the Adair equation. It is for this reason that experimental data are usually fitted by multivariate regression analysis to the Adair equation. The parameters (usually in reduced form) that characterize any particular model can then be calculated from the Adair parameters. We now proceed to consider models of cooperativity in which specific molecular mechanisms are proposed.

The Monod, Wyman and Changeux model

This model is based upon the assumption that a multimeric protein may assume *two* separate conformations in the *absence* of ligand. These two states are envisaged as either R (relaxed) or T (tensed) states which have different affinities for the ligand. Each site in a particular conformation binds ligand identically, so all the binding events to a single conformation may be described by a single equilibrium constant. Cooperativity is achieved as a result of the differential binding affinity of the R and T states and allows the *lower* affinity T state to be pulled over to the R *high* affinity state as the ligand concentration is increased. This has the effect of increasing the overall affinity for ligand as the concentration increases and leads to a sigmoid binding curve. The model is depicted diagrammatically below:

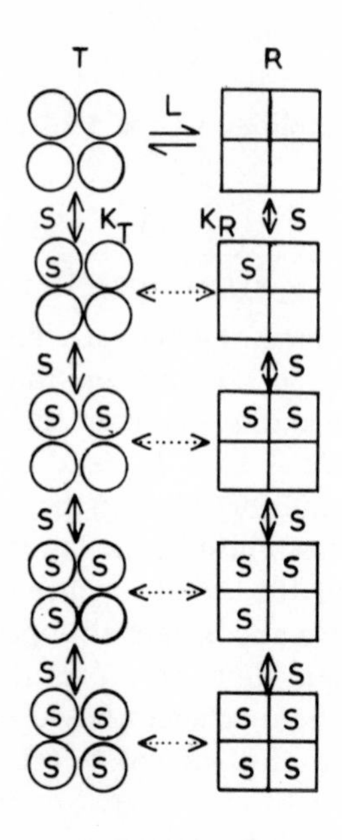

(The reversible processes shown as dotted lines are not an essential feature of the model.)

A most important feature of the model is that *all* the subunits change their conformation coincidentally. The model is defined in terms of three parameters: L, which is the ratio of the concentration of non-liganded T-state to that of non-liganded R-state, K_T which is the intrinsic dissociation constant for ligand-T-state interaction and K_R, the equivalent term for the R-state. In the derivation of the MWC binding equation statistical factors must be taken into account in the same way as they were during the derivation of the Adair equation. The fractional saturation is evaluated using simple algebra as for the Adair equation and yields the equation (7.37) for four sites.

$$Y = \frac{(1+(\mathrm{S})/K_R)^3(\mathrm{S})/K_R + Lc(1+c(\mathrm{S})/K_R)^3(\mathrm{S})/K_R}{(1+(\mathrm{S})/K_R)^4 + L(1+c(\mathrm{S})/K_R)^4} \qquad (7.37)$$

where $c = K_R/K_T$.

Not surprisingly, if $L = 0$ or ∞ equation (7.37) reduces to the hyperbolic forms (7.38) and (7.39) respectively.

$$Y = \frac{(S)}{K_R + (S)} \tag{7.38}$$

$$Y = \frac{(S)}{K_T + (S)} \tag{7.39}$$

Accordingly, the observation of cooperative behaviour relies upon the presence of both the R- and T-forms of the protein. Equation (7.37) also reduces to hyperbolic form if $K_R = K_T$, i.e. $c = 1$ when the ligand affinity is the same for both R- and T-forms of the protein. If $c = 0$, i.e. the ligand binds only to the R-form, then cooperativity results, the degree being dependent upon the value of L.

As mentioned earlier, the Adair equation represents a general form of cooperativity equation so it must be possible to express the Adair parameters in terms of the MWC parameters. The result of such a comparison made by multiplying out the MWC equation yields:

$$K_1 = \frac{K_R(1+L)}{1+Lc}$$

$$K_2 = \frac{(1+Lc)K_R}{1+Lc^2}$$

$$K_3 = \frac{(1+Lc^2)K_R}{1+Lc^3}$$

$$K_4 = \frac{(1+Lc^3)K_R}{1+Lc^4}$$

Taking the ratio of two of the Adair constants we get

$$\frac{K_2}{K_1} = \frac{(1+Lc)K_R}{(1+Lc^2)} \cdot \frac{1+Lc}{(1+L)K_R}$$

which for $K_1 > K_2$

$$(1+L)(1+Lc^2) \geqslant (1+Lc)^2$$

$$1+L+Lc^2+L^2c^2 \geqslant 1+2Lc+L^2c^2$$

which yields on reduction

$$L(1-c)^2 \geqslant 0$$

Thus $K_1 > K_2$ for all positive values of L and c. Since this relationship applies to all pairs of Adair constants such that $K_n > K_{n+i}$, it is apparent that the MWC model applies only to *positive* cooperativity. The fundamental assumption of the MWC model, namely that of the existence of two forms of the protein (R and T) in the absence of ligand seems to many to be rather implausible. However, in view of the known conformational mobility of some proteins in solution (also "fuzzy" regions in X-ray crystallographic structures) the existence of R- and T-states in the absence of ligand may not be so unreasonable. We must remember that induced fit can be interpreted in terms of either induced deformation or conformation selectivity. Recent advances in the observation and computer simulation of the conformational dynamics of proteins favour the latter interpretation. Since the forms are supposed to be in equilibrium at room temperature the activation energy for their interconversion cannot be very large. The MWC theory perhaps has more relevance to binding systems such as haemoglobin rather than to catalytic systems where one expects either induced fit or transition state complementarity (see chapter 9). An important factor in favour of the MWC model rests in its ability to provide relatively simple explanations for *heterotropic* or allosteric effects. Heterotropic effects involve modification of binding or catalysis by an effector other than the substrate acting at a site other than the active site. For example 2, 3-diphosphoglycerate is a heterotropic effector of haemoglobin and binds at a site distinct from the oxygen-binding sites. Heterotropic interactions will be considered in more detail after the main alternative model for cooperativity has been described.

The Koshland, Nemethy and Filmer model

The KNF model is based upon the idea of induced fit which Koshland had previously suggested was relevant to the mechanisms of enzymic catalyses. The inability of hexokinase to hydrolyse ATP to any extent in the absence of glucose is an oft-quoted example of the importance of induced fit.

Thus the KNF model involves the assumption that subunit conformational changes are induced specifically by interaction with ligand. Each subunit whose conformation is perturbed by ligand interaction has an effect upon neighbouring subunits such that their ability to bind ligand may be perturbed. A rather obvious example of this may be visualized as an attempt by the system to maintain symmetry. A subunit perturbed as a result of ligand interaction may interact with a ligand-free neighbour in such a way as to optimize the interaction between the neighbours. Since in an initially symmetric system this is most likely to be achieved by the preservation of symmetry, the ligand-free neighbour will deform towards the structure of the ligand-bound state. Since energy must be expended to produce the deformation that results from induced fit, the ligand-free neighbour will bind ligand more tightly than did the subunit that loaded ligand in an earlier step. Thus the KNF model provides a relatively simple mechanistic explanation of positive cooperativity. It is a simple matter to extend the above rationale to cover negative cooperativity although it has to be admitted the rationale is not so pleasing since a distortion away from the ligand-induced conformation must be proposed. The scheme for the KNF model is shown below.

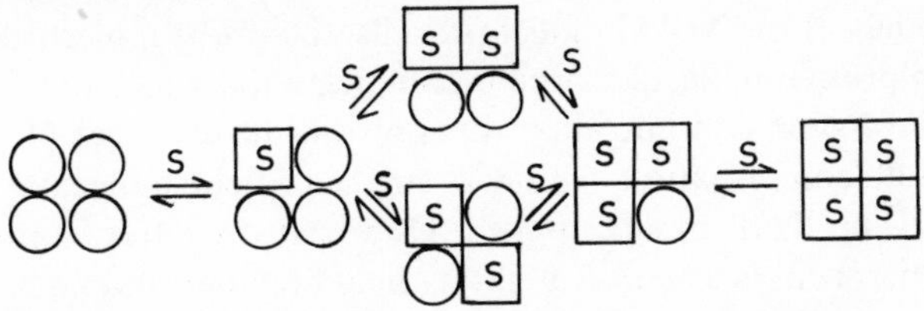

The dependence upon the geometric array of the subunits in the KNF model is simply explained on the basis that an immediately adjacent subunit will be subject to a larger perturbation by a ligand-bound subunit than will a subunit further removed from the source of the perturbation. In fact, in the usual derivation of the KNF scheme it is assumed that only adjacent subunits are affected. This means that in the "square" geometry shown above any one subunit may influence two other subunits but not the one diagonally opposite it. In a tetrahedral geometry all subunits interact, whilst in a linear geometry the end subunits each interact with one subunit while the others interact with the two adjacent subunits. The KNF model is usually analysed in terms of interface stability constants which describe the three permutations possible for a given interface, i.e. AA-type, AB-type or BB-type, where A represents a ligand-free subunit and B a liganded subunit. The derivation of the KNF equation using the interface concept together with dissociation constants for ligandation of A- and B-type subunits is relatively straightforward but undoubtedly rather cumbersome. For this reason the quantitative derivation of the equations for the KNF model will not be presented here. The coefficients of the equations may be related to the Adair constants as with the MWC model. The KNF model differs however from the MWC model in that it provides a fit to negatively cooperative data and is sensitive to the geometry of the subunit assembly.

Distinction between the MWC and KNF models

As mentioned above, the MWC model is not consistent with negative cooperativity; it is difficult to distinguish between the models in cases of positive cooperativity. It has been claimed that the data for haemoglobin oxygenation is better fitted by the MWC model than by the KNF model but the evidence for this claim is weak. The measurement of rate constants in principle allows distinction between the models: the MWC model should give rise to fewer relaxation processes since only two conformations are invoked. Fluorescence and e.s.r. experiments in which conformational probe molecules have been attached to each of the subunits of an oligomeric protein have been performed in order to attempt the distinction of the models. If the MWC model prevails, the probe molecules attached to a given protein molecule should all indicate the same conformation at a particular ligand concentration. In contrast, the existence of more than one subunit conformation within a single protein molecule would be expected if the KNF model applied. However, the interpretation of this type of experiment is usually fraught with ambiguities, and none has as yet provided unequivocal support for one model against the other. It is however quite popularly supposed that the MWC model may be more applicable to haemoglobin than to enzymes and vice-versa for the KNF model.

We end this section by noting that it is very important to appreciate that the MWC and KNF models of cooperativity represent special forms of a more general "square" model. The MWC model is represented by two sides of the "square" and the KNF model by a diagonal. Unfortunately the algebra involved in the derivation of the equation for the general scheme becomes hopelessly complicated, so it cannot be used in practice.

Heterotropic (allosteric) interactions

By allosterism is meant modification of a binding or catalytic function by an "other shape". Thus allosterism does not imply competitive or product inhibition—the important point is that interaction takes place at an allosteric site distinct from the main catalytic or binding site. Some examples are 2, 3-diphosphoglycerate, an allosteric effector of haemoglobin; AMP, an effector of isocitrate dehydrogenase; and fructose diphosphate, an effector of pyruvate kinase.

It is frequently found that allosteric effectors reduce or abolish homotropic cooperativity, and the MWC model in contrast to the KNF model lends itself to relatively simple explanations of these allosteric effects. An effector is supposed to act by binding preferentially to either the R- or the T-form, and thus one form will be stabilized with respect to the other, which results in modification of the cooperativity. An activator is proposed to bind to the R-form and, if it binds sufficiently strongly, will cause the R-form to predominate, leading to abolition of homotropic cooperativity as L tends to zero. The effectors AMP and fructose diphosphate mentioned above are both of this type. At low substrate concentration (i.e. in the "lag" phase of the v_i vs. S profile) an allosteric activator can produce a dramatic increase in enzymic activity, as shown in figure 7.7.

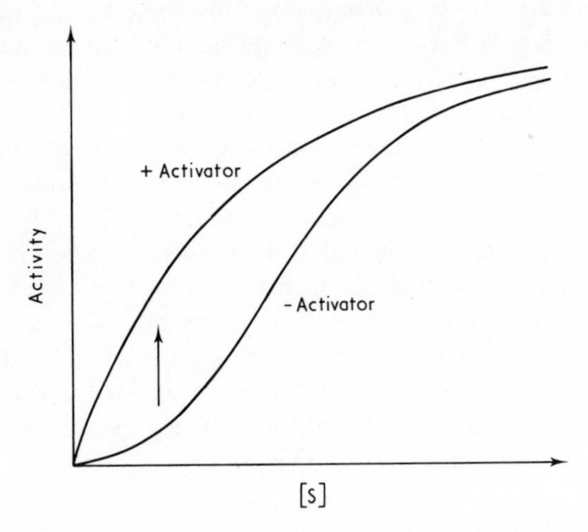

Figure 7.7 A diagram which illustrates the large increase in activity that may be achieved by an allosteric activator acting under conditions of low substrate concentration. The arrow indicates the increase in activity.

Inhibitors are supposed to bind to the T-form which has the lower affinity for substrate. The concentration of the T-form will increase, leading to a lower concentration of enzyme–substrate complex and hence a lower activity. Again cooperativity will be abolished if the interaction between the inhibitor and the T-form is strong since L will tend to infinity. Thus the modification of the catalytic activity is seen as being due to a change in the apparent affinity of the enzyme for its substrate; such systems are known as *K systems*. The explanation is delightfully simple, which is an important point in favour of the MWC model. However, it is naïve to expect the effect to be explained entirely on the basis of binding perturbation, since it seems improbable that the substrate complexes of the R- and T-forms have identical catalytic activities.

Some enzymes show quite a different pattern of response to effectors. Pyruvate carboxylase, which does not show homotropic cooperativity in substrate, is allosterically activated by acetyl-CoA. The response of the initial velocity to increasing acetyl-CoA is sigmoid rather than hyperbolic; in other words, coöperativity is shown in the allosteric response. The activator increases the maximum velocity of the enzyme reaction so the effect of the activator must be to alter the catalytic activity of the enzyme rather than the substrate binding affinity. AMP acts as an allosteric inhibitor of fructose 1,6-diphosphatase by lowering the V_{max} value. Systems such as these are known as V-systems in contrast to the K-systems described above. In general, mixed V–K systems would be expected and several enzymes seem to follow this type of behaviour.

The KNF model, although appealing in terms of the molecular mechanism which is involved in the definition of the model, is not easily adapted for the consideration of heterotropic effects. Many different assumptions are possible concerning the number of conformations available to the system as well as the identity of the site of interaction of the effector. This leads to many possible explanations of heterotropic effects with no simple way of distinguishing between them. Thus the arguably naïve simplicity of the MWC model wins the day as far as interpretation of heterotropic effects is concerned, particularly in circumstances where little "molecular" information is available.

Half-of-the-sites reactivity

An extreme example of negative cooperativity is one in which only half of the active sites in an oligomer are active at any particular time. Such a situation has supposedly been detected in the mechanism of several enzymes, for example, glyceraldehyde-3-phosphate dehydrogenase and aldolase. Half-of-the-sites reactivity is characterized by rapid reaction of substrates or irreversible inhibitors at half the total active sites. The phenomenon has been explained on the basis of pre-existing asymmetry which results from the asymmetric assembly of identical subunits. The reliable observation of half-of-the-sites reactivity requires an accurate knowledge of the active site concentration of the enzyme; that this information has not always been of a high quality casts doubt on many of the reported observations of this phenomenon.

Other models of cooperativity—subunit depolymerization

Cooperativity can arise as a result of subunit dissociation which gives rise to a mixture of oligomers. A wide variety of detailed models may be proposed. Two-state models (in which two forms of the enzyme exist in equilibrium) give rise to equations that bear close resemblance to that of the MWC model, except that the enzyme concentration occurs in the equation. Since the degree of cooperativity is predicted as dependent upon the enzyme concentration, this type of cooperativity should be easier to detect than those discussed earlier. Cooperative nucleotide binding to glutamate dehydrogenase has been explained on the basis of reversible dissociation of the enzyme.

Kinetic models of cooperativity

Consider the random order bisubstrate mechanism

$$\begin{array}{ccccc} & \overset{A}{\nearrow\!\!\swarrow} & EA & \overset{B}{\searrow\!\!\nwarrow} & \\ E & & & & EAB \rightleftharpoons EPQ------ \\ & \underset{B}{\searrow\!\!\nwarrow} & EB & \underset{A}{\nearrow\!\!\swarrow} & \end{array}$$

The equilibrium derivation of the equation for the initial velocity leads to equation (7.40)

$$v_0 = \frac{V}{1 + \dfrac{K_m^A}{(A)} + \dfrac{K_m^B}{(B)} + \dfrac{K_S^A K_m^B}{(A)(A)}} \tag{7.40}$$

At a fixed concentration of B this equation predicts simple hyperbolic kinetic behaviour of v_0 as (A) is changed. However, the *steady-state* derivation leads to a much more complex equation which includes squared terms in A and B. At a fixed concentration of B the equation may be simplified to give equation (7.41)

$$v_0 = \frac{\alpha_1(A) + \alpha_2(A)^2}{\beta_0 + \beta_1(A) + \beta_2(A)^2} \tag{7.41}$$

This equation gives rise to a *sigmoid* response in v_0 to variation of (A) if $\beta_0\alpha_2 > \beta_1\alpha_1$. Other complex responses are possible including curves which possess maxima. Equation (7.41) may be related to the Adair equation as may all the other models. Thus apparent cooperativity may occur in situations that do not involve subunit interaction.

The Rabin mechanism

This mechanism involves enzyme isomerization as shown in the scheme below.

$$\mathrm{E+S \rightleftharpoons ES \rightleftharpoons E'S \rightarrow E'+P}$$

(with $-\mathrm{S}$ and $+\mathrm{S}$ arrows linking E′S and E′, and an arrow linking E′ back to E)

in which E and E′ represent isomeric forms of the enzyme. If the step E′S → E′ + P is fast compared with the steps E′ → E and ES → E′S, then at high (S) the enzyme will not have time to revert to the E form but will form E′S instead. The reaction will be rapid since it bypasses both the slow steps (E′ → E, ES → E′S). At low (S) the enzyme will have time to relax from E′ to E so the reaction will proceed via the slow step ES → E′S. Thus the isomerization steps have the effect of altering the ratio of the pathways as the substrate concentration is changed. Since the slower pathway is favoured at low (S) a lag phase will be introduced into the v_0 vs. (S) plot giving rise to a sigmoid response.

The mnemonical mechanism

The mnemonical mechanism requires two isomeric forms of the enzyme in the absence of substrate and represents an extension of Rabin's mechanism. It has been proposed that wheatgerm hexokinase and rat liver glucokinase follow this mechanism, which is described in chapter 9. Rat liver glucokinase has been established as a single polypeptide chain so the possibility of subunit interaction may be excluded.

Other situations in which apparent cooperativity may be observed

Before sigmoid kinetics may be interpreted in terms of genuine cooperativity, three important sources of apparent cooperativity must be eliminated. These are:

(1) The enzyme is subject to compulsory substrate activation. In this mechanism *two* substrate molecules must bind to the enzyme before catalytic activity occurs.
(2) More than one enzyme present in the assay mixture is capable of catalysing the reaction.
(3) The substrate is impure: (a) it contains an activator which will be added in constant proportion—sigmoid kinetics result; or (b) it contains more than one species which may be catalytically transformed by the enzyme—this leads to apparent negative cooperativity.

Readers are advised to convince themselves that the mechanisms mentioned above do indeed give rise to apparent cooperativity.

Molecular mechanisms of cooperativity

The molecular mechanism has not been established for cooperativity in enzymes; perhaps the main reason for this is the great complexity (and potential ambiguity) of the experiments which will be required to establish this. Studies which make use of substrate analogues and perhaps transition state analogues should prove useful in that they eliminate the catalytic elements. However, since the catalytic (as against binding) aspect is likely to be of considerable importance these methods will not be completely adequate. Cryogenic techniques would seem to hold out the most hope for progress in this field. It is possible to essentially "stop" enzyme reactions by cooling them to suitably low temperatures in special solvents (see chapter 8). This allows direct structural study of the true enzyme–substrate complex(es) provided that the reaction proceeds in the cryogenic solvent as it would in an aqueous environment. Thus the conformational mobility of the protein must not be much affected by the solvent (up to 60% dimethyl sulphoxide) otherwise the cooperative properties of the enzyme may be modified or lost in the solvent.

It may well be that the cooperative properties *are* either lost or seriously modified when the enzyme is treated as above, in which case progress will have to rely upon the more traditional forms of study (such as NMR and X-ray crystallography). For instance the conformational change induced by glucose binding to hexokinase has recently been determined by X-ray crystallography. This is possible because hexokinase is a bisubstrate enzyme. Such an enzyme may be crystallized in the presence of the substrate for which it shows cooperative kinetics without the problem of catalytic turnover. If an enzyme-catalysed reaction has an equilibrium constant which strongly favours substrates or products it should be possible to produce crystals of the thermodynamically stable enzyme–substrates (products) complex from a mother liquor of these substrates. Analysis of the X-ray structures in terms of conformational changes which result in alteration of hydrophobic contact, salt bridges, hydrogen bonding and substrate binding should lead to a detailed rationale of the cooperativity process, as has been possible in the case of haemoglobin (see bibliography).

CHAPTER EIGHT

THE ANALYSIS OF ENZYME MECHANISMS BY MEANS OF FAST REACTION TECHNIQUES

IN CHAPTERS 4 AND 7 WE HAVE SEEN THAT THE INFORMATION OBTAINABLE from steady-state kinetic studies is extensive but limited in certain important respects. Steady-state kinetic studies usually allow the definition of the stoichiometry and minimal kinetic mechanism of a given enzyme reaction. The quantitative results arise in the form of substrate (or product) binding equilibria which may be kinetically perturbed and catalytic rate constants which may define a unique rate-limiting step or may be a function of several rate constants which contribute to the rate-limiting step. Whether or not the apparent equilibria and catalytic rate constants relate to a single well-defined step of the reaction depends upon the relative magnitudes of the component rate constants that comprise the expressions for the apparent equilibria and rate constants. For instance in chapter 4 the Michaelis constant for a simple enzyme reaction is defined as:

$$K_m = \left(\frac{k_{-1} + k_2}{k_1}\right) \tag{8.1}$$

Clearly $K_m = K_S$, the true dissociation constant of the enzyme–substrate complex, only if $k_{-1} \gg k_2$. Similarly the apparent k_{cat} for the chymotrypsin acyl enzyme mechanism (see chapter 9) is given by

$$k_{cat} = \frac{k_2 k_3}{k_2 + k_3} \tag{8.2}$$

where k_2 represents acylation of the enzyme to yield the first product and acyl-enzyme, and k_3 represents the rate constant for deacylation. Clearly if $k_2 \gg k_3$, $k_{cat} = k_3$ and if $k_3 \gg k_2$, $k_{cat} = k_2$, but if $k_2 \simeq k_3$ then $k_{cat} \simeq \frac{1}{2}k_2$

or $\frac{1}{2}k_3$. Many special tricks have been developed that enable steady-state kinetic methods to be applied to the problem of resolving the component rate constants of expressions of this type. The greatest success has been achieved in relation to studies of hydrolytic enzymes, particularly those which catalyse polymer hydrolysis, since such enzymes will tolerate considerable chemical variation in the structure of their substrates (see chapter 9).

Enzymes that are highly substrate-specific (such as lactate dehydrogenase) will not tolerate extensive chemical interference with their substrates and thus the techniques outlined above could not be applied even if they were relevant. An alternative and preferably more direct method for the kinetic dissection of the minimal mechanism defined by steady-state kinetics must be used. Methods that have proved successful in this respect depend upon unambiguous observations of well-defined single catalytic steps and to achieve this it is necessary to make observations in the time domain before the steady state is achieved.

Analysis of the simple mechanism

We shall now consider the nature and quality of information that can be gained from the study and analysis of the pre-steady-state kinetics of the simple two-step enzymic reaction:

$$\mathrm{E+S} \underset{k_{-1}}{\overset{k_1}{\rightleftharpoons}} \mathrm{ES} \xrightarrow{k_2} \mathrm{E+P}$$

Assuming only that no product is present initially we need conservation equations for E and S_0 and rate equations for ES and P, which are given by (concentration brackets omitted for clarity):

$$\mathrm{S_0 = S+ES+P} \tag{8.3}$$

$$\mathrm{E_0 = E+ES} \tag{8.4}$$

$$\frac{d\mathrm{ES}}{dt} = k_1\mathrm{E \cdot S} - (k_{-1}+k_2)\mathrm{ES} \tag{8.5}$$

$$\frac{d\mathrm{P}}{dt} = k_2\mathrm{ES} \tag{8.6}$$

$$\frac{d^2\mathrm{P}}{dt^2} = k_2\frac{d\mathrm{ES}}{dt} \tag{8.7}$$

Simple algebraic combination of these equations leads to the intractable non-linear second order differential equation:

$$\frac{d^2P}{dt^2} + \frac{k_1}{k_2}\left(\frac{dP}{dt}\right)^2 + (k_1(E_0+S_0+P)+k_{-1}+k_2)\frac{dP}{dt} + k_1k_2E_0P = k_1k_2E_0S_0 \quad (8.8)$$

It is necessary to introduce some simplifying assumption(s) in order to improve the tractability of the system since equation (8.8) can be solved only by computer simulation (see later). If we introduce the assumption that $S_0 \gg E_0$ and proceed as above, neglecting the conservation equation in substrate (8.3) we find that the solution is given by the non-linear second order differential equation:

$$\frac{d^2P}{dt^2} - k_1P\frac{dP}{dt} + (k_1S_0+k_{-1}+k_2)\frac{dP}{dt} + k_1k_2E_0P = k_1k_2E_0S_0 \quad (8.9)$$

which is of the same form as equation (8.8), cannot be linearized by substitution and thus is not amenable to analytical solution. If conditions are arranged such that $E_0 \gg S_0$, a situation that can often be achieved when a stopped-flow spectrophotometer is used, a linear differential equation describes the kinetics since the conservation equation in enzyme (8.4) may be deleted.

$$\frac{d^2P}{dt^2} + (k_1E_0+k_{-1}+k_2)\frac{dP}{dt} + k_1k_2E_0P = k_1k_2E_0S_0 \quad (8.10)$$

This equation has the solution:

$$P = \frac{S_0}{m_1-m_2}\{(m_1-m_2)+m_2\exp(m_1t)+m_1\exp(m_2t)\} \quad (8.11)$$

where m_1 and m_2 are the roots of the auxiliary quadratic:

$$m^2+(k_1E_0+k_{-1}+k_2)m+k_1k_2E_0 = 0 \quad (8.12)$$

Since the quadratic (8.12) cannot conveniently be solved in the absence of numerical values of the coefficients, the analytical solution is rather cumbersome. However, the sum and product of the roots of (8.12) can be used to obtain relatively simple relationships between the relaxation times and the individual rate constants.

$$m_1+m_2 = k_1E_0+k_{-1}+k_2 \quad (8.13)$$

$$m_1m_2 = k_1k_2E_0 \quad (8.14)$$

If $m_1 \gg m_2$ then (8.13) and (8.14) can be simplified to give:

$$m_1 = k_1 E_0 + k_{-1} \tag{8.15}$$

and

$$m_2 = \frac{k_1 k_2 E_0}{k_1 E_0 + k_{-1}} \tag{8.16}$$

Values of k_1 and k_{-1} can thus in principle be determined by plotting m_1 against $[E_0]$ which gives a slope of k_1 and an intercept of k_{-1}. k_2 is obtained as the intercept from a plot of $1/m_2$ against $1/[E_0]$. The value of m_1 may however be too large to measure on a stopped-flow instrument, while k_2 can be obtained from steady-state measurements.

It is both surprising and educational to note that neither of the assumptions introduced above leads to equations whose analytical solution is simple and which may easily be applied to practical problems. In order to achieve a relatively simple solution which is readily amenable to analysis, we must try a different set of assumptions. Let us arrange that $S_0 \gg E_0$ as before and in addition assume that $S = S_0$, i.e. that in the time interval we shall consider the substrate concentration will remain essentially unchanged. This is a reasonable assumption since we are concerned here with the pre-steady-state region. The second order differential equation is:

$$\frac{d^2P}{dt^2} + (k_1 S_0 + k_{-1} + k_2)\frac{dP}{dt} = k_1 k_2 E_0 S_0 \tag{8.17}$$

which may be integrated (by substitution for dP/dt) to give:

$$\frac{dP}{dt} = k_1 k_2 E_0 S_0 \frac{\{(1 - \exp(-(k_1 S_0 + k_{-1} + k_2)t)\}}{(k_1 S_0 + k_{-1} + k_2)} \tag{8.18}$$

The rate is initially zero but accelerates to the steady-state value, which is the Michaelis–Menten equation, as the exponential decays as in figure 8.1(a). Integration of this equation to give an equation describing product accumulation yields:

$$P = \frac{k_1 k_2 E_0 S_0 \cdot t}{k_1 S_0 + k_{-1} + k_2} \cdot \frac{k_1 k_2 E_0 S_0 \{1 - \exp(-(k_1 S_0 + k_{-1} + k_2)t)\}}{(k_1 S_0 + k_{-1} + k_2)^2} \tag{8.19}$$

The first term of this equation is the simple Michaelis–Menten equation integrated with respect to time, assuming constant substrate, whilst the second term will be negative initially, decaying to a constant value as the steady state is achieved. The net result is a lag phase in product production

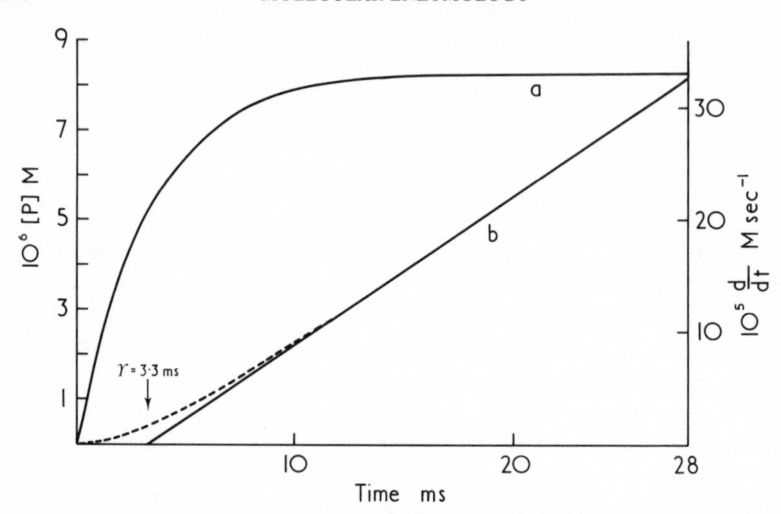

Figure 8.1 The pre-steady state rate and product accumulation of the simple enzyme reaction $E+S \rightleftharpoons ES \rightarrow P$. Calculated using equation 8.19 for line *a* and equation 8.20 for line *b*. The parameters used are: $k_1 = 10^6\ M^{-1}\ sec^{-1}$; $k_{-1} = 10^2\ sec^{-1}$; $k_2 = 10^2\ sec^{-1}$; $S_0 = 10^{-4}\ M$; $E_0 = 10^{-5}\ M$.

as shown in figure 8.1(b). After the exponential has decayed the product concentration is given by:

$$P = \frac{k_2 E_0 S_0}{K_m + S}\left\{t - \frac{1}{k_1 S_0 + k_{-1} + k_2}\right\} \tag{8.20}$$

Thus extrapolation of this steady-state rate of product generation to zero product concentration will give a finite intercept on the time axis of $1/(k_1 S_0 + k_{-1} + k_2)$. If the lag phase is determined at several values of S_0 and the reciprocal of the time axis intercept at $P = 0$ is plotted against S_0 the slope is k_1 and the intercept is $(k_{-1} + k_2)$. Knowledge of k_2 from the steady-state region allows determination of k_1, k_{-1} and k_2 which represents a complete kinetic description of this reaction scheme.

Flow techniques of following fast reactions

It is necessary at this stage to consider the apparatus that may be used to follow product generation in the pre-steady-state region of enzyme-catalysed reactions. Flow techniques which rely on a continuous mixing of enzyme and substrate at a constant rate, and observation further down the

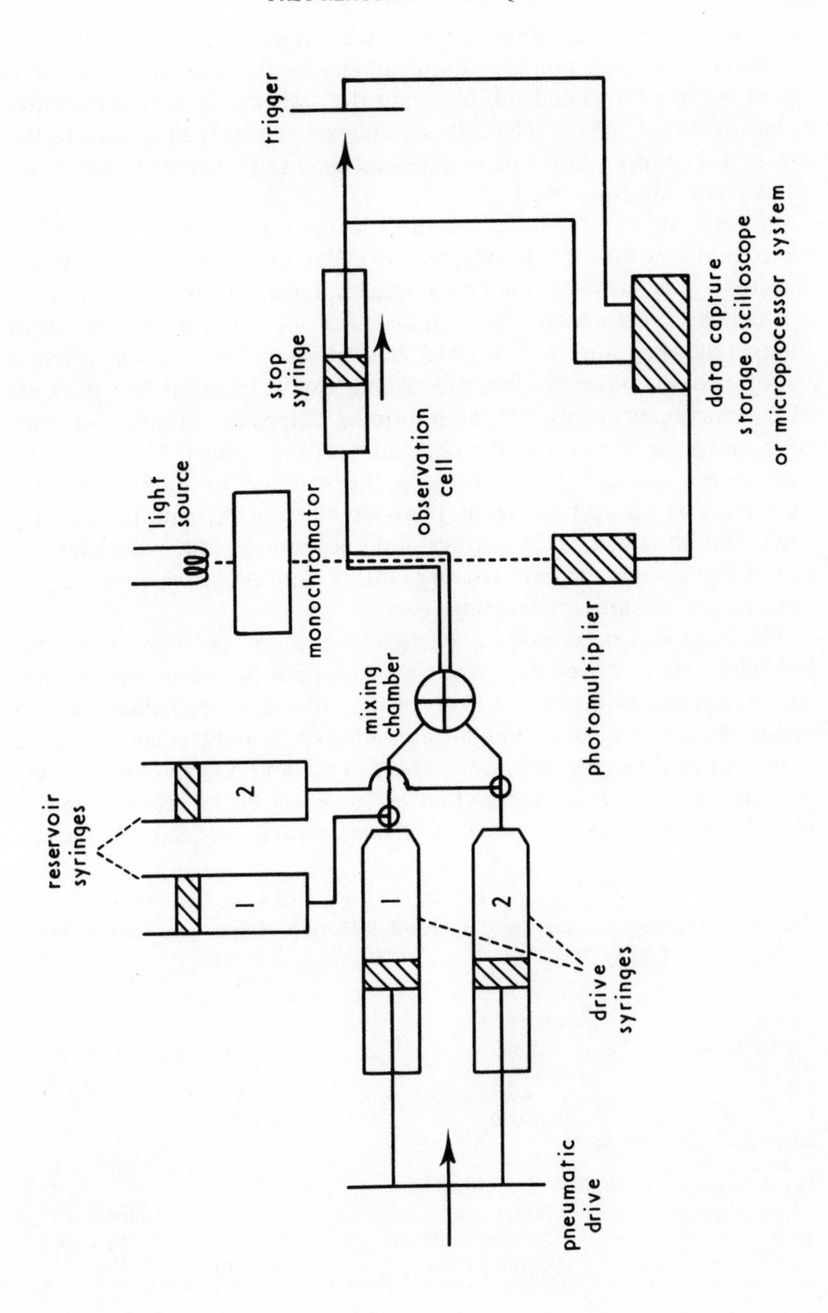
trigger
stop syringe
light source
monochromator
observation cell
data capture
storage oscilloscope
or microprocessor system
mixing chamber
photomultiplier
reservoir syringes
1
2
1
2
drive syringes
pneumatic drive

flow line of product accumulation are excellent in principle but little used in practice. The main practical disadvantages are the requirement for large quantities of enzyme and substrate and the necessity for a high flow rate to ensure turbulent flow. These disadvantages are largely eliminated by the use of the stopped-flow spectrophotometer, the elements of which are shown on the previous page.

The enzyme and substrate solutions in the drive syringes are rapidly mixed and flow through the observation cell to fill the stop syringe. When the stop syringe is filled, the flow is discontinued and observation of the cell contents commences. The minimal time for complete mixing (dead time) of the apparatus is of the order of a millisecond and so events having half-lives of the order of a few milliseconds can be followed. The progress of the reaction is observed by monitoring a change in optical absorbance or fluorescence. A very useful modification of the stopped-flow technique that allows quenching (i.e. stopping the reaction by addition of, for example, a strong acid) at various times after mixing has come to be widely used. This allows chemical analysis of the reaction products after various time intervals and permits reactions to be studied in which no useful absorbance or fluorescence change occurs.

The stopped-flow method can be used to detect the lag phase during the pre-steady-state acceleration of enzyme reactions provided that the time axis intercept $1/(k_1S_0+k_{-1}+k_2)$ is of the order of a few milliseconds or greater. In practice this technique has limited applicability, mainly because of the large values commonly found for k_1 (see table 8.1) and the necessity of using sufficient substrate so that the progress of the reaction can be reliably observed (i.e. that the signal to noise ratio is satisfactory). The rate

Table 8.1 Binding and dissociation rate constants for various enzyme–ligand interactions

	Ligand	$k_1(\text{sec}^{-1}\,\text{M}^{-1})$	$k_{-1}(\text{sec}^{-1})$
Catalase	H_2O_2	5×10^6	—
Chymotrypsin	Proflavin	1.2×10^8	8.3×10^3
	Acetyl-L-trp *p*-nitrophenylester	6×10^7	6×10^4
Lysozyme	$(NAG)_2$	4×10^7	10^5
Glyceraldehyde 3-phosphate dehydrogenase	NAD	1.9×10^7	10^3
Muscle lactate dehydrogenase	NADH	10^9	10^4
Trypsin	Pancreatic inhibitor	10^6	6.6×10^{-8}
Chymotrypsin	Chymotrypsin	3.7×10^3	0.68

constant k_1 represents the second order diffusion-limited constant for enzyme–substrate complex formation and typically varies from 10^6–$10^8\ M^{-1}\ sec^{-1}$. Choosing a value of 10^{-4} M for K_m and 100 sec^{-1} for k_{cat}, the lag phase intercept on the time axis will be approximately 0.5 msec at a substrate concentration of 10^{-4} M. If, however, the value of K_m is taken as 10^{-6} M and k_{cat} as 5 sec^{-1}, then at a substrate concentration of 10^{-6} M the intercept occurs at 50 msec. The latter situation is less typical than the former and many enzyme reactions will give rise to time axis intercepts that are difficult or impossible to measure using stopped-flow methods. Fluorescent measurement techniques that allow the use of very low substrate concentrations and judicious variation of the pH can sometimes be used to bring the pre-steady-state reaction time scale into a measurable region. Radioactive tracer methods can be used with quenched stopped-flow in order to achieve extremely sensitive assays.

Burst kinetics

When α-chymotrypsin is mixed with an excess of *p*-nitrophenyl acetate at pH 7.8 the absorbance at 400 nm increases due to release of *p*-nitrophenolate anion. If mixing is carried out by the usual manual method and the increase in absorbance at 400 nm monitored from (for example) thirty seconds after mixing for a period of minutes, the recorded trace (or graph) is of the form shown in figure 8.2(a). If the straight line, which apparently represents the steady-state reaction, is extrapolated to zero time the intercept on the 400 nm absorbance (product) axis is finite and positive. Repetition of the reaction at various enzyme concentrations produces intercepts which are proportional to the enzyme concentration. Calculation of the molarity of the *p*-nitrophenolate anion released in the "burst" phase reveals that at sufficiently high substrate concentrations the molarity of the "burst" is equal to that of the active enzyme. This situation may best be rationalized in terms of the "acyl-enzyme" kinetic scheme:

$$E + S \underset{k_{-1}}{\overset{k_{+1}}{\rightleftharpoons}} ES \xrightarrow{k_2} ES^1 + P_1 \xrightarrow{k_3} E + P_2$$

where S represents nitrophenyl acetate; ES, the Michaelis complex; ES^1, acetyl-chymotrypsin; P_1, *p*-nitrophenolate anion and P_2, acetate. Thus the stoichiometric burst of *p*-nitrophenolate anion released on admixture of E and S is coincident with formation of ES^1. The steady-state rate is due to the hydrolytic breakdown of ES^1 yielding E which is reacylated by S. Such a "burst", followed by a steady turnover rate, indicates that $k_2 > k_3$. If the

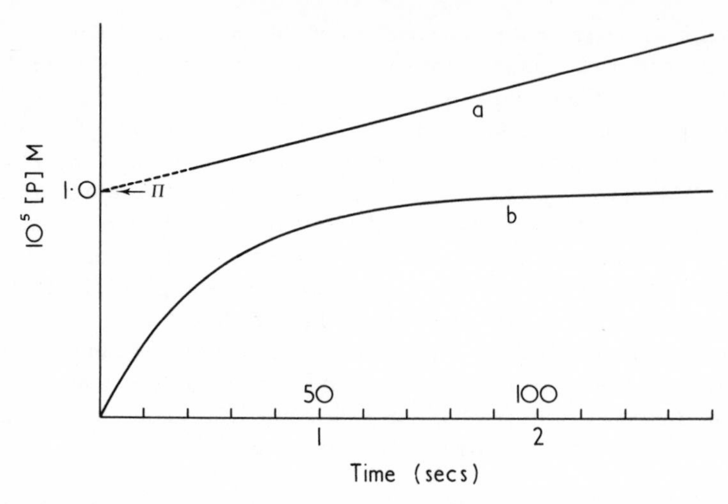

Figure 8.2 Burst kinetics of the α-chymotrypsin-catalysed hydrolysis of *p*-nitrophenyl acetate. Line *a* (upper time axis) represents the reaction as measured by a typical manual method, line *b* as measured in the stopped-flow spectrophotometer. The lines were calculated using equations 8.21 and 8.23 with the parameter values: $K_S = 10^{-3}$ M, $S_0 = 10^{-3}$ M, $E_0 = 10^{-5}$ M_1, $k_2 = 4$ sec^{-1}, $k_3 = 5 \times 10^3$ sec^{-1}, representing the reaction at pH 7.8, 25°.

production of acetate were to be followed, a lag phase would be evident during the period in which ES^1 accumulates to its steady-state value. If we assume that $S \gg E_0$ and that $K_m = K_S$ (i.e. that $k_{-1} \gg k_2$) we can solve the equations relating to the above scheme to yield an equation which relates the product concentration with the time after mixing:

$$P = \frac{k_2 k_3 E_0}{k_2 + \gamma k_3} t + \frac{k_2^2 E_0}{(k_2 + \gamma k_3)^2} \{1 - \exp(-(k_2/\gamma + k_3)t)\} \qquad (8.21)$$

where

$$\gamma = 1 + K'_S/S \quad \text{and} \quad K'_S = \left(\frac{k_{-1} + k_2}{k_1}\right)^* \qquad (8.22)$$

After the exponential term has died out the equation becomes:

$$P = \frac{k_2 k_3 t}{k_2 + \gamma k_3} + \frac{k_2^2 \cdot E_0}{(k_2 + \gamma k_3)^2} \qquad (8.23)$$

* Note that $K_m = \left(\frac{k_{-1} + k_2}{k_1}\right)\left(\frac{k_3}{k_2 + k_3}\right)$.

which on extrapolation to zero time gives:

$$\pi = \frac{k_2^2 \cdot E_0}{(k_2+\gamma k_3)^2} = \frac{E_0}{\left(1+\dfrac{\gamma k_3}{k_2}\right)^2} \tag{8.24}$$

where π is the magnitude of the burst. The "burst" will be equivalent to the enzyme concentration if the denominator is equal to unity, i.e. $k_2 \gg \gamma k_3$. If the enzyme concentration and K'_S are known, the value of π can be used to estimate k_3/k_2. If $k_2 \gg k_3$ then $k_{cat} = k_2k_3/k_2+k_3 = k_3$ and so the value of k_2 can be obtained. In the case of the α-chymotrypsin-catalysed hydrolysis of *p*-nitrophenyl acetate, $k_2 \gg k_3$ since the burst is near stoichiometric, the ratio k_2/k_3 being 750 at pH 7.8.

This type of method can be used to "titrate" the operational* molarity of an enzyme solution provided that a suitable active site titrant (e.g. *N*-*trans*-cinnamoyl-imidazole for α-chymotrypsin) is available. Also, a pH value has to be found such that $k_3 \simeq 0$ while k_2 is as large as possible consistent with *specific* reaction at the enzyme active site.

It is perhaps obvious that a stopped-flow spectrophotometric study of the reaction of α-chymotrypsin with *p*-nitrophenyl acetate will yield more direct information. The pre-steady-state portion of the reaction (as shown in figure 8.2(b)) is available for detailed analysis. If the pre-steady-state P values are subtracted from the extrapolated steady-state values a simple first order curve results, the rate constant being given by the exponential factor of equation (8.21):

$$\frac{(k_2+k_3)S+k_3K'_S}{S+K'_S} \tag{8.25}$$

which enables calculation of (k_2+k_3) and K_S if, as is the case with *p*-nitrophenylacetate, the term $k_3K'_S$ can be neglected. Much the most direct measurement of k_2 can be achieved under conditions where $[E_0] \gg [S_0]$. This is known as a "one turnover" reaction and deacylation does not need to be included in the kinetic analysis of *p*-nitrophenolate anion release. The reaction is first order with a rate constant of

$$k_{app} = \frac{k_2E_0}{K_S+E_0} \tag{8.26}$$

Clearly determination of the apparent value of the rate constant at several values of $[E_0]$ followed by use of a suitable graphical plot allow the

* That proportion of the enzyme protein that is enzymically active.

determination of k_2 and K'_S. This method can be applied effectively to the study of hydrolase-catalysed hydrolyses of relatively non-specific substrates. However, the half-times for reaction of specific reactive substrates (e.g. acyl-L-amino acid *p*-nitrophenyl esters with α-chymotrypsin) are likely to be too short to measure using the stopped-flow apparatus. The hydrolysis of less reactive specific substrates (e.g. methyl esters) which cannot easily be observed spectrophotometrically can be studied using the quenched stopped-flow method.

Thus, despite the several disadvantages that limit the stopped-flow method, it has found wide applicability, particularly in the study of enzymes whose catalyses proceed via one or more intermediates which are kinetically significant (see chapter 1).

Rapid scanning spectrophotometry

Recently spectrophotometric equipment has been developed capable of the repeated scanning of spectra over a range of 200 nm in a millisecond, immediately (3 ms) after mixing in a stopped-flow apparatus. This has added another dimension to the observation of fast reactions. Although in principle spectral information can be obtained by repetition of stopped-flow experiments in the simple apparatus at a variety of wavelengths, this is, in practice, so time-consuming as to be virtually prohibitive. The rapid scanning apparatus is very useful for the examination of enzyme reactions with a view to the detection and characterization of intermediates on the reaction pathway. An example of such an experiment is shown in figure 8.3.

Perturbation methods

As noted in the section on stopped-flow methods, many enzyme reactions reach steady-state conditions in a time period that is too short for accurate measurement with the stopped-flow apparatus. Very fast substrate binding and desorption processes can however be studied in detail by means of perturbation methods. In such methods the factor that causes displacement from equilibrium or steady state is applied to the system for a very brief time interval (of the order of 1 μsec). Much the most widely-used method in enzyme kinetics employs a very rapid temperature change—approximately 10°C in 1 μsec—the heat pulse being applied to the cell contents by means of an electric current produced by the discharge of

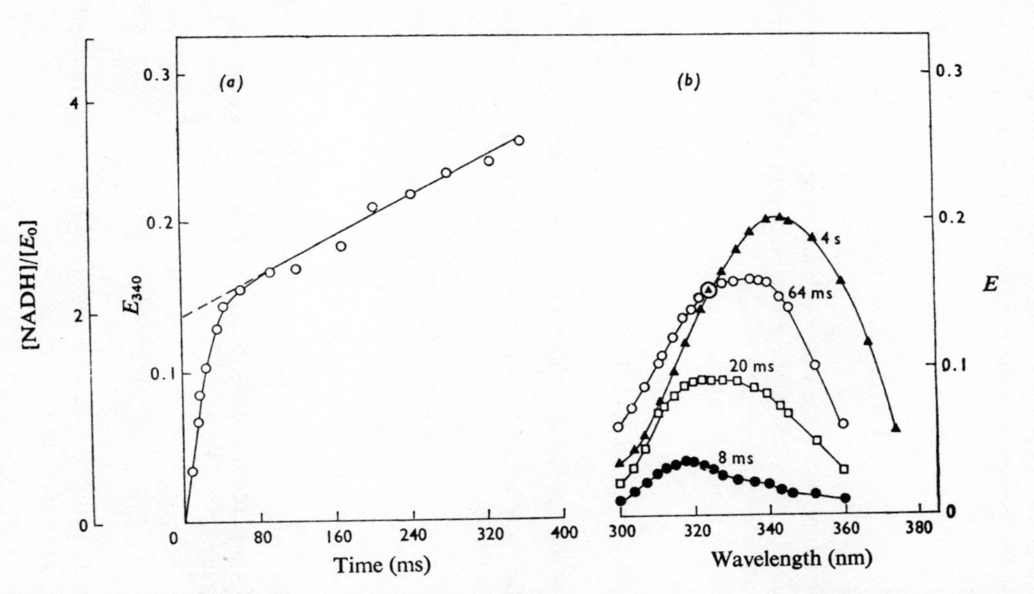

Figure 8.3 Rapid scanning spectrophotometry. Horse liver alcohol dehydrogenase was mixed with NAD and ethanol in a stopped-flow system. The burst of NADH followed by steady state NADH-production is seen in (a). In (b) the spectrum of bound NADH (λ_{max} 320 nm) is seen in the early spectra followed by the accumulation of free NADH (λ_{max} 340 nm). The stoichiometry of the burst is shown in (a) and represents approximately one mole of NADH per subunit of the enzyme. Taken from Holloway and White (1975).

capacitors. The cell contents must, of course, contain sufficient electrolyte to allow the passage of the current pulse. Equilibria (where $\Delta H \neq 0$) and steady-state processes will change their values upon such a temperature change. The method relies upon spectroscopic observation of the process of relaxation of the perturbed system towards the new equilibrium or steady state.

We shall first examine the analysis of a simple dissociation equilibrium:

$$\mathrm{A+B} \underset{k_{-1}}{\overset{k_1}{\rightleftharpoons}} \mathrm{AB}$$

Let the concentrations of the various species after re-equilibration be a_e, b_e and ab_e, and Δa, Δb, Δab be the changes in concentration of these species. The heat pulse displaces the system Δab_0 to the new equilibrium value ab_e and at any time before this is reached the rate of change of ab is given by:

$$\frac{dab}{dt} = k_1(a_e+\Delta a)(b_e+\Delta b) - k_{-1}(ab_e - \Delta ab)$$

$$= k_1(a_e b_e + b_e\Delta a + a_e\Delta b + \Delta a \cdot \Delta b) - k_{-1}(ab_e - \Delta b) \qquad (8.27)$$

If the perturbation is small we can ignore $\Delta a \cdot \Delta b$, and since at equilibrium $dab/dt = 0$ we can write

$$k_1 a_e b_e = k_{-1} ab_e \qquad (8.28)$$

$$\therefore \quad a_e b_e = (k_{-1}/k_1) ab_e \qquad (8.29)$$

Also, conservation of mass requires that

$$\Delta a = \Delta b = -\Delta ab \qquad (8.30)$$

Substitution of these relationships, together with the relation $dab/dt = d\Delta ab/dt$ into equation (8.27) gives:

$$\frac{d\Delta ab}{dt} = -(k_1(a_e+b_e)+k_{-1})\Delta ab \qquad (8.31)$$

This equation may be integrated, making use of the boundary condition that $\Delta ab = \Delta ab_0$ at $t = 0$, to give

$$\Delta ab = \Delta ab_0 \exp[-k_1(a_e+b_e)+k_{-1}]t \qquad (8.32)$$

Thus the relaxation is a simple first order process regardless of the relative concentrations of a and b. The values of k_1 and k_{-1} can be calculated from the relaxation time, τ, of the process as follows:

$$\frac{1}{\tau} = (k_1(a_e + b_e) + k_{-1})$$

The relaxation time τ of an exponential process is the time taken for it to decay to $1/e$ (37%) of the initial value, and is equal to the reciprocal of the rate constant. The values of a_e, b_e and $Ke_a = k_{-1}/k_1$ are independently determined. Alternatively, plotting data obtained at different values of $(a_e + b_e)$ in the form of $1/\tau$ against $(a_e + b_e)$ yields a straight line of slope k_1 and intercept k_{-1}. The method outlined above can be extended to cover mechanisms involving more than one step and the two mechanisms shown below can, in principle, be characterized and distinguished:

$$\mathrm{E} \underset{k_{-1}}{\overset{k_1}{\rightleftharpoons}} \mathrm{E'} \underset{k_{-2}}{\overset{k_2\mathrm{S}}{\rightleftharpoons}} \mathrm{ES'}$$

and

$$\mathrm{E} \underset{k_{-1}}{\overset{k_1\mathrm{S}}{\rightleftharpoons}} \mathrm{ES} \underset{k_{-2}}{\overset{k_2}{\rightleftharpoons}} \mathrm{ES'}$$

Each of these mechanisms is characterized by two relaxation times, and provided conditions suitable for the resolution of the two relaxation times can be achieved by manipulation of E and S, then it is possible to characterize and distinguish the mechanisms by making use of the different concentration dependencies of the relaxation times as shown in (8.13) *et seq.* As wide a range of concentration of E and S as possible should always be used in order to ensure that these two-step mechanisms are properly distinguished from the single-step mechanism.

As mentioned above, the temperature jump method can be applied to the analysis of steady-state systems. The temperature jump alters the steady-state rate and the reaction rate relaxes to the new steady state at the higher temperature. The analysis proceeds along similar lines to that of the equilibrium situation. The simple two-step reaction scheme is considered:

$$\mathrm{E + S} \underset{k_{-1}}{\overset{k_1}{\rightleftharpoons}} \mathrm{ES} \xrightarrow{k_2} \mathrm{E + P}$$

where $S_0 \gg E_0$ and S_0 is regarded as being constant. If E_{SS} and ES_{SS} are the new steady-state values then

$$\frac{d\mathrm{ES}}{dt} = k_1\,(\mathrm{E_{SS}} + \Delta\mathrm{E})\mathrm{S_0} - (k_{-1} + k_2)\,(\mathrm{ES_{SS}} + \Delta\mathrm{ES}) \tag{8.33}$$

where $\Delta E = -\Delta ES$ and $d\Delta ES/dt = dES/dt$. The steady-state condition

$$k_1 E_{SS} S_0 = (k_{-1} + k_2) ES_{SS}$$

is introduced leading to:

$$\frac{d\Delta ES}{dt} = -(k_1 S_0 + k_{-1} + k_2)\Delta ES \tag{8.34}$$

which after integration has the relaxation time

$$\frac{1}{\tau} = (k_1 S + k_{-1} + k_2) = k_{obs} \tag{8.35}$$

which, not surprisingly, is the same as that which applies to the stopped-flow method for the same mechanism.

As before, measurement of the relaxation time at various values of S_0 enables the calculation of k_1 and k_{-1} when the value of k_2 is available. Since the temperature jump takes place in approximately 10^{-6} sec it is possible to measure relaxation times up to approximately 1000-fold faster than those that can be measured in the stopped-flow apparatus. Thermal instability makes long-term recording difficult and the temperature jump apparatus is restricted to the range from $\sim 10^{-6}$ sec to $\sim 10^{-2}$ sec. Thus the stopped-flow method, since it is effective in the range 10^{-2}–10^{2} secs, is not usurped by the temperature jump method. Temperature jump apparatus in which the heat pulse is provided by an infrared laser can be used to measure relaxation times of a few tens of nanoseconds. Thus the temperature jump technique is in many respects ideal for the measurement of the kinetics of binding and dissociation processes, whether the ligand be a proton or a protein. In table 8.1 are collected some values of binding and dissociation rate constants for various enzyme–ligand interactions.

Finally a word of caution is in order. It is essential that the observed relaxation time(s) be properly identified in terms of real events upon the reaction pathway. The event must be identified in physico-chemical terms and the kinetic characteristics of the event must be consistent with the overall kinetic pathway. Failure to achieve these objectives renders the study rather pointless and certainly the results cannot be used to discuss mechanistic criteria.

Cryogenic techniques

Quite recently it has proved possible to measure some enzyme reaction rates in a specially-adapted stopped-flow spectrophotometer at temperatures down to $-80°C$. This remarkable achievement has been made possible by the observation that many enzymes are not denatured in a solvent comprising 65% dimethyl sulphoxide, provided that the temperature does not rise above 5°C. It has been shown for example in the cases of α-chymotrypsin- and papain-catalysed hydrolyses that the overall reaction pathway is not *qualitatively* affected, as compared with the reaction in aqueous solution at 25°, by transfer to 65% aqueous dimethyl sulphoxide at sub-zero temperatures. The overall *rate* of reaction at a given substrate concentration is however very much slower, partly due to the temperature effect and partly due to an increase in K_m which results from the high organic solvent content of the reaction medium. The great reduction of the various rate constants combined with the rapid mixing available with the stopped-flow apparatus allows the detailed examination of the transformations which occur along the reaction pathway. For instance it is possible to study non-rate limiting conformational changes which occur before the rate-limiting step provided that a suitable signal is available that has a direct relationship with the conformational change. Measurements designed to elucidate the mechanism of a particular enzyme reaction can be initiated at a low temperature (e.g. $-50°$), and subsequently the temperature can be raised to allow the convenient observation of slower processes. It is thus possible to observe processes with a very wide range of relaxation times by this method of temperature variation.

Fast reaction techniques have steadily gained in importance over recent years primarily because they allow such a detailed dissection of enzyme kinetic mechanisms. It is probably wise to preface rapid reaction studies with carefully designed steady-state studies and to use these techniques to expand upon the minimal definition of the kinetic mechanism provided by the steady-state studies. In circumstances where a particular aspect of a mechanism can be studied by both steady-state and rapid reaction techniques then both should be used, but it is likely that the rapid reaction studies will provide the more pertinent information.

CHAPTER NINE

ENZYME MECHANISMS

IN THIS CHAPTER WE SHALL CONSIDER THE CURRENT STATUS OF OUR knowledge and understanding of the mechanisms of several types of enzymes. We outline some experimental results that, in our opinion, have contributed most effectively to the deduction of the chemical or kinetic mechanism of the enzyme under discussion. An attempt will be made to draw out the common features of enzyme mechanism. We shall also consider catalytic criteria in terms of the evolution of enzyme catalytic power.

The serine proteases

The serine proteases are characterized by the presence at their active sites of a reactive serine residue which is essential for enzymic activity. The serine proteases are rapidly and completely inactivated by diisopropyl fluorophosphate (DFP) which reacts with the active centre serine residue to yield a phosphate ester. Use of ^{32}P-labelled DFP allows the position of this serine in the sequence of amino acids to be determined. The mammalian enzymes that will be considered in this chapter are chymotrypsin, trypsin and elastase, which are all secreted from the pancreas. The corresponding bacterial enzymes are subtilisin (from *B. subtilis*) and α-lytic protease from myxobacter.

The pancreatic enzymes are secreted as inactive zymogens and activated by limited proteolysis in the small intestine. The zymogens are single polypeptide chains which are extensively cross-linked with disulphide bonds. These enzymes have rather rugged globular structures with molecular weights in the region of 25 000. They are endopeptidases and have a specificity which is reflected in a preference for the amino acid on the acyl side of the cleavage point. Chymotrypsin has a preference for hydrophobic aromatic amino acids at this position while trypsin prefers amino acids which carry a positive charge on the side chain. Elastase has a more general specificity. That the specificity is relative rather than absolute is demonstrated by consideration of the kinetic parameters given in table 9.1.

Table 9.1 α-Chymotrypsin- and trypsin-catalysed hydrolysis of ester substrates, 25°

Enzyme	Substrate	k_{cat} (sec^{-1})	K_m (M)	k_{cat}/K_m ($M^{-1}\,sec^{-1}$)
α-Chymotrypsin	*N*-acetyl-L-tyrosine ethyl ester	173[a]	3.2×10^{-3}	5.4×10^{4}
Trypsin	*N*-acetyl-L-tyrosine ethyl ester	14.5[a]	4.2×10^{-2}	3.45×10^{2}
Trypsin	*N*-benzoyl-L-arginine ethyl ester	8.4[a]	1.0×10^{-5}	8.4×10^{5}
α-Chymotrypsin	*N*-benzoyl-L-arginine ethyl ester	1.4[a]	1.5×10^{-2}	9.3×10^{1}
Trypsin	*N*-acetylglycine ethyl ester	0.028[b]	0.79	3.5×10^{-2}
α-Chymotrypsin	*N*-acetylglycine ethyl ester	0.066[b]	0.41	1.6×10^{-1}
α-Chymotrypsin	*N*-benzoylglycine ethyl ester	0.10[c]	2.3×10^{-3}	4.3×10^{1}

[a] Parameters measured at pH 8.0.
[b] pH-independent values.
[c] pH 7.0.

The value of k_{cat}/K_m, the best measure of specificity, varies over a range of 10^7-fold for the trypsin-catalysed reactions and 10^5-fold for the chymotrypsin-catalysed reactions. However, each enzyme catalyses the hydrolysis of the most specific substrate of the other, and the hydrolyses are characterized by k_{cat}/K_m values similar to that for the α-chymotrypsin-catalysed hydrolysis of benzoyl glycine ethyl ester.

We have discussed relative specificity in terms of simple synthetic substrates rather than the proteins upon which the enzymes work in natural circumstances. It is not possible to study the detailed kinetics of protease-catalysed hydrolysis of proteins since there are likely to be multiple cleavage sites which leads to a dynamically changing substrate identity as a result of secondary cleavage. Nearly all hydrolases whose natural substrates are polymeric will tolerate considerable truncation of the length of the substrate as well as variation in the leaving group. Indeed in this case the chemical nature of the substrate has been changed from the naturally encountered peptide amide to a (non-reactive) ester.

The chymotrypsin-catalysed hydrolysis of p-nitrophenyl acetate

The reactive ester *p*-nitrophenyl acetate (PNPA) (see chapter 2) possesses few if any of the elements of specificity of α-chymotrypsin. The ester grouping is tolerated and the phenyl ring of the leaving group is found in the *side chain* of specific amino acid derivatives. The presence in the leaving group of a phenyl ring with none in the acyl portion leads to the expectation that PNPA will be non-productively bound. Nonetheless the hydrolysis of PNPA is catalysed by α-chymotrypsin. The reaction is unusual in that a rapid "burst" of absorbance at 400 nm appears when the enzyme and substrate are mixed. This is followed by a steady linear increase in the absorbance at 400 nm (see figure 8.2). The absorbance at 400 nm is due to release of *p*-nitrophenolate anion. The steady-state reaction subsequent to the burst obeys Michaelis–Menten kinetics and the burst of *p*-nitrophenolate is stoichiometric with the active enzyme concentration when the substrate is at saturating concentration. These observations are consistent with the kinetic mechanism:

$$E + S \underset{k_{-1}}{\overset{k_1}{\rightleftharpoons}} ES \xrightarrow{k_2} \underset{P_1}{\overset{}{ES' + }} \xrightarrow{k_3} E + P_2$$

where E is free enzyme, S is PNPA, ES is the Michaelis absorptive complex, ES′ is acetyl enzyme, P_1 is *p*-nitrophenolate and P_2 is acetate.

The rate constants $k_1$and k_{-1} have their usual meaning, k_2 is the first order rate constant for acylation of the enzyme and k_3 is the rate constant for deacylation. The apparent Michaelis parameters are given by

$$K_m^{app} = \frac{k_{-1}+k_2}{k_1} \cdot \frac{k_3}{k_2+k_3}$$

$$k_{cat}^{app} = \frac{k_2 k_3}{k_2+k_3}$$

The second order overall reaction constant is given by k_{cat}^{app}/K_m^{app} which is equal to k_2/K_S if $k_{-1} \gg k_2$. The results obtained with PNPA as substrate fit the kinetic scheme above if $k_2 \gg k_3$. Thus $k_{cat} = k_3$, the deacylation rate constant which is rate-limiting in the steady-state portion of the A_{400nm} vs. time curve. Note that K_m^{app} is much less than $K_S = (k_{-1}/k_1)$ since K_S is multiplied by k_3/k_2 to give K_m. At saturating substrate concentration the enzyme is almost entirely in the form of the acyl-enzyme which can be isolated by lowering the pH followed by gel chromatography to remove excess PNPA. The pH-dependence of the deacylation reaction shows a requirement for a single group which must be in the base form of $pK \simeq 7$. If ^{14}C-labelled PNPA is used the acetyl group of acetyl-chymotrypsin can be shown to be esterified with the side-chain oxygen of Ser-195 as was the diisopropylphosphate group mentioned earlier. Thus there is evidence for an acetyl-enzyme intermediate upon the reaction pathway and that a basic group of $pK \simeq 7$ is required for its hydrolysis. Based upon the pK value we might suspect this to be a histidine residue.

Active site directed inhibition

Tosyl-L-phenylalanine chloromethyl ketone (TPCK)

$$\mathrm{Me{-}C_6H_4{-}SO_2NH\,CH(CH_2C_6H_5)\,CO\,CH_2Cl}$$

is an active site directed inhibitor of chymotrypsin that possesses obvious elements of specificity. The lysine analogue (9.13) is an effective inhibitor of trypsin. These reagents react with chymotrypsin or trypsin at the chloromethyl group, which is activated by the presence of the carbonyl group in a manner that is subject to competitive inhibition of the reaction by substrates. Use of suitably radioactively labelled inhibitor allows the mode and position of the attachment in the amino acid sequence to be established. The inhibition reaction shows saturation in inhibitor. TPCK is found to alkylate His-57 of chymotrypsin and to cause complete loss of enzymic activity. Note that the point of attack is one carbon atom removed from that normally attacked in a specific substrate and this may be the reason histidine rather than serine-195 is alkylated. The above results provide direct evidence for the involvement of histidine at the active site, probably in a position immediately adjacent to serine-195.

Active site titration

N-transcinnamoyl imidazole (NTCI) (9.1) is an effective active site titrant for chymotrypsin.

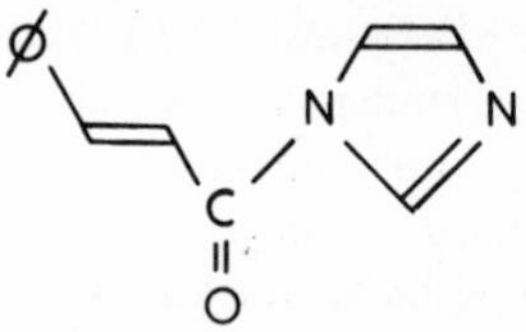

$$\phi\text{-CH=CH-C(=O)-N(imidazole)} \quad (9.1)$$

NTCI possesses some elements of the specificity of chymotrypsin. The phenyl group is the same number of carbon atoms removed from the reactive centre, as in a phenylalanine-containing substrate, although it is not isosteric due to the presence of the double bond. It possesses a form of amide bond albeit a highly reactive version. NTCI reacts rapidly with chymotrypsin at pH 5.5, and the reaction can be followed (at 310 nm or 335 nm) by loss of the NTCI absorbance (λ_{max} 306 nm). The products of the reaction are imidazole and the acyl-enzyme transcinnamoyl-chymotrypsin. This acyl-enzyme has a λ_{max} of 292 nm which is red-shifted 9 nm compared with the λ_{max} of 281 nm of the model compound *O*-cinnamoyl-*N*-acetyl serineamide. The model compound is relevant since it has been established that the cinnamoyl moiety is esterified with serine-195. Both the acylation *and* deacylation reactions have been shown to be dependent upon a group required in the base form of p$K \simeq 7$. The acylation reaction is very much faster than the deacylation reaction at pH 5.5 and this is why NTCI can be used as an active site titrant. The change in absorbance upon acylation can be used to calculate the concentration of enzyme that is enzymically active (usually $\simeq 80\%$) in a solution of α-chymotrypsin. Table 9.2 gives the rate constants for the various steps of the reactions of NTCI with trypsin and α-chymotrypsin.

Examination of table 9.2 reveals that the difference in specificity of these two enzymes towards NTCI is expressed entirely in the acylation step. The transcinnamoyl-enzymes are much more stable towards denaturation than are the free enzymes. This is an example of the general finding that enzyme–substrate complexes and/or intermediates tend to be more

Table 9.2 The reaction of trypsin and α-chymotrypsin with NTCI

Enzyme	$k_2^{(a)}$ (M^{-1} sec^{-1})	$k_3^{(b)}$ (sec^{-1})	k_{OH^-} (M^{-1} sec^{-1})
α-Chymotrypsin	1.2×10^4	1.3×10^{-2}	4.1×10^{-2}
Trypsin	63	1.45×10^{-2}	4.5×10^{-2}

[a] Note that k_2 is a second order constant, i.e. binding was not detected.
[b] pH-independent value.

stable towards denaturation than are free enzymes. The deacylation rate constants are of the order of 10^{-5} of those for specific substrates and the value of k_2 for the trypsin-NTCI reaction is low due to the lack of side chain specificity. The high reactivity of NTCI towards these enzymes is the result of the reactive nature of acyl-imidazoles (see chapter 2). The k_2 values are much larger than typical values of k_{cat}/K_m for highly specific non-reactive amide substrates (e.g. in the chymotrypsin-catalysed hydrolysis of *N*-acetyl-L-Phe amide, where $k_{cat}/K_m = 1\ M^{-1}\ sec^{-1}$).

The study of the reactions of trypsin and chymotrypsin with NTCI again implicate the involvement in the catalytic mechanism of serine and histidine. Histidine involvement is implied by the pH-dependence of both acylation and deacylation. The lack of specificity in deacylation is notable and is presumably due to the presence of very similar non-productive modes in both acyl-enzymes.

The X-ray structure of indolylacryloyl-chymotrypsin has been determined and shows that the indolylacryloyl group is located in a non-productive mode. The water molecule which partakes in deacylation is not properly aligned with the carbonyl carbon but is hydrogen bonded to the carbonyl oxygen.

The catalysis of the hydrolysis of specific substrates

Table 9.3 gives some rate constants for the α-chymotrypsin-catalysed hydrolysis of specific substrates.

Since for each group of substrates the acyl portion is constant, the value of k_{cat} should be constant if deacylation is rate-limiting but variable if acylation is rate-limiting. This may be rationalized in terms of the

Table 9.3 Kinetic constants of the α-chymotrypsin-catalysed hydrolysis of specific substrates[a]

Substrate	k_{cat} (sec^{-1})	K_m (mM)	k_{cat}/K_m ($M^{-1}\ sec^{-1}$)
Acetyl-L-Trp amide	0.03	7	4.3
Acetyl-L-Trp ethyl ester	27	0.1	2.7×10^5
Acetyl-L-Trp *p*-nitrophenyl ester	30	0.002	1.5×10^7
Acetyl-L-Phe amide	0.04	37	1.1
Acetyl-L-Phe ethyl ester	63	0.1	6.3×10^5
Acetyl-L-Phe *p*-nitrophenyl ester	77	0.02	4×10^6
Benzoyl glycine amide			
Benzoyl glycine ethyl ester	0.1	2.3	43
Benzoyl glycine *p*-nitrophenyl ester	0.5	0.03	1.7×10^4

[a] Measured at pH 7.0, 25°.

intermediacy of a common acyl-enzyme whose breakdown does not involve the original leaving group. The chemical reactivity of the ester substrates varies widely and this would be reflected in the k_{cat} values if acylation were rate-limiting. The amide substrates have much lower k_{cat} values indicating that acylation is rate-limiting. Clearly the acylation step is the difficult one in the case of natural amide substrates; it is at this step that specificity will be expected to be of prime importance.

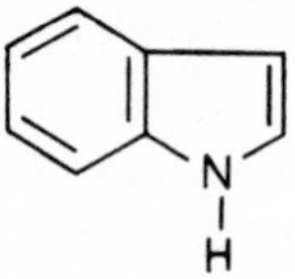

N
H
(9.2)

Indole (9.2), which is equivalent to the side chain portion of tryptophan, accelerates the steady-state rate of the α-chymotrypsin-catalysed hydrolysis of PNPA, albeit by a small factor, about 3-fold. The magnitude of the burst is reduced in the presence of indole, probably as a result of competitive inhibition with PNPA, while the deacylation rate is enhanced as a result of "stereo population control" or induced fit. Presumably the binding position of the acetyl group of acetylchymotrypsin is rather ill-defined or non-productive relative to that of a specific acyl-enzyme (e.g. acetyl-L-trp-chymotrypsin). This is reflected in the approximately 2000-fold larger deacylation rate constant of the specific acyl-enzyme. Rate enhancement by "stereo population control" involves the supposition that the presence of indole restricts the acetyl group to more productive orientations for deacylation. An explanation based upon the concept of induced fit supposes that a conformational change of the active site induced by the presence of indole results in better organization of the catalytic apparatus relative to the acetyl group which allows an enhancement of deacylation. The authors prefer the former hypothesis, since conformational changes in the structure of α-chymotrypsin which might be regarded as characteristic of induced fit have not been observed at *neutral* pH. Indole does induce a conformational change at higher pH but the enzyme at neutral pH (particularly the acyl-enzyme which is more stable towards denaturation and conformational changes than the free enzyme) is already in the conformation that is *induced* by indole at higher pH. A similar effect is seen in the trypsin-catalysed hydrolysis of acetyl glycine ethyl ester, in which case *N*-alkyl ammonium ions accelerate the rate of the reaction. The alkyl portion of these modifiers plays an important role which indicates the presence of a hydrophobic region in the

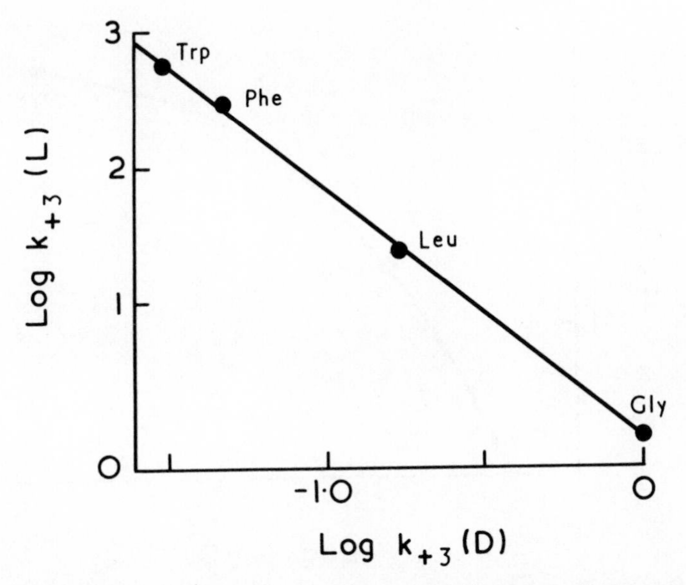

Figure 9.1 The stereospecificity of chymotrypsin. The log of the normalized rate constants for the deacylation of *N*-acetyl-L-aminoacyl-chymotrypsins plotted against values for the D-enantiomers. The substrates used were *p*-nitrophenyl esters of the *N*-acetyl amino acids, deacylation being rate-limiting. The normalization, relative to the *N*-acetylglycine ester eliminates any contribution due to intrinsic reactivity. Redrawn from Ingles and Knowles (1967).

trypsin binding site which interacts with the alkyl portions of arginine and lysine substrates.

α-Chymotrypsin catalyses the hydrolysis of both L- and D-enantiomers of *N*-acetyl amino acid *p*-nitrophenyl esters. If the side chain is specific (i.e. bulky, aromatic) for the chymotrypsin binding pocket, the L-enantiomer is a very good substrate but the D-enantiomer is very poor. Figure 9.1 demonstrates this relationship—as the specificity of the side chain is increased, the L-enantiomer becomes a better substrate but the D-enantiomer becomes a worse substrate. Note that deacylation is rate-limiting in all cases and that the rate constants have been corrected for variation in the intrinsic reactivity. Figure 9.2 demonstrates an effect that can be described as "specificity saturation", i.e. there is a limit to the "correctness" of a productive enzyme–substrate interaction. An increase in the number of amino acids in the substrate makes very little difference to the specificity in chymotrypsin-catalysed hydrolyses in contrast to those of elastase and papain (see later).

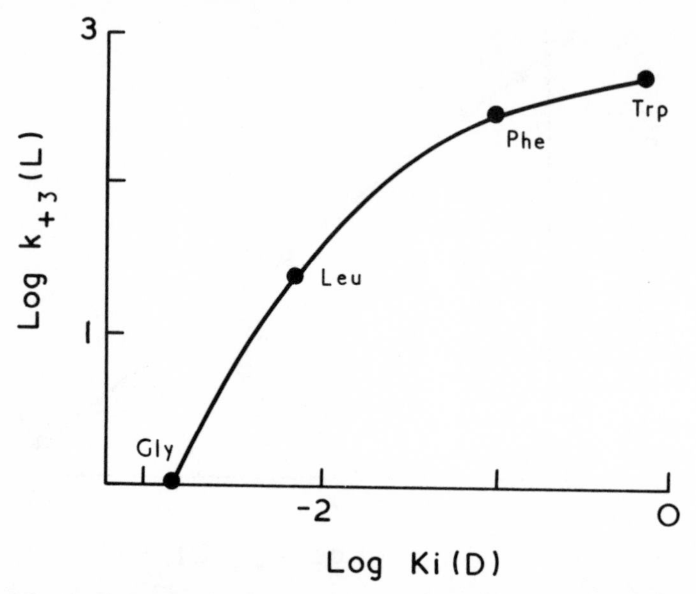

Figure 9.2 Specificity saturation in chymotrypsin-catalysed hydrolyses. The log of the normalized deacylation rate constants for the *N*-acetyl-L-aminoacyl-chymotrypsins are plotted against the logarithms of the competitive inhibition constants (pK_i) for the *N*-acetyl-D-amino acid amides. The amides were assayed as competitive inhibitors of *N*-acetyl-L-leucine *p*-nitrophenyl ester. The normalization of the k_3 values was as in figure 9.1. Redrawn from Ingles and Knowles (1967).

Nucleophile partitioning

If an enzyme reaction proceeds via an acyl-enzyme it ought to be possible to trap the ester intermediate by reaction with a nucleophile other than water. The relevant kinetic scheme is shown in (9.3).

$$E + S \underset{}{\overset{K_S}{\rightleftharpoons}} ES \xrightarrow{K_2} ES' \xrightarrow{K_3(H_2O)} E + P_2$$
$$+ P_1 \qquad ES' \xrightarrow{K_4(N)} E + P_2{-}N \qquad (9.3)$$

In the case of chymotrypsin-catalysed hydrolyses amino acid amides such as glycinamide, alaninamide and tryptophanamide are effective nucleophiles and give rise to significant quantities of P_2—N. Such small molecules as methanol and methylamine also compete but less effectively than the amino acid amides, presumably because the binding interaction of the small molecules is minimal. The lack of a Brønsted correlation in nucleophile partitioning experiments indicates that the predominant factor in the reactivity of nucleophiles is their binding ability.

Analysis of product ratios in the hydrolysis of various substrates by α-chymotrypsin reveals that the ratio remains constant as the nature of the substrate changes as shown in table 9.4; this is good evidence for the existence of an acyl-enzyme intermediate. The rate-limiting step in the chymotrypsin-catalysed hydrolysis of esters is deacylation and this step can be accelerated by the addition of nucleophiles. When the rate of reaction is measured it is seen to saturate in nucleophile concentration and to increase in value to above that found in the absence of nucleophile. That this is *not* due to saturation of the nucleophile binding interaction is indicated by the observation that the saturating rate is different for various ester substrates that have a common acyl group.

Table 9.4 Product ratio analysis in the α-chymotrypsin-catalysed hydrolysis of various substrates[a]

	Product ratio $\frac{[P_2{-}N]}{[P_2][N]}$ (M^{-1})		
Substrate	Ala NH_2	Gly NH_2	H_2NNH_2
Ac-Phe-OMe	43	13	2.2
Ac-Phe-NH ϕNMe_3	45	11	1.8
Ac-Phe-Ala NH_2	43	9	—

[a] pH 9.3.

These results must be interpreted in terms of a change in rate-limiting step from deacylation to acylation as the rate of deacylation is accelerated. The rate-limiting step in the chymotrypsin-catalysed hydrolysis of amides is acylation and the acyl-enzyme does not accumulate. The observation of a constant product ratio in nucleophile partition experiments with several amide substrates strongly indicates that these hydrolyses proceed via an acyl-enzyme. The rate of the reaction is unchanged but the fraction of P_2—N increases with increasing nucleophile concentration. Caution must be applied in the interpretation of experiments in which rates are measured since the added nucleophiles can have non-specific effects upon the kinetic parameters—product ratio *and* rate measurements should be combined to form the basis of a definitive study.

The pH-*dependence of α-chymotrypsin-catalysed hydrolyses*

The pH-dependence of k_{cat}/K_m for chymotrypsin-catalysed hydrolyses is bell-shaped, and characterized by pK values of approximately 7 and 8.5–9. The pK of 7 is likely to be associated with the active site histidine but the pK of $\simeq 9$ has not been encountered in any of the results quoted so far. It is known that nitrous acid-treated chymotrypsin is inactive, the loss of activity being ascribed to the loss of the amino group of N-terminal Ileu-16. Chymotrypsin may be acetylated so that all amino groups are blocked and then activated with trypsin to give acetylated γ-chymotrypsin (the chymotryptic cleavage in α-chymotrypsin is absent) which has 80% of the enzymic activity of α-chymotrypsin. It has been demonstrated, by difference titration and N-terminal analysis, that acetylated γ-chymotrypsin has a single Ileu N-terminal amino group of pK 8.3. The optical rotatory dispersion at 313 nm of acetylated forms of chymotrypsinogen, γ-chymotrypsin and diisopropylphosphoryl-γ-chymotrypsin over a range of pH is shown in figure 9.3. The acetylated forms of the zymogen and the phosphorylated enzyme show pH-independent optical rotatory dispersion between pH values of 6 and 11. The active form shows an increase in this

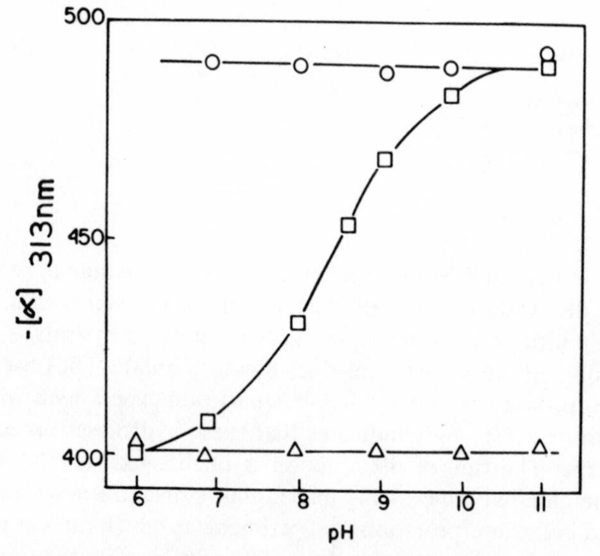

Figure 9.3 The pH-dependence of the optical rotation of various forms of chymotrypsin. ○, acetylated chymotrypsinogen; □, acetylated δ-chymotrypsin and △, acetylated diisopropylphosphoryl-δ-chymotrypsin. Redrawn from Oppenheimer *et al.* (1966).

pH range from the value of the phosphorylated enzyme to that of the zymogen. The increase is characteristic of a single ionization process associated with a pK value of 8.3.

It is proposed that the phosphorylated form of the enzyme is analogous to an acyl-enzyme in which the enzyme is stabilized in the active conformation. The difference titration experiments show that neither the zymogen nor the phosphorylated form of the enzyme show an ionization with a pK of 8.3. Accordingly it may be deduced that Ileu-16 is stabilized in the cationic form in the phosphorylated form of the enzyme. For such stabilization to occur, Ileu-16 must be buried (see chapter 3) in a hydrophobic region which in turn requires charge pairing if the energy required for the burial is not to be excessive.

Thus the pH-dependence of k_{cat}/K_m may be interpreted on the basis of these and earlier results. The acidic limb (low pH) characterized by a pK of 7 may be ascribed to the requirement for a histidine residue which must be in the base form for catalytic activity. Since $k_{cat}/K_m = k_2/K_S$ the acidic limb represents k_2. The basic limb may be ascribed to an effect upon K_S. The enzyme requires Ileu-16 to be in the acidic form in order that the active conformation of the enzyme may be achieved. This assignment is supported by the observation that the catalytic competence of the zymogen is significant but that it is incapable of binding substrate. We note that it has been possible to assign the acidic and basic limbs of the bell-shaped pH-profile to specific processes requiring specific ionic forms by means of independent (i.e. non-kinetic) experiments. In general it is not possible to assign the ionic forms of bell-shaped pH-dependencies without independent evidence since the *intrinsic* reactivity of the system is not known and "crossing-over" of the pK values is possible (see chapter 6).

The acid limb of the pH-dependence of k_{cat}/K_m for the δ-chymotrypsin-catalysed hydrolysis of N-acetyl-L-tryptophan *p*-nitrophenyl ester is characterized by a pK value of 6.5. This value is abnormally low compared with the normal value for other substrates of 6.8 ± 0.03. The pH-independent value of k_{cat}/K_m (when the data are fitted to a sigmoid curve) has a value of 3×10^7 M^{-1} sec^{-1}. This value approaches that expected for enzyme and substrate diffusional encounter and suggests that k_{cat}/K_m may approximate to k_1, the rate constant for the formation of the absorptive complex. k_1 will be expected to be pH-independent in the region pH 5–7.5, but k_2 will increase over this range. Thus if a partial or complete change in rate-limiting step from $k_2 \rightarrow k_1$ occurs as the pH is increased, then the pH-profile will level off at a lower value of k_{cat}/K_m than predicted on the basis of a pK of 6.8. This results in a lowered apparent value of the pK as is seen in the case described above. All of the rate constants which describe the simple acyl-enzyme mechanism (i.e. excluding possible conformational changes) for this catalysis have been determined as a result of steady-state kinetic analysis. It is rather unusual to be able to dissect the kinetic mechanism to this extent without recourse to rapid reaction techniques; the rate constants are shown in (9.4).

$$\text{Ac-Trp ONp} + \delta\text{-CHy} \underset{6 \times 10^{4}\ \text{sec}^{-1}}{\overset{6 \times 10^{4}\ \text{m}^{-1}\ \text{sec}^{-1}}{\rightleftharpoons}} \text{E-S} \overset{\text{pKa} = 6{\cdot}8}{\rightleftharpoons} \overset{+}{\text{E}}\text{-S} \not\rightarrow$$

$$\text{E-S} \xrightarrow{7 \times 10^{4}\ \text{sec}^{-1}} \text{E-Ac-Trp} + {}^{-}\text{ONp} \xrightarrow{65\ \text{sec}^{-1}} \text{E} + \text{Ac-Trp} \qquad (9.4)$$

We note since $k_{-1} \simeq k_2$ that Briggs–Haldane (steady-state) rather than Michaelis–Menten (equilibrium) kinetics apply to this system.

Transition state analogues

Boric acid is known to ionize in an unusual fashion by taking on a hydroxide ion rather than by losing a proton. The anionic tetrahedral structure that results bears some resemblance to the structure of a tetrahedral intermediate in ester hydrolysis. Substitution of a hydroxyl group by a specificity-conferring benzyl group to give a boronic acid as in 2-phenylethane boronic acid (9.5) might be expected to give rise to a transition state analogue upon interaction with chymotrypsin.

$$C_6H_5\text{-}CH_2\text{-}CH_2\text{-}B(OH)_2 \qquad (9.5)$$

The active site serine-195 might be expected to insert at the boron atom to yield a tetrahedral adduct (9.6) essentially isosteric with the tetrahedral intermediate derived from insertion at a carbonyl centre in the course of chymotrypsin-catalysed ester hydrolysis.

$$\overset{\oplus}{\text{ImH}} \quad \overset{\text{R}}{\underset{\text{HO}\ \ \text{OH}}{\overset{|}{\overset{\ominus}{\text{B}}}}}\text{—O—CH}_2 \tag{9.6}$$

In fact 2-phenyl ethane boronic acid binds *only* to the kinetically active form of the enzyme, with a binding constant some 200-fold larger than that which characterizes hydrocinnamide (9.7) binding.

$$C_6H_5\text{—}CH_2-CH_2-C\overset{\displaystyle =O}{\underset{\displaystyle \diagdown NH_2}{}} \tag{9.7}$$

The binding of 2-phenyl ethane boronic acid is dependent upon pK values of 6.4 and 8.9, the pH value of optimum binding being around 7.6. The pK values may be ascribed, as above, to the active site His-57 and N-terminal Ileu-16. Hydrocinnamide binding is dependent only upon the pK of 8.9 and is insensitive to the state of ionization of His-57, whereas 2-phenylethane sulphonic acid, which is anionic at all pH values above zero, binds most strongly to the enzyme form at low pH where His-57 is cationic. It binds less strongly at neutral pH and not at all to the form in which Ileu-16 is deprotonated. The nature of the pH-dependence and the relative tightness of the binding of 2-phenylethane boronic acid strongly suggest that the adduct does form and can be regarded as a transition state analogue.

Aldehydes readily form adducts with water to give hydrates, an extreme example being chloral hydrate. Thus it might be supposed that suitably specific aldehydes would be transition state analogues for the serine (and thiol) proteases. Elastase, in contrast to chymotrypsin, has an extended active site and interacts favourably with several amino acid residues in

Table 9.5 Binding and kinetic constants for the interaction of oligopeptide derivatives with elastase

Substrate		K_i(mM)	K_m (mM)	k_{cat} (sec^{-1})	k_{cat}/K_m (M^{-1} sec^{-1})
Acetyl · Ala · Pro · Ala	$CONH_2$		4.2	0.09	21
Acetyl · Ala · Pro · Ala	CH_2OH	7.0			
Acetyl · Ala · Pro · Ala	CHO	0.062			
Acetyl · Pro · Ala · Pro · Ala	$CONH_2$		3.9	8.5	2200
Acetyl · Pro · Ala · Pro · Ala	CH_2OH	0.6			
Acetyl · Pro · Ala · Pro · Ala	CHO	0.0008			

oligopeptide substrates. Table 9.5 gives some binding and kinetic constants for the interaction of oligopeptide derivatives with elastase.

The results shown in table 9.5 demonstrate several important features of the binding and catalytic properties of elastase. Firstly we note that the aldehyde derivatives bind very much more tightly to the enzyme than do the amide and alcohol derivatives. This indicates that the enzyme probably forms a hemi-acetal adduct with the aldehyde to give a tetrahedral structure which will resemble to a considerable extent the transition state structure for acylation. Since the aldehyde is likely to be extensively hydrated in solution the overall reaction is given by (9.8).

$$\text{E-OH} + \text{R—CH(OH)}_2 \rightleftharpoons [\text{E-OH} \cdots \text{R—CH(OH)}_2] \rightleftharpoons \text{E—O—CH(R)(OH)} + H_2O \quad (9.8)$$

Extension of the chain length by one amino acid increases the binding efficiency of the aldehydes some 800-fold and that of the alcohols some 11-fold, but does not alter the K_m value for the elastase-catalysed hydrolysis of the amides. However, the k_{cat} value increases by about 100-fold and so the increased interaction that presumably occurs on chain lengthening of the amide is entirely expressed in terms of an increased acylation rate. This may be explained in terms of the introduction of strain into the susceptible amide bond upon binding of the amide to the enzyme. The susceptible amide bond of the longer substrate will be more strained than that of the shorter substrate. The strain that occurs must be *towards* the transition state structure in which state the strain must be relieved for optimal catalysis. These results represent an elegant demonstration of the concept of optimal transition state binding which is achieved as a result of relief of strain in a ground state complex.

It is however important to note that, despite these considerations, the major factor responsible for the catalytic power of elastase has been identified as the reduction in entropy of activation that results from non-covalent complex formation (i.e. intramolecularity, see chapter 2). Only a relatively small role is assigned to the strain effect. The results of theoretical computations upon the rigidity of protein structures have been interpreted in terms of a rather easily deformable structure over the small distances involved in substrate strain. It therefore seems likely that the enzyme rather than the substrate will be distorted by the energy available from the enzyme–substrate "solvation" interaction. Consequently it was suggested that chemical models of enzyme catalysis which involve the introduction of a high degree of ground state strain (e.g. the "trialkyl lock", chapter 2) during synthesis are not realistic models of enzymic catalysis. Even if this is so, their value as model systems is not much reduced since they clearly demonstrate (albeit in an exaggerated fashion) a factor that makes some contribution to catalysis.

X-ray crystallographic studies

As a result of X-ray crystallographic studies, several structural features of the enzyme, which chemical experiments (see earlier) have implicated as being intimately involved in the mechanism, have been identified.

(1) At the active centre there exists a "catalytic triad", which comprises the hydrogen-bonded side chains of Asp-102, His-57 and Ser-195 (9.9).

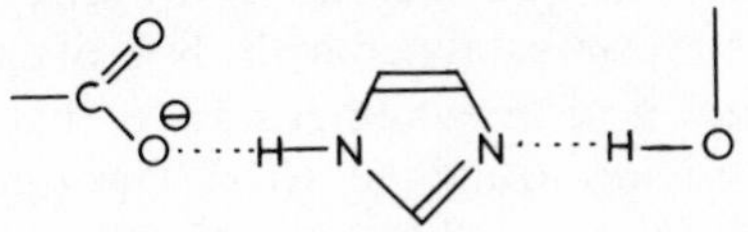

(9.9)

The histidine and serine residues identified (as a result of chemical studies) as associated with active-centre reactivity are thus seen to be in intimate hydrogen-bonded contact. It is perhaps surprising that the triad system should show a kinetic pK value of about 7, which is the normal imidazole pK value. However, firstly it is important to think in terms of *system* ionizations rather than individual group ionizations, and secondly to consider likely environmental perturbations. Asp-102 is buried in a hydrophobic pocket and this would, neglecting the effect of adjacent His-57, tend to raise its pK value. However, an adjacent cationic imidazolium ion would tend to lower the pK of Asp-102, particularly if the dielectric constant in the intervening region is low. When His-57 is protonated it cannot *accept* a hydrogen bond from Ser-195 but may donate one to it as in (9.10). This form will not be catalytically active.

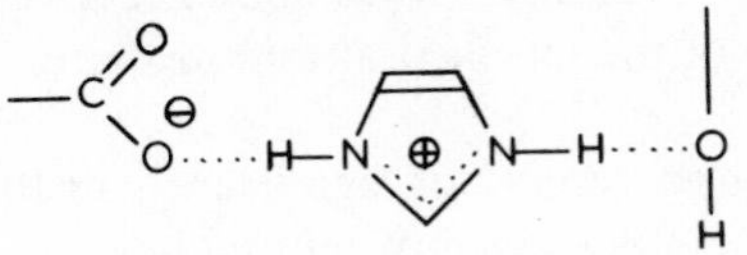

(9.10)

A hydrogen bond from Ser-195 to neutral His-57 will result only if the imidazole proton is localized upon the nitrogen atom which is adjacent to the Asp-102 carboxyl group. Since the model compound N-acetyl-L-serine amide has been shown to have a pK of ~ 13.6, it seems most unlikely that the $\gamma - O$ of Ser-195 could act as an (unassisted) nucleophile towards non-reactive substrates (but see later). This consideration immediately leads to the suggestion that His-57 may act as a general *base* upon Ser-195, rendering it sufficiently nucleophilic to insert at poorly electrophilic carbonyl centres. Such insertion probably results in tetrahedral intermediate formation whose breakdown to form acyl-enzyme will be subject to acid catalysis by His-57, i.e. protonation of the leaving group. There seems however to be no direct experimental evidence for the action of His-57 as a general *acid*. This may be explained on the basis of fast tetrahedral intermediate breakdown, so that this step does not contribute to the rate-limiting process. Alternatively, lack of equilibrium of His-57 with solvent in the enzyme-tetrahedral intermediate complex may be proposed as an explanation, if tetrahedral intermediate breakdown is rate-limiting. It is notable that there is no correlation between σ^- and $\log k_{cat}$ in the chymotrypsin-catalysed hydrolysis of anilides. The carboxyl group of Asp-102 serves to orientate His-57 so that these functions are optimized, as well as localizing the His-57 proton. Alternatively, the carboxylate may act as a general base upon His-57 which in turn acts upon Ser-195. In this case a proton will be transferred to the carboxylate which will be protonated consequent upon tetrahedral intermediate formation. Leaving-group protonation, concerted with or prior to tetrahedral intermediate breakdown, will occur by general acid catalysis by the Asp-102 carboxyl group mediated by His-57 or directly by protonated His-57. The various possibilities will be further discussed when the putative mechanism of action is presented below.

(2) Adjacent to the triad there is a hydrophobic pocket of dimensions suitable for the inclusion of a hydrophobic group of the size of an indole ring system. This was demonstrated as a result of X-ray crystallographic examination of tosyl-chymotrypsin, which is an acyl enzyme analogue.

(3) There is an "oxyanion" hole that accommodates the carbonyl oxygen of the tosyl group in hydrogen-bonding contact with the amide protons of Ser-195 and Gly-195. This serves to stabilize negative charge development upon the oxyanion of the tetrahedral intermediates that occur upon the reaction pathway.

(4) The enzyme is apparently not significantly distorted upon substrate binding and thus does not show any induced fit.

(5) The bound substrate is probably slightly distorted towards a configuration that resembles that of the tetrahedral intermediate. This is deduced from X-ray crystallographic studies on the very strong interaction of trypsin with its natural pancreatic inhibitor.
(6) Ileu-16, in the positively charged form, interacts with the carboxylate of Asp-194 as a buried ion pair. Ileu-16 is not N-terminal in the zymogen and Asp-194 is not in the same conformation as in the active enzyme. The movement of Asp-194 consequent upon zymogen activation with ion-pair formation causes a conformational change that creates the substrate binding site. The catalytic triad seems to be largely correctly organized in the zymogen although some small changes occur on zymogen activation. These findings are supported by the observation that of the 10^6-fold increase in catalytic activity that results upon activation, 10^4-fold is due to improved binding and 10^2-fold due to optimization of the catalytic apparatus.

The chemical mechanism

The observations above, together with the results of chemical experiments, allow a mechanism of action to be proposed (9.11).

The role of Gly-193 is seen as allowing the provision of a strong hydrogen bond in the tetrahedral intermediate and/or transition state. The hydrogen bond from the carbonyl oxygen increases in strength as the double bond character is lost consequent upon Ser-195 nucleophilic attack, and negative charge accumulates on the oxygen. This interaction represents an excellent example of transition state stabilization achieved as a result of optimum complementarity between enzyme and transition state.

The mechanism (9.11) is presented in a form in which Asp-102 does not change its ionic state during the catalytic cycle although there is evidence from NMR studies that it may do so. Asp-102, as mentioned earlier, thus serves to orient His-57 and to localize the proton on the nitrogen atom adjacent to Asp-102. His-57 now becomes cationic as a result of Ser-195 nucleophilic insertion, since the proton is not fully relayed to Asp-102. Asp-102 and His-57 therefore form a charge pair at the tetrahedral intermediate stages of the mechanism, in contrast to the neutral pair which would result if the proton were relayed to Asp-102. Asp-102 is buried in a hydrophobic area, and this will tend to raise its pK value. However, the negatively-charged form will be well-stabilized in the presence of cationic His-57.

α-Lytic protease (produced by myxobacter) has a catalytic triad very

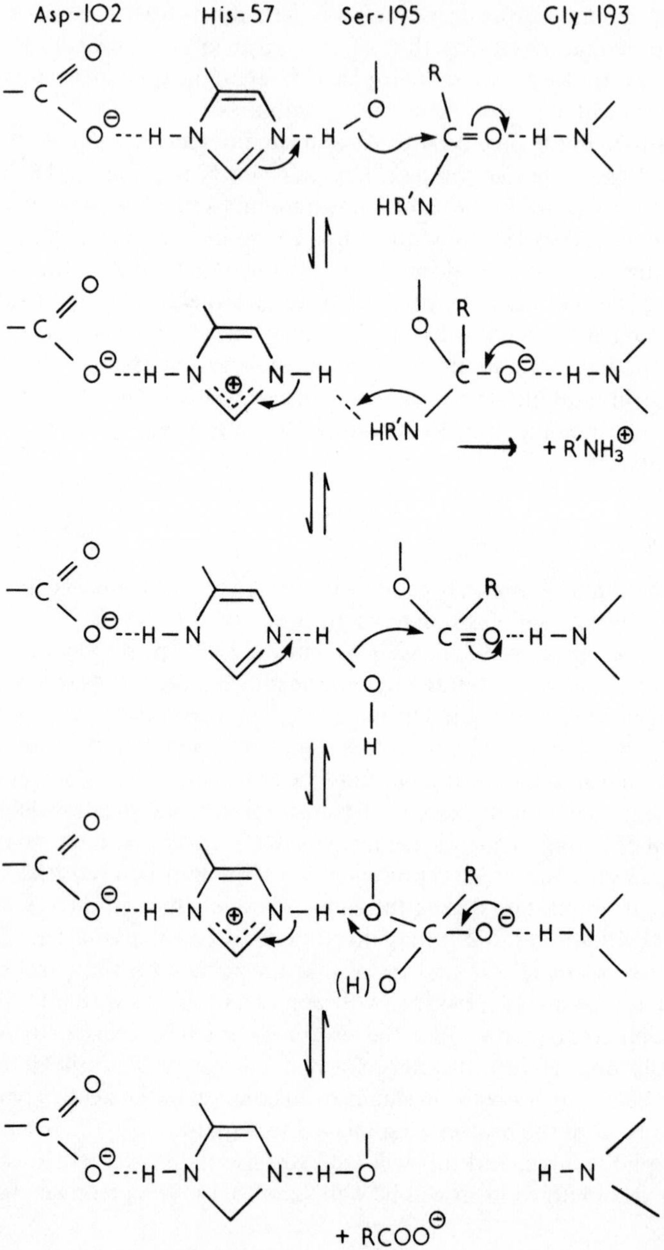

Asp-102 His-57 Ser-195 Gly-193

(9.11)

similar to that of chymotrypsin, but possesses only a single histidine residue. An α-lytic protease which is extensively isotopically labelled in this single histidine residue has been isolated by growing the bacterium in the presence of histidine enriched in ^{13}C in the C-2 position or ^{15}N in the N-1 and/or N-3 positions. The pH-dependence of the coupling constants and/or the chemical shifts of the isotopically labelled histidine residue in the enzyme have been measured by NMR spectroscopy. The results of the two series of experiments (^{13}C and ^{15}N) give rise to contrasting interpretations of the protonic equilibria of the triad. In the ^{13}C experiments, the chemical shift of the C-2 carbon titrates with a p*K* value of about 7. The C—H coupling constant, however, does not alter in this pH region but changes to a value characteristic of protonated histidine only at a much lower pH value. These observations have been interpreted in terms of a "p*K* reversal" between Asp-102* and His-57.

The ^{15}N-NMR experiments involved the measurement of the pH-dependence of the chemical shifts of both the N-1 and N-3 nitrogens of the active site histidine. The chemical shifts of both nitrogen atoms alter in a fashion that may be characterized by p*K* values close to 7. These results can be interpreted to mean that His-57 has a p*K* value close to 7. NMR studies of variously substituted imidazole compounds have allowed the deduction, based upon the enzyme spectra, that the proton is located on the nitrogen atom adjacent to the aspartate residue.

The magnetic resonance of the C-2 *proton* of His-57 of chymotrypsin can be resolved from the other histidine C-2 proton resonances when a high-field (470 MHz) instrument is used. The pH-dependence of the chemical shift of this proton clearly supports the assignment of the group in the triad which has a p*K* of ca. 7 as His-57.

In the opinion of the authors the evidence for the latter interpretation is stronger than that for the former—in addition, simple chemical reasoning suggests that the role of Asp-102 should be as depicted in (9.11).

A considerable quantity of indirect experimental evidence has accumulated that suggests that the tetrahedral intermediate in the acylation reaction does not accumulate in circumstances where the leaving group of the substrate is not *p*-nitroaniline. The ^{15}N kinetic isotope effect (1.01 at pH 7) indicates that partial C—N bond cleavage has occurred in the rate-limiting transition state. Evidence derived from extensive studies of the chemical mechanisms of amide hydrolysis and ester aminolysis may be combined with the results of enzymological studies to suggest that, at low

* The amino acids are numbered as for chymotrypsin to avoid confusion.

Table 9.6 The kinetic parameters of some elastase-catalysed hydrolyses

Substrate	k_{cat} (sec^{-1})	K_m (M)	k_{cat}/K_m (M^{-1} sec^{-1})	Fast phase (sec^{-1})
Ac-Ala-Pro-Ala-PNA[a]	2.8	1.6×10^{-3}	1750	17
Ac-Ala-Pro-Ala-PNA OMe[b]	300	1.0×10^{-4}	3.0×10^{6}	—
Ac-Ala-Pro-Ala-PNA PNPE[c]	300	1.0×10^{-4}	3.0×10^{6}	—

[a] *p*-nitroanilide; [b] methyl ester; [c] *p*-nitrophenyl ester.

pH, tetrahedral intermediate formation is rate-limiting. At low pH, amine expulsion from the tetrahedral intermediate will be optimal and the concentration of the reactive protonic form of the triad will be minimal. At high pH general base-catalysed nucleophilic attack should be rapid, but amine expulsion may be rate-limiting if His-57 is in protonic equilibrium with the solvent. If, however, it is not, then acid catalysis will be undiminished at high pH. The ^{15}N kinetic isotope effect suggests that both formation and breakdown may be partially rate-limiting in the region of neutral pH, which fits in well with this reasoning.

Direct evidence for the accumulation of a tetrahedral intermediate has been obtained in the elastase-catalysed hydrolysis of Ac-Ala-Pro-Ala-*p*-nitroanilide. The data in table 9.6 clearly demonstrate that acylation is rate-limiting in the hydrolysis of the substrate.

Figure 9.4 shows the change in absorbance at 410 nm when elastase and the *p*-nitroanalide substrate are mixed. A fast "burst" phase is seen which, it has been proposed, represents tetrahedral intermediate accumulation to the extent of 80% of the maximum possible value. The rationale is that the electron-withdrawing nitro group will enhance nucleophilic attack upon the anilide, but that acid catalytic protonation of the tetrahedral intermediate nitrogen will be unfavourable because of the relative acidity of the aniline. Accordingly, tetrahedral intermediate breakdown will be rate-limiting. We note that if longer substrates were used the rate-limiting step might well become deacylation, as remote interactions are able to enhance elastase catalysis greatly (see earlier). Similar results have been obtained for α-lytic protease.

Proton inventories

The measurement of the deuterium kinetic isotope effect in aqueous solvents having a deuterium atom fraction content ranging from zero to unity allows the compilation of a *proton inventory*. From this may be

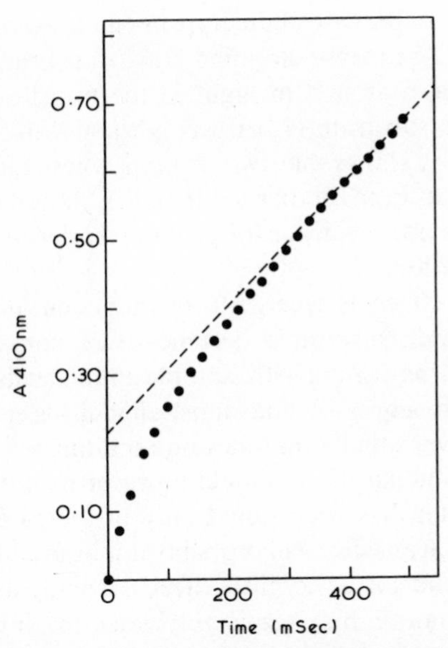

Figure 9.4 The burst of absorbance at 410 nm consequent upon mixing elastase with Ac-L-Ala-L-Pro-L-Ala *p*-nitroamilide at 25°. $[S_0]=1.5\times10^{-3}$ M; (a) $[E_0]=2.13\times10^{-5}$ M; (b) $[E_0]=3.49\times10^{-5}$ M. Redrawn from Hunkapiller *et al.* (1976).

deduced the protonic structure of the transition state relative to that of the ground state, and an estimate can be made of the number and nature of the protons "in flight" in the transition state. If the measured kinetic isotope effect is linearly related to the atom fraction of deuterium present, it may be deduced that a single proton is "in flight". A linear relation between the root of the kinetic isotope effect and the atom fraction of deuterium indicates two protons are "in flight". Some information concerning the degree of coupling of the motion of the two protons may also be obtained. This method does, of course, rely upon the presence of a primary isotope effect.

Application of the method to the hydrolysis of acetyl-chymotrypsin shows that, as in all the chymotrypsin reactions so far studied, a single proton (presumably that of water) is in flight in the rate-limiting transition state. In contrast, the α-lytic protease-catalysed hydrolysis of Ac-Ala-Pro-Ala-PNA is apparently characterized by the coupled movement of *two* protons in the rate-limiting transition state, which has been shown to be the transition state for acylation. This movement therefore provides evidence for protonic relay to the residue equivalent to Asp-102 in α-lytic protease acylation. In the case of the trypsin-catalysed hydrolysis of simple substrates (e.g. α-N-benzoyl-L-arginine ethyl ester) the proton inventory shows that a single proton is in flight in the transition state. However, when the extended substrate N-benzoyl-L-Phe-L-Val-L-Arg-PNA is used the proton inventory shows that two protons whose motions are coupled are in flight. As it has been established that deacylation is the rate-limiting step, this result provides evidence for proton transfer to the Asp residue of the triad in deacylation.

Two important concepts emerge from these considerations. Firstly it seems that as amide substrates become more completely specific by appropriate chain lengthening with suitable amino acids, so acylation (the step that the enzyme normally finds most difficult) is enhanced relative to deacylation which eventually becomes rate-limiting when the substrate is near perfect. Two qualifications should however be noted. The change in rate-determining step has been noted only in the case of *p*-nitroanilide substrates; also, such an effect will probably not be found for chymotrypsin since chain-lengthening has very little effect. Secondly it appears that dual concerted proton motion may occur only when the substrate is properly and completely specific for the enzyme. Binding of a suitably specific substrate may achieve a "zipping-up" of the relay system causing a change from a diad to a triad mechanism. This surmise is supported by the observation that the serine enzymes asparaginase and glutaminase both show *dual* proton motion when offered asparagine and glutamine respectively as substrates but only *single* proton motion when the substrates are interchanged. All the proton inventory experiments must, however, be interpreted with caution since the results may be interpreted on the basis of complex compensating effects perhaps due to several superimposed factors. Nonetheless the proton inventory results do suggest that the relay system operates as a triad in some circumstances (9.12) in contrast to the mechanism (9.11).

$$-C(=O)-O^{\ominus}\cdots H-N\langle\text{imidazole}\rangle N\cdots H-O- \qquad (9.12)$$

Re-examination of the detailed geometry of the His-Ser interaction in subtilisin by X-ray crystallography has led to the suggestion that these residues are too far apart to form an effective hydrogen-bonded interaction, at least in the *free* enzyme. If this is so, the mechanism probably involves electrophilic enhancement of the substrate. If, on binding, the substrate is extensively deformed towards the structure of the transition state then it may become sufficiently electrophilic to encourage *unassisted* attack by protonated serine oxygen. Histidine will, as before, function as a general acid but the nature of the proton dynamics of such a mechanism is not clear, since the histidine must acquire the oxonium proton subsequent to nucleophilic insertion. We note that extensive deformation of the substrate may be precluded by the flexibility of the protein, as was mentioned earlier.

These last observations "throw back into the melting pot" the discussion of the serine protease mechanisms and may require a detailed reappraisal of all previous evidence. It is possible that some degree of induced fit is involved if the serine and histidine residues assume hydrogen-bonding contact consequent upon substrate binding. Thermal agitation may allow the hydrogen-bonded configuration to be transiently achieved, after which reaction would be rapid.

The origin of the catalytic rate enhancement

The catalytic rate enhancement achieved by the serine proteases arises from three sources: approximation, strain and chemical catalysis. The nucleophilic catalysis (see chapter 2) of acylation can have a very high (up to 10^8 M) effective concentration, whilst deacylation is rendered pseudo-intramolecular as a result of the acyl-linkage and strong hydrogen bonding of the bound water molecule. Under these conditions, the general base-catalysed deacylation will have an effective concentration approaching that of a nucleophilic reaction as compared with the very low values that characterize simple general-base catalysed reactions. Here we can see the symmetry of the mechanism and the way in which evolution has balanced the effectiveness of the two steps of the reaction.

We have described the evidence and putative mechanism of the serine proteases in very considerable detail, because we consider that the development of the ideas concerning the mechanism of these enzymes provides the clearest and most wide-ranging insight into the methods and concepts that may be employed for the solution of enzyme mechanisms in general. We end this section on an evolutionary note by observing that subtilisin and α-lytic protease, which have little or no sequence homology with the mammalian serine proteases, must have evolved to a common mechanism by *convergent evolution*. In contrast it seems highly likely that the mammalian serine proteases, which do have considerable sequence homology, must have evolved their differing specificities by means of *divergent evolution.*

The thiol proteases

Some plants and bacteria synthesize proteolytic enzymes that require unblocked thiol groups for their catalytic activity. Papain, ficin and bromelain are found in papaya, fig and pineapple tissues respectively. The natural function of these enzymes is unknown, although it has been proposed that they may represent a defensive system against attack by insects. Papain has been studied to a much greater extent than the others and will be primarily considered here.

Inhibition and titration of the thiol proteases

All the thiol proteases are inhibited by heavy metals that react with the thiol group. The thiol group is also readily alkylated by simple reagents such as chloroacetamide and iodoacetic acid. Sulphur is a very soft polarizable nucleophile and readily partakes in S_N2 displacement reactions at saturated sp^3 carbon centres associated with good leaving groups (e.g. halides). The transition state of these reactions is stabilized by the ability of the easily-polarized sulphur to overlap with the electrons of the leaving group. The active-site directed inhibitor α-N-tosyl-L-lysine chloromethylketone (9.13) reacts with the single active centre thiol group of papain at the alkylating centre of the inhibitor, with expulsion of chloride.

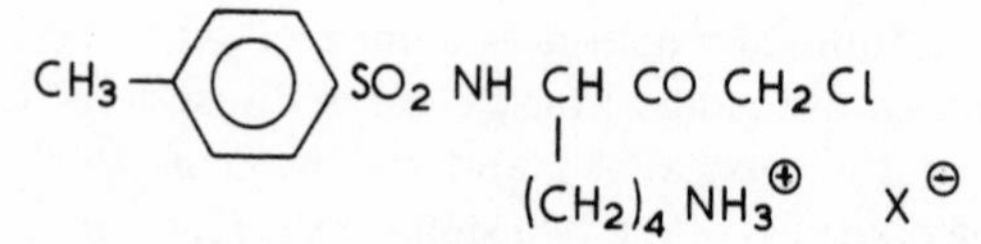

$$CH_3-C_6H_4-SO_2\,NH\,\underset{\displaystyle (CH_2)_4\,NH_3^{\oplus}}{\underset{|}{CH}}\,CO\,CH_2Cl \quad X^{\ominus} \tag{9.13}$$

The active centre thiol group of the enzyme reacts readily with aromatic disulphides in an exchange reaction, e.g. (9.14).

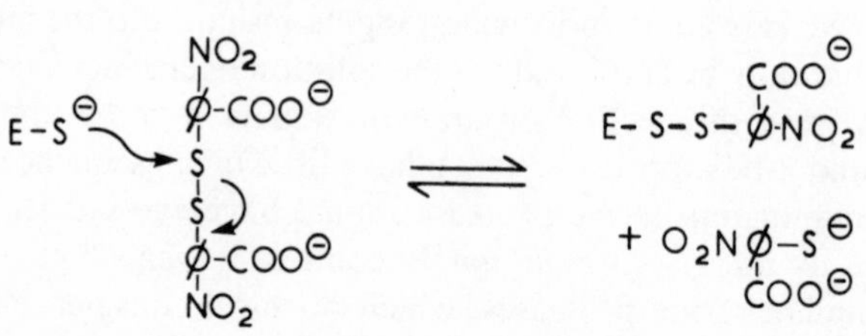

$$E-S^{\ominus} + (NO_2)\phi(COO^{\ominus})-S-S-\phi(COO^{\ominus})(NO_2) \rightleftharpoons E-S-S-\phi(COO^{\ominus})-NO_2 + O_2N\phi(COO^{\ominus})-S^{\ominus} \tag{9.14}$$

The enzyme is inactivated and the aromatic thiolate anion that is released absorbs strongly, with a λ_{max} value of 412 nm. This reaction (which is completely specific for thiol groups) can be used to titrate the active-centre thiol of the enzyme. When the concentration of active-site thiol is compared with an estimate of the enzyme concentration (obtained by independent measurement of the protein concentration) the thiol concentration is usually found to have a lower value. The amount of titratable thiol can usually be increased by treatment with the reducing agents cysteine or dithiothreitol, but not to the extent of one thiol per mole of protein. These observations are rationalized in terms of the enzyme forms shown in (9.15).

$$
\begin{aligned}
&1.\ \mathrm{E{-}SH} \\
&2.\ \mathrm{E{-}SS{-}R} \\
&3.\ \mathrm{E{-}SO_x^-} \quad (x = 2, 3)
\end{aligned}
\tag{9.15}
$$

Only the form (1) is active while (2) can be activated to give (1) in the presence of a reducing agent, but (3) is in an irreversible oxidized state. The reagent 2,2′-dipyridyldisulphide (2PDS) has a particular advantage: it may be used to titrate the enzyme in the presence of other simple thiol compounds. Figure 9.5 shows the pH-profile for the reaction of 2PDS with

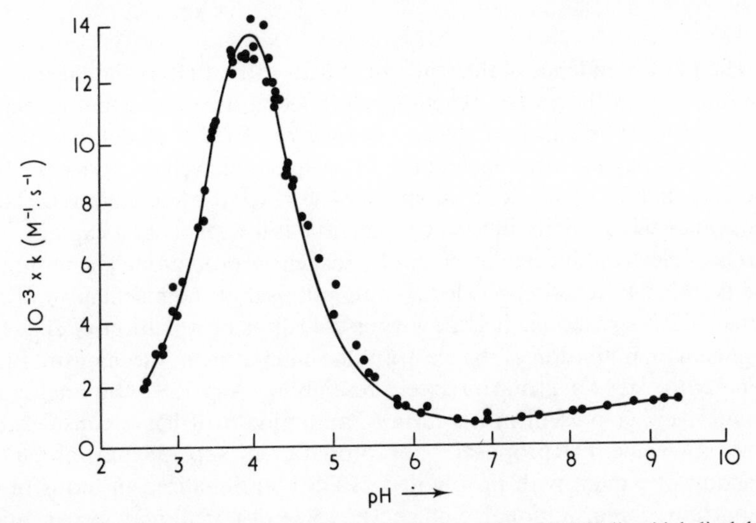

Figure 9.5 The pH-dependence of the reaction of papain with 2,2′-dipyridyl disulphide (2PDS). The second order rate constants were measured at 25°, $I = 0.1$. Taken from Shipton and Brocklehurst, 1978.

papain. The profile shows unexpectedly high reactivity around pH 4. Although this could be due to electrophilic enhancement as a result of protonation of the reagent, such an explanation may be discounted since the highest pK of the reagent has a value of 2.45. The pK values which characterize the bell around pH 4 both have a value of 3.9, and the shape of the bell indicates that there is significant positive cooperativity. These two pK_a values have been ascribed to deprotonation of the active centre thiol system (see later) which generates nucleophilic reactivity and acid catalysis by Asp-158. Acid catalysis by Asp-158 would be effected by interaction of this residue with the nitrogen of the pyridine ring distal to the sulphur atom being attacked by the enzyme (9.16).

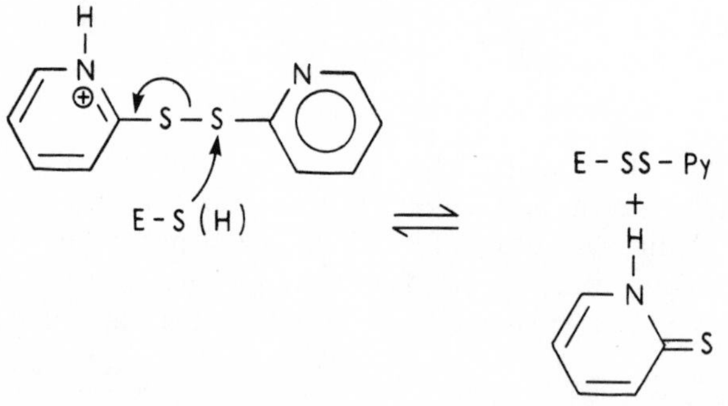

(9.16)

The pH-dependence of the reaction of ficin with 2PDS is similar in form compared with the papain reaction but differs in two important respects: the rate of the reaction in the bell-shaped region is considerably higher than in the papain reaction, and the pK values which characterize the bell are 2.42 and 3.82 so it is apparent that the reagent pK is involved. Thus the enhanced reactivity in the ficin reaction at low pH is best explained in terms of electrophilic enhancement by reagent protonation. The nature of the ficin-2PDS reaction has led to a reappraisal of the mechanism of the papain-2PDS reaction. It is now proposed that protonation of Asp-158 results in an inhibition of the electrophilic enhancement mechanism. Ficin apparently lacks a group precisely resembling Asp-158—the analogous group is either present in the form of an amide or it has a considerably lower pK value. The proposal that protonation of Asp-158 may inhibit the reaction of papain with protonated 2PDS is rationalized in terms of an important conformational change (DOWN ↔ UP) that may occur upon Asp-158 protonation (see later). An important result of the study of the 2PDS reactions is that the thiol group at the active centre is seen to have nucleophilic activity towards simple reagents at neutral pH, well below the pK values of simple thiol compounds which show reactivity *only* in the deprotonated form.

It has been shown that 1,3-dibromoacetone cross-links the active centre thiol with an imidazole group of histidine residue that is within 5Å of the thiol group.

The specificity of papain

Papain is an endopeptidase and, like elastase, has an extended binding site. This is able to recognize up to five amino acids at the N-terminal side of the cleavage site as well as at least one residue on the C-terminal side. Although papain has rather general specificity, it is characterized by a preference for the location of an amino acid with an aromatic side chain in a position one amino acid removed, on the N-terminal side, from the point of cleavage. N-acyl amino acid derivatives are substrates for papain, the most effective having Arg or Lys as the amino acid moiety. Thus an excellent substrate for papain would have the form (9.17)

$$\text{X—Phe—Lys—COR} \tag{9.17}$$

where X is an acyl group (e.g. benzyloxycarbonyl–) or further amino acids, and R may be a phenol, alcohol, ammonia, an aniline or one or more amino acids, provided that if a single amino acid is present it is blocked as an amide.

The kinetic parameters given in table 9.7 can be interpreted as with the serine proteases to indicate the presence of an intermediate acyl-enzyme upon the reaction pathway. Nucleophile partitioning experiments, al-

Table 9.7 The papain-catalysed hydrolysis of some simple synthetic substrates

Substrate	k_{cat} (sec^{-1})	K_m (mM)	k_{cat}/K_m (m^{-1} sec^{-1})
N-benzoylglycine methyl ester[a]	2.72	20.5	132
N-benzoylglycine amide[b]	0.6	158	3.8
N-benzoyl-L-arginine ethyl ester[c]	16	11.8	1.35×10^3
N-benzoyl-L-arginine amide[d]	8.9	34.5	258
N-acetyl-L-Phe-Gly methyl ester[a]	5.4	0.032	1.68×10^4
N-acetyl-L-Phe-Gly *p*-nitrophenyl ester[a]	6.6	3.9×10^{-4}	1.7×10^7
N-acetyl-L-Phe-Gly *p*-nitroanilide[a]	1.3	0.88	1.48×10^3

[a] $I = 0.3$ (NaCl) pH 6.0, 35°; [b] $I = 0.3$, pH 6.0, 38°; [c] $I = 0.1$, pH 6.5, 25°; [d] $I = 0.1$, pH 6.68, 25°. Note that acylation is partially rate-limiting in papain-catalysed hydrolyses of amides.

though these have been less rigorously performed compared with the similar experiments on chymotrypsin, also indicate that an acyl-enzyme intervenes. A most ingenious direct demonstration of the intermediacy of an acyl-enzyme was achieved by the spectroscopic examination of the papain-catalysed hydrolysis of methylthionohippurate. Dithioesters are known to absorb in the UV with a λ_{max} value around 305 nm. When papain and methylthionohippurate were mixed, a steady-state value of absorbance having $\lambda_{max} = 313$ nm developed, which decayed only when the substrate was exhausted. These observations can be interpreted in terms of the scheme shown in (9.18).

$$\text{E-S} \frown\!\!\rightarrow \underset{\text{OMe}}{\overset{\text{R}}{\underset{|}{\overset{|}{\text{C}}}}}\text{=S} \longrightarrow \underset{+\ \text{MeOH}}{\text{E-S-}\overset{\text{R}}{\overset{|}{\text{C}}}\text{=S}} \longrightarrow \underset{+\ \text{RCOS}^{\ominus}}{\text{E-S}} \qquad (9.18)$$

The pH-*dependence of papain-catalysed hydrolyses*

NTCI reacts with papain, albeit much more slowly than with chymotrypsin, the cinnamoyl-enzyme hydrolysis being characterized by a pK of (base form required) 4.7, a value similar to that found for the pH-dependence of k_{cat} for specific substrates. The plot of k_{cat}/K_m for the papain-catalysed hydrolysis of specific substrates is bell-shaped, having pK values near to 4 and 8. The pK of ca. 8 is characteristic (if rather low) of that of a thiol group but the value of ca. 4 is very low for a histidine residue and is more characteristic of a carboxyl group.

The X-ray structure of papain

The enzyme, which has a molecular weight of 23 400, is composed of two lobes. The active site lies between these lobes in a deep cleft. The thiol of Cys-25 is provided by one lobe and the imidazole of His-159 by the other; these two residues are very close to one another (within hydrogen bonding distance) and are reminiscent of the chymotrypsin triad system. The proton upon the nitrogen of His-159 disposed *away* from Cys-25 is hydrogen bonded to the carbonyl oxygen of the side chain of an asparagine residue, Asn-175. The carboxyl group of Asp-158 is in the active centre area but is not sufficiently close to His-159 or Cys-25 to be able to form a hydrogen bond to either *unless a conformational change occurs.*

The reactive protonic form of the enzyme

pH-dependence studies cannot distinguish which of two or more stoichiometrically equivalent systems is the reactive form. The Michaelis constant K_m is often approximately pH-independent and therefore the pH-dependence of k_{cat}/K_m is that of k_2, acylation. Thus, assuming for the time being that the pK values of k_2 refer to His-159 and Cys-25, two possibilities for the reactive form exist (9.19).

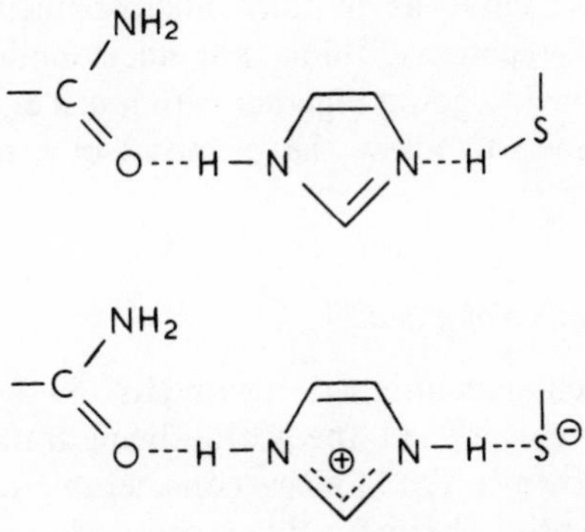

(9.19)

Several lines of evidence have been presented, and it is proposed that these imply that the zwitterionic form is predominant at neutral pH in the free enzyme. This evidence is provided (i) by the double sigmoid pH-dependence of the alkylation of the enzyme by chloroacetamide having pK_a values of 4.0 and 8.4, (ii) the absorbance change at 250 nm consequent upon alkylation which is also double sigmoid, and (iii) direct potentiometric difference titrations. The potentiometric titrations performed upon free enzyme and enzyme in which Cys-25 was blocked in the form of a methyl disulphide indicate strong interaction between His-159 and Cys-25. The ionization scheme is shown in (9.20).

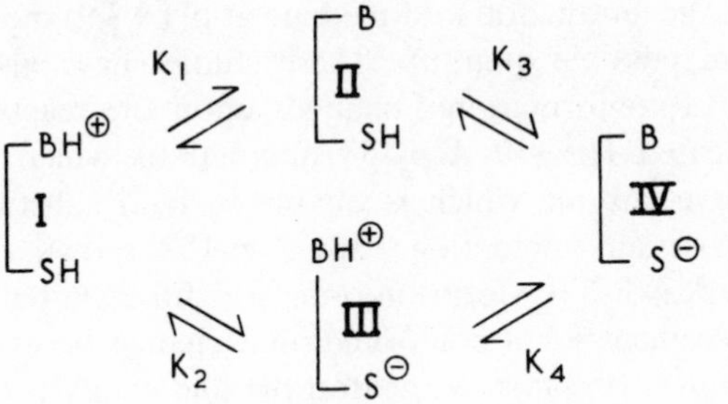

$pK_1 = 4.2$
$pK_2 = 3.1$
$pK_3 = 7.9$
$pK_4 = 9.0$

(9.20)

Thus the thiol pK changes from 3.1 to 7.9 consequent upon the ionization of His-159 and the His-159 pK changes from 4.2 to 9 when the thiol ionizes at 15°, $I = 0.05$. It seems reasonable to assume that the enzyme form which is predominant at the pH of maximal activity is the catalytically active form. The evidence from the alkylation experiments strongly suggests that II is the reactive form in alkylation reactions. However, we must bear in mind that the catalytically active form *may* be the neutral form (III) as the evidence quoted above does not constitute proof. General base catalysis of thiol nucleophilicity has never been observed in model reactions; thiols are nucleophilic only in the deprotonated form. The foregoing together with much evidence not included here can be combined to allow the proposal of a mechanism for this enzyme (9.21).

The catalytic mechanism of papain

If Asp-158 is involved in an interaction with His-159 then the Asp-His-Cys system could operate as a triad, the proton being transferred to Asp-158. The mechanism shown in (9.21) bears considerable resemblance to that given for chymotrypsin: the main difference is the zwitterionic form of the reactive diad. The lower pK of the thiol group as compared with that of the serine hydroxyl group renders zwitterion formation relatively favourable.

The acid limb ($ImH^+SH \rightarrow ImH^+S^-$) of the bell-shaped pH-profile for k_2 probably represents tetrahedral intermediate formation and the alkaline limb ($ImH^+S^- \rightarrow ImS^-$) rate-limiting acid catalysed breakdown of the tetrahedral intermediate. The pH-independent ^{15}N isotope effect of 1.024 in the papain-catalysed hydrolysis of benzoyl-L-arginine amide indicates almost complete breakage of the C—N bond in the rate-limiting transition state.

Cryoenzymological studies have allowed the direct observation of the accumulation of the tetrahedral intermediate at pH 9.3 to more than 95% of the maximum possible quantity. These studies have also suggested the importance of conformational changes upon the reaction pathway and have implicated His-159–Asp-158 interactions which give rise to catalytically active enzyme which is unable to bind substrate. His-159 must change from an interaction with Asp-158 (DOWN form) to an interaction with Asn-175 (UP form) in order for substrate binding to take place. The requirement for a conformational change between substrate binding to the minor (UP) form at neutral pH and catalysis (in the DOWN

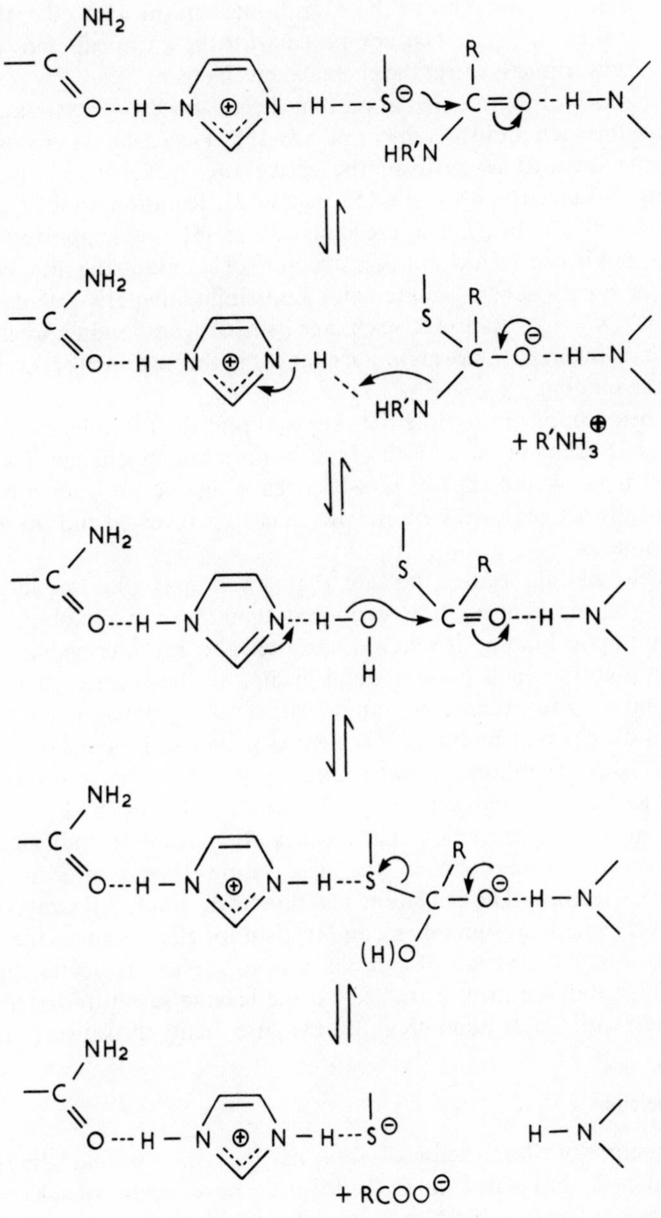

NH_2 — C = O---H — N (⊕) N — H---$S^{\ominus}$ → C(R)(HR′N) = O---H — N

⇌

NH_2 — C = O---H — N (⊕) N — H ... S — C(R)(HR′N) — $O^{\ominus}$---H — N

$+ R'NH_3^{\oplus}$

⇌

NH_2 — C = O---H — N N---H — O(H) → C(S)(R) = O---H — N

⇌

NH_2 — C = O---H — N (⊕) N — H---S — C(R)((H)O) — $O^{\ominus}$---H — N

⇌

NH_2 — C = O---H — N (⊕) N — H---$S^{\ominus}$ H — N

$+ RCOO^{\ominus}$

(9.21)

form) requires elaboration of the overall mechanism. However, the conformational changes are fast compared with the chemical steps and do not otherwise directly affect the chemical mechanism.

It is important to note that at present there is no X-ray crystallographic evidence for such an involvement of Asp-158. The carboxyl group of this residue is directed away from the active site, well beyond hydrogen-bonding distance to either His-159 or Cys-25. Rotation about C_α–C_β of Asp-158 will not bring it appreciably closer. This is supported by the recently published 1.7Å resolution structure of actinidin, the thiol protease from Chinese gooseberries. It remains a possibility that His-159 may move by C_α–C_β, C_β–C_γ rotation to exchange its hydrogen-bonding interaction. Modest main chain conformational changes in this region may occur upon substrate binding.

It is also important to note the proposal that the pK values of His-159 and Cys-25 exchange as a result of the conformational change. The pK of His-159 is ca. 4 and that of Cys-25 is ca. 8 in the UP form but in the catalytically active DOWN form the pK values are reversed and the ion pair predominates.

Model-building studies indicate that strain may play an important part in the mechanism. It is proposed that the strain induced in the substrate upon binding is relieved upon tetrahedral intermediate formation. Inhibitors which present small groups at the reactive centre bind much more tightly than compounds with bulky groups in this position which indicates that binding of the former may avoid the introduction of strain. Aldehyde inhibitors have been proposed to be transition state analogues and to form hemithioacetal adducts with Cys-25.

Proton inventory studies of the deacylation reaction indicate that a single proton is in flight in the deacylation transition state for all substrates studied. The acylation reaction with amide substrates shows only a very small, presumably secondary, isotope effect, as does the reverse of acylation. The absence of a primary isotope effect indicates that acid (His-159) catalysed proton transfer to the leaving group nitrogen is not concerted with C—N bond breaking, the latter being the slowest step.

Ribonuclease

The structure of ribonuclease (RNase) has been discussed at some length in chapter 3 and some catalytic features have been considered. The chemical mechanism of RNase was determined as a result of ingenious chemical experimentation long before the X-ray structure became available. The X-ray structure has contributed relatively little to our understanding of the mechanism of this enzyme but has given important information concerning the groups which surround the active site and the polarity of the catalytic environment. The enzyme hydrolyses polyribonucleotides and has a strong specificity preference for a pyrimidine residue on the 3′ side of the cleavage point. Cytidyl-cytidine (CpC) (9.22) is an excellent simple substrate for the enzyme.

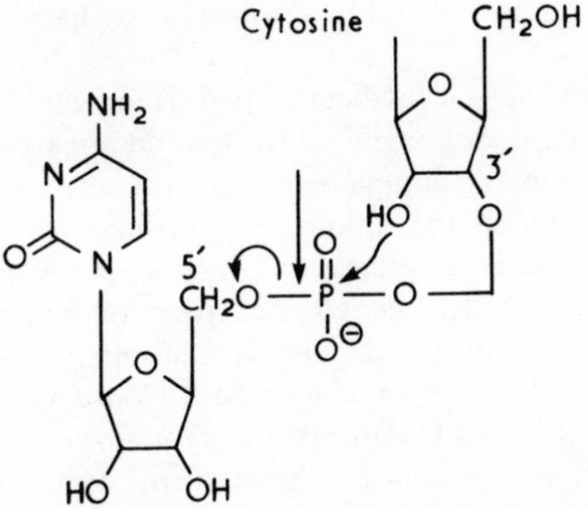

Cytosine CH_2OH NH_2 N 3′ HO O O 5′ CH_2O—P—O O $O^{\ominus}$ N O HO OH (9.22)

Chromatographic and spectrophotometric examination of the RNase-catalysed hydrolysis of CpC revealed that an intermediate accumulated during the course of the hydrolysis but disappeared when the substrate was exhausted. This intermediate (which was not enzyme-bound) was characterized by the usual methods and found to be cytidine 2′, 3′ cyclic phosphate (9.23). Such a cyclic intermediate was shown to be obligatory—the reaction proceeds wholly via this route.

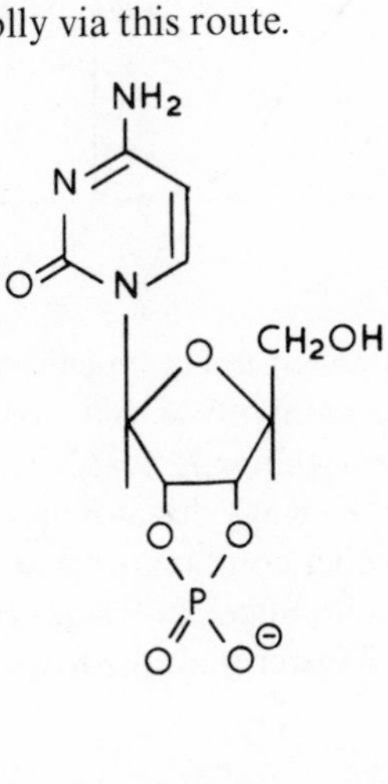

NH_2 N O N O CH_2OH O O P O $O^{\ominus}$ (9.23)

The pH-*dependence of RNase catalysis*

The pH-dependence of k_{cat}/K_m for the RNase-catalysed hydrolysis of simple substrates is bell-shaped with pK values of 5.2 and 6.8 which represent free enzyme ionizations. The pH-dependence of k_{cat} is also bell-shaped, and characterized by pK values in the enzyme–substrate complex of 6.3 and 8.1, both values being significantly perturbed compared with the free enzyme values. The magnitude of these pK values suggests that they could represent a pair of histidine residues or perhaps a carboxyl and a histidine.

Organic solvents have a predictable perturbing effect upon pK values. Neutral acids, such as carboxylic acids, show an apparently increased pK value in the presence of organic solvents as a result of less favourable differential solvation of the anion in a medium of reduced dielectric constant. Cationic acids are subject to a decrease in pK but the magnitude of the effect is much smaller since charge separation is not involved. RNase is a particularly rugged enzyme and is not conformationally affected by the presence of organic solvents in assay media. Consequently it has proved possible to determine the pH-dependence of the enzyme in organic solvent. (Plainly the buffer system used to control the pH of the assays will also be affected by the presence of the organic solvent and this needs to be taken into account.) The results obtained are shown in schematic form in (9.24). The pH_{app} values are those that would pertain if the organic solvent were absent (solid line) or present (broken line). Neutral buffer is shown on the left and cationic buffer on the right.

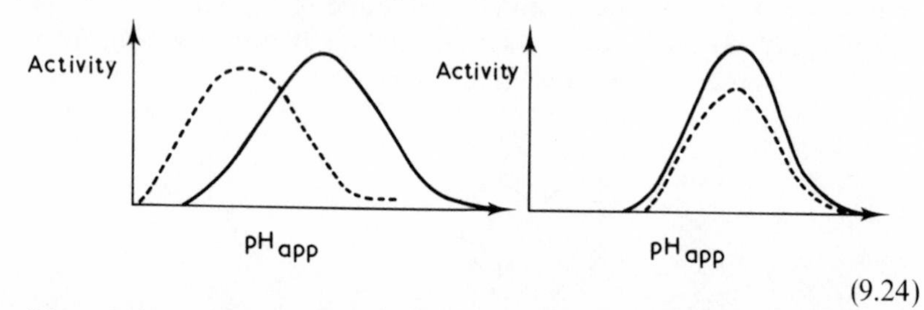

(9.24)

The results can be interpreted unequivocally to show that two cationic acids are involved in the catalytic activity of RNase. The bell-shaped nature of the profiles means that one acts as an acid and the other as a base. We leave it to readers to work out as a useful exercise the pH-profiles that would be expected if other combinations of groups were present.

The cyclic phosphate accumulates in RNase catalysed hydrolyses, since the cyclization step is much faster than the hydrolytic step. Representative

Michaelis parameters are (for the cyclization step of CpC) $k_{cat} = 240\,sec^{-1}$, $K_m = 4\,mM$, and (for the hydrolytic step) $k_{cat} = 5.5\,sec^{-1}$, $K_m = 3\,mM$.

Inhibition of RNase

Photoxidation in the presence of methylene blue results in the oxidation of three of the four histidine residues in the enzyme with complete loss of catalytic activity. Iodoacetic acid also causes loss of activity. This reagent reacts with one of two histidine residues at neutral pH, the reactivity of the enzyme towards this reagent being characterized by a bell-shaped profile. At pH 2.8 iodoacetate reacts with methionine, and at pH 8–10 with a lysine residue. Thus in the pH range in which the enzyme has catalytic activity the reaction is exclusively with histidine. Apparently one of the histidines is required in the base form and the other in the acid form for reaction to take place. Analysis of the enzyme when inhibited with ^{14}C-iodoacetate revealed that 7/8 of the enzyme molecules were labelled at N-1 of His-119 and 1/8 at N-3 of His-12. This suggests that the predominantly reactive form has His-11 in the basic form and His-12 in the cationic form. Iodoacetamide did not inhibit the enzyme or react with histidine at pH 5.5 which implies that the charge interaction between cationic histidine and the carboxylate anion of iodoacetate represents an essential part of the interaction process. Interaction of His-119 and Asp-121 may be important since if Asp-121 is converted to an amide by chemical modification His-119 is not alkylated by iodoacetate. Alkylation of Lys-41 at pH 8.5 also causes loss of enzymic activity.

When freeze-dried from 50% acetic acid, RNase forms dimers which are fully enzymically active. Elegant inhibition experiments with iodoacetate have demonstrated a remarkable property of these dimers. The reaction of the histidines in the dimer is as found for the monomer (i.e. mutually exclusive) but the pair of histidines comprising the reactive duo are provided by each of the two enzyme molecules in the dimer! This remarkable interaction can occur since His-12 is near the N-terminal and this portion of the molecule is able to unfold from the protein and interact with the His-119 region of the other molecule. Thus dissociation of the inactivated dimers leads to the recovery of some enzymic activity due to the mutually exclusive alkylation reaction of the dimer.

The X-ray structure of the enzyme with the inhibitor UpcA (uridyladenine with a CH_2 in place of the 5′-oxygen of the adenine residue) bound at the active site shows that His-12 is disposed close to the 2′-oxygen of the uracil residue and that His-119 is disposed close to the 5′ position of the adenine residue. Thus examination of the X-ray structure of the inhibitor complex answers the question of which of the histidine residues acts as a base in the cyclization step. We note that the assignment is the *opposite* of that which appears to allow optimum reactivity in the alkylation reaction.

All of the above information may be combined to allow the proposal of a mechanism for RNase catalysis which is given in (9.25).

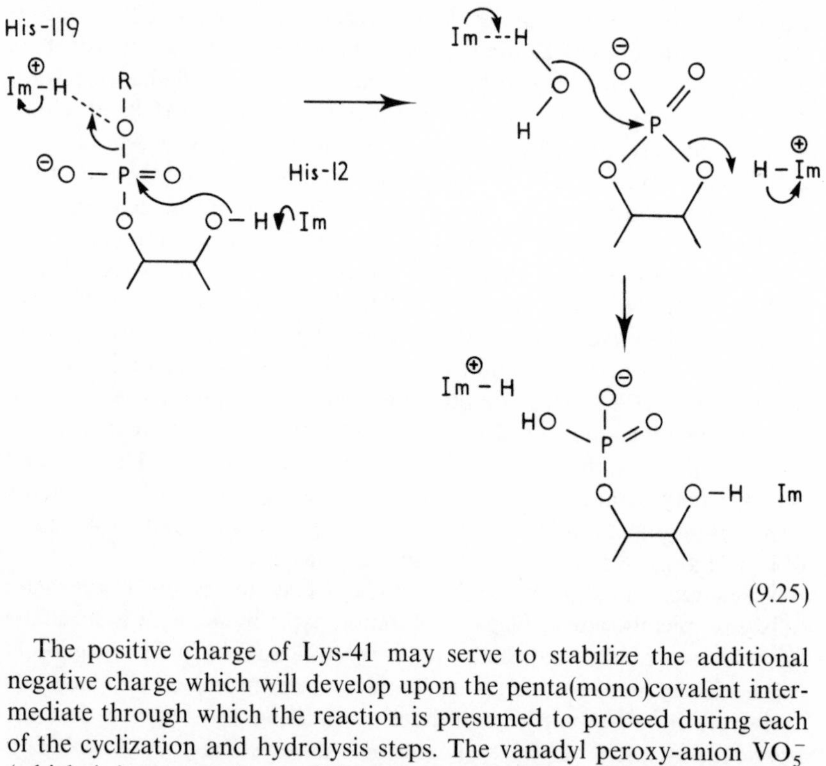

(9.25)

The positive charge of Lys-41 may serve to stabilize the additional negative charge which will develop upon the penta(mono)covalent intermediate through which the reaction is presumed to proceed during each of the cyclization and hydrolysis steps. The vanadyl peroxy-anion VO_5^- (which, it is proposed, resembles the structure of the pentacovalent intermediate) binds very tightly to RNase in the presence of uridine and may thus represent a transition state analogue.

The stereochemical course of the enzyme-catalysed reaction may proceed by either an *in-line* mechanism or an *adjacent* mechanism. The in-line mechanism is similar to an S_N2 displacement where the incoming group attacks the opposite side of the phosphorus atom from the leaving group. In the adjacent mechanism the attack occurs from the same side as the leaving group and a geometric reorganization of the pentacovalent intermediate (pseudo-rotation) is required before leaving group expulsion can occur. Model-building studies suggest that an in-line mechanism is most reasonable. Incorporation of a sulphur atom into uridine 2′, 3′ cyclic phosphate to give chiral cyclic phosphorothioate (9.26) and determination of its absolute configuration by X-ray crystallography has permitted definitive experiments that are capable of distinguishing these mechanisms.

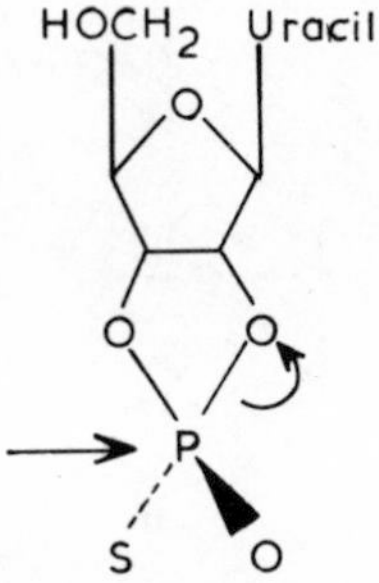

HOCH$_2$ Uracil

(9.26)

The methanolysis of this compound in the presence of RNase gave a methyl ester whose geometry confirmed the in-line mechanism.

We note finally that the 2′-hydroxyl group of the substrates is an essential feature and implies that the enzyme will be totally inactive towards DNA; a corollary is that DNase must hydrolyse DNA by quite a different mechanism.

Glyceraldehyde 3-phosphate dehydrogenase (G3PDH)

This enzyme, which plays a key role in glycolysis, catalyses the production of NADH and 1, 3-diphosphoglycerate which in turn yields ATP in the subsequent reaction. G3PDH has a reactive active centre thiol that is essential for activity, and hydrolyses *p*-nitrophenylacetate in much the same way as does chymotrypsin, albeit via a thioacetyl enzyme intermediate. X-ray crystallographic studies implicate the involvement of His-176 as well as Cys-149 and analogy has been drawn between this system and that found in papain, i.e. that the His–Cys pair occurs in the form of a zwitterionic system. All of the recent studies upon this enzyme support the mechanism originally proposed in the 1950's (9.27).

His-176 has been included in the mechanism where it performs a role equivalent to that of His-195 of lactate dehydrogenase in the hydride transfer step of the reaction. His-176 may also assist phosphate attack by stabilizing the developing negative charge upon the carbonyl oxygen and

Cys 149 His 176

$-S^{\ominus}$ C(H)=O···H—$Im^{\oplus}$ / CHOH / CH_2O(P)

⇌ $NAD^{\oplus}$

$NAD^{\oplus}$ —S—C(H)(R)—O—H Im

⇌ (9.27)

NADH —S—C(R)(O-PO(OH)$O^{\ominus}$)=O····H—$Im^{\oplus}$

⇌

$-S^{\ominus}$ (R)C(O-PO(OH)$O^{\ominus}$)=O····H—$Im^{\oplus}$ + NADH

the charge upon the tetrahedral intermediate. G3PDH, which is a tetramer, shows cooperativity in the binding of NAD^+ but not in G3P binding. It has been proposed that the enzyme may be a "dimer of dimers" and exhibit "half of the sites" reactivity (i.e. a "flip-flop" mechanism) but this is the subject of some controversy.

A comparison of the mechanisms of lactate and alcohol dehydrogenases

The structure of lactate dehydrogenase has been described in chapter 3 and the general chemical reactivity of NAD^+ in chapter 5. We saw that the fundamental requirement for catalysis of hydride transfer is rather simple. When the reduced form of the substrate is oxidized, a general basic group is required to remove the hydroxylic proton from the substrate. The general basic species in lactate dehydrogenase is His-195 which conversely acts as a general acid in the reduction of the ketonic form of the substrate.

Horse liver alcohol dehydrogenase is a symmetrical dimer of 40 000 molecular weight, and has two bound zinc atoms and one NAD^+ binding site per subunit. The yeast enzyme, which is a tetramer of 145 000 molecular weight, has a single bound zinc atom and one NAD site per subunit.

X-ray crystallographic studies on the liver enzyme indicate that one of the two zinc atoms per subunit is situated at the junction between the NAD and substrate binding sites. It has been deduced that the zinc atom is directly involved in the catalysis since there are no other groups capable of acting as a general acid/base catalyst which are adequately positioned with respect to the bound substrates. Proton release upon NAD^+ binding to the apo-enzyme indicates that the holoenzyme has a functional group with a pK of ca. 7.6 which is characteristic of the ionization of a zinc-bound water molecule. The pH-dependence of k_{cat} shows a sigmoid relationship being characterized by a group of pK 8.25 required in the base form; again characteristic of the ionization of a zinc-bound water molecule.

One proposal for the mechanism supposes that the substrate displaces the bound water molecule and becomes liganded to the zinc (9.28).

$$\begin{matrix}\text{Cys-46}\\ \text{Cys-174}\\ \text{His-67}\end{matrix}\cdots Zn^{2+}\cdots {}^{-}O-\underset{|}{\overset{|}{C}}-H \quad \overset{+}{NAD} \rightleftharpoons \cdots Zn^{2+}\cdots O{=}C\langle \quad H-NAD \qquad (9.28)$$

An alternative proposal is that the water molecule is not displaced but acts in the zinc-bound hydroxide ion form as a base upon the substrate (9.29).

$$\cdots Zn^{2+}\cdots \underset{\substack{|\\ H}}{\overset{-}{O}}\cdots H-O-\underset{|}{\overset{|}{C}}-H \quad \overset{+}{NAD} \rightleftharpoons \cdots Zn^{2+}\cdots \underset{\substack{|\\ H}}{O}-H \quad O{=}C\langle \quad H-NAD \qquad (9.29)$$

Apparently this arrangement is not consistent with the geometry of the active site. A variation of this scheme, where the zinc atom becomes pentacoordinate by *adding* the substrate to the ligand set (which is consistent with the active site geometry) is shown in (9.30).

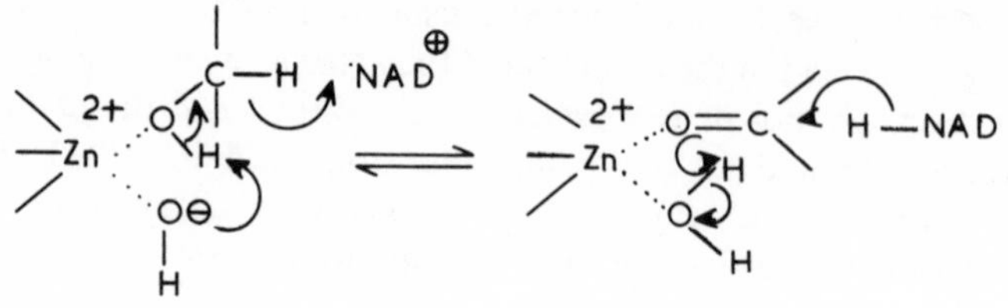

(9.30)

The comparison of these enzymes demonstrates the apparently similar role played by histidine and zinc-bound water.

An appealing scheme known as the "oil–water histidine" mechanism has been proposed to explain the catalytic power of lactate dehydrogenase. NAD^+ binds to a conformation of the enzyme in which a surface loop is disposed away from the active site, thus allowing solvent and substrate access. After lactate has bound, a conformational change occurs, in which the loop closes down upon the active site rendering the pyridinium area of the NAD^+ binding domain hydrophobic. Solvent and counterions are expelled, which results in destabilization of the pyridinium positive charge. This in turn encourages electron flow through the bound substrate from His-195, which is in a more polar environment. This results in hydride transfer to the C-4 atom of the pyridine ring with neutralization of the positive charge in the hydrophobic region. The closure of the loop over the active site provides the driving force for overcoming the very unfavourable equilibrium for the formation of NADH and pyruvate. Although the reverse reaction will involve transfer of a positive charge *into* a hydrophobic region, the energy can be provided by the very favourable equilibrium in this direction.

The enzyme *carbonic anhydrase* has been proposed to have a mechanism relying upon a zinc-bound hydroxide ion. The zinc-bound water molecule, which has a pK of 7, acts as a nucleophilic hydroxide ion when deprotonated and so we can see that the role of zinc is directly comparable with that in alcohol dehydrogenase.

X-ray crystallographic studies of *carboxypeptidase* have shown that the zinc atom in this case acts as a Lewis acid. The oxygen atom of the carbonyl group of the susceptible peptide bond is coordinated to the zinc, which renders the carbonyl carbon more electrophilic. This interaction will also help to orient the peptide bond for the attack of water, probably general base-catalysed by Glu-270 (a nucleophilic mechanism involving Glu-270 has not been excluded). The role of zinc in carboxypeptidase-catalysed reactions is comparable in some respects to that played by zinc in the reduction of acetaldehyde by alcohol dehydrogenase.

Triose phosphate isomerase

The muscle enzyme is a dimer of molecular weight 53 000 that shows no cooperativity. It catalyses the reaction (9.31).

$$\begin{array}{ccc} HC{=}O & & CH_2OH \\ | & & | \\ CHOH & \rightleftharpoons & C{=}O \\ | & & | \\ CH_2O\text{Ⓟ} & & CH_2O\text{Ⓟ} \\ \text{Glyceraldehyde 3Ⓟ (G3P)} & & \text{Dihydroxyacetone Ⓟ (DHAP)} \end{array} \qquad (9.31)$$

The equilibrium of the reaction is strongly in favour of DHAP and this has allowed the determination of the difference X-ray map between the free enzyme and the enzyme–DHAP complex. Glu-165 is positioned so that it may act as a proton shuttle between C-1 and C-2, and His-95 so that it could perform a similar function between the oxygen atoms attached to C-1 and C-2. The pH-dependence of k_{cat}/K_m is bell-shaped with pK values of 6 and 9 whilst that of k_{cat} is sigmoid with a pK of ca. 6. The pK of 6 represents the substrate phosphate group. The pH-dependence of the chemical modification by chloroacetol sulphate of the carboxyl group of Glu-165 is characterized by a pK value of 3.9. The pK of 9 in the pH-dependence of k_{cat}/K_m might be due to ionization of His-95 but the pK is rather high unless the cation is stabilized by an adjacent anion.

Extensive and sophisticated kinetic studies involving both isotopically labelled substrate(s) and solvent have proved extremely important in determination of the mechanism of triose phosphate isomerase. Incorporation of deuterium from the solvent on to C-1 of DHAP indicates that the reaction proceeds by proton transfer on to C-1 and not by hydride ion addition. That hydride ions do not exchange with solvent is seen for example in the isotopic studies of dehydrogenase stereospecificity (chapter 4). If G3P labelled at C-2 with tritium is mixed with enzyme, some of the tritium is transferred to C-1 in the DHAP product. This indicates that the C-2 → C-1 transfer is mediated by a single base. Assuming that G3P/DHAP must be proton-deficient on the carbon atoms *during* the transfer the most plausible intermediate is an enediol (9.32).

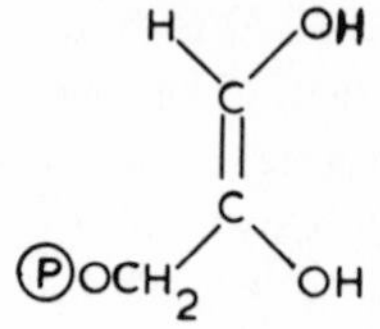

(9.32)

The stereochemistry of the enediol has been established to be cisoid. The information provided above allows the mechanism of the enzyme to be postulated as in (9.33).

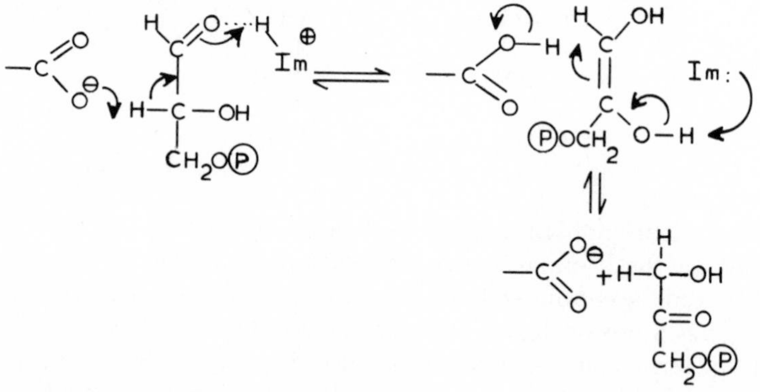

(9.33)

The complete Gibbs free energy diagram for triosephosphate isomerase catalysis has been evaluated as a result of isotope experiments and is shown in chapter 1 (1.4). This enzyme will be discussed further in the section on evolution since it seems to be subject to diffusional encounter rate limitation.

Glucokinase

Rat liver glucokinase is a *monomer* enzyme of 48 000 molecular weight. It catalyses the phosphorylation of glucose at the 6-position using $MgATP^{2-}$. A surprising property of this enzyme is its cooperative kinetics in glucose concentration. The degree of cooperativity is positively dependent upon the $MgATP^{2-}$ concentration, but $MgATP^{2-}$ is *not* itself homotropically cooperative. These observations cannot easily be reconciled with any straightforward kinetic scheme in which rate processes directly reflect equilibrium binding events. An alternative form of kinetic mechanism, known as a *mnemonic model*, which involves a kinetically significant conformational change of the free enzyme has been proposed. This is shown in (9.34) where MT = $MgATP^{2-}$, MD = $MgADP^{-}$, G = glucose and G6P = glucose-6-phosphate.

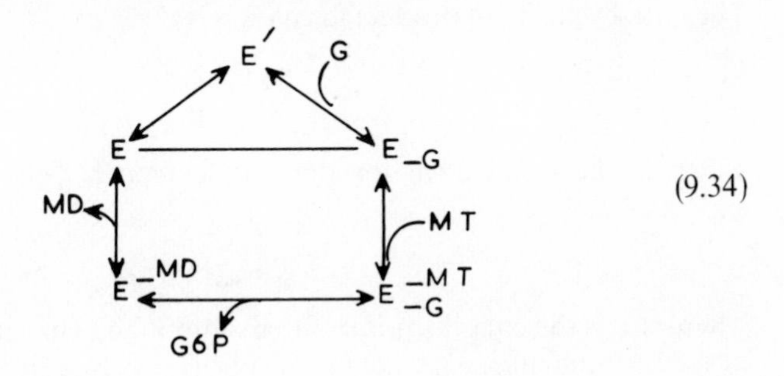

(9.34)

At very low $MgATP^{2-}$ concentration the free enzyme forms will be in equilibrium with E-glucose and no cooperativity will be apparent. At high $MgATP^{2-}$ concentration the equilibria will be perturbed and cooperativity can occur to an extent dependent upon the perturbation from equilibrium. Thus this enzyme represents an interesting example of an enzyme that shows cooperativity without having interacting active sites.

The origin and evolution of enzyme catalytic power

The considerable progress in our understanding of these concepts made in recent years is primarily a result of contributions from Fersht, and Albery and Knowles. It has been stated elsewhere that optimal catalysis results from optimal transition state stabilization. This concept has formed the background to our thinking throughout this chapter. We shall now consider the veracity of this concept within a logical qualitative rather than quantitative format.

We first recall that the theory of absolute reaction rates assumes pseudo-equilibrium between ground and transition states. The rate of a reaction is equal to the concentration of transition state multiplied by a constant (kT/h). Thus the effect of free energy changes upon rates and equilibria can be considered in terms of the Gibbs equation $-\Delta G = RT \ln K(K^{\ddagger})$; the important consequence of this is that changes in free energy of states of a system will have the same *relative* effect whether the state be thermodynamically or kinetically defined.

The kinetic model upon which our considerations will in the first instance be based is the simple two-step form (see chapter 4)

$$E+S \underset{}{\overset{K_m}{\rightleftharpoons}} ES \xrightarrow{k_{cat}} E+P$$

The initial velocity of product formation is given by

$$v_0 = \frac{k_{cat}(E_0)(S)}{K_m + (S)}$$

which is the usual form for the simple model. This equation can be expressed in the form

$$v_0 = k_{cat}/K_m(E_f)(S)$$

where (E_f) is the concentration of free or unbound enzyme.

We assume our enzyme to be partly evolved in that it interacts favourably with both substrate and transition state but that it is capable of further evolution of its catalytic power. Three modes by which further evolution could occur must be considered.

(1) *The enzyme achieves tighter interaction with the substrate*, so that the free energy of the ES complex is decreased but the free energy of the enzyme-bound transition state is unchanged (i.e. the additional interaction is lost upon activation). This situation is shown in figure 9.6(a). The result is that K_m is decreased, k_{cat} is decreased, and since ES is more stable, (E_f) is decreased. The net effect is a reduction in rate since the effects upon K_m and k_{cat} compensate.
(2) *The interactions between enzyme and substrate and enzyme and transition state are* equally *enhanced* (figure 9.6(b)). K_m decreases, k_{cat} is unchanged and (E_f) decreases. The rate of the reaction is unchanged if initially (i.e. before reduction) $K_m \ll (S)$ but increases somewhat if initially $K_m > (S)$, i.e. a 5.5-fold increase if K_m is reduced by a factor of 10 to $(S) = K_m$ from an initial value $10 \times (S)$.
(3) *The transition state is bound more tightly but the enzyme–substrate interaction remains unchanged.* In this instance k_{cat} increases, K_m is unchanged and (E_f) is unchanged. Clearly the improvement in interaction with the transition state is directly reflected in an increase in k_{cat} and this will be the preferable mode of evolution. This situation is shown in figure 9.6(c).

Tighter binding of a transition state implies an improvement in the complementarity of enzyme and transition state relative to that between enzyme and substrate. An excellent example of this is the formation, resulting from tetrahedral intermediate formation, of a strong hydrogen

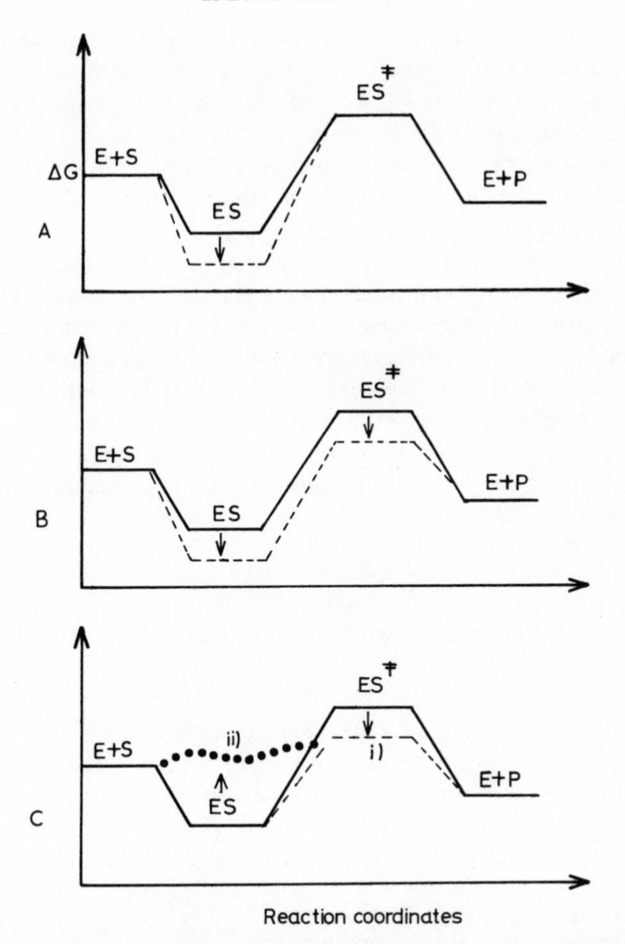

Figure 9.6 Free energy profiles for the evolution of a simple enzyme reaction. (a) The free energy of the Michaelis complex (ES) is decreased (from ——— to -----) but the extra binding energy is lost consequent upon activation to $ES^{\ddagger}$; (b) The free energy of both ES and $ES^{\ddagger}$ are decreased equally, the extra binding interaction in ES being preserved upon activation to $ES^{\ddagger}$; (c) (i) The free energy of ES is unaffected but the free energy of $ES^{\ddagger}$ is decreased, an additional binding interaction is achieved upon transformation from $ES \rightarrow ES^{\ddagger}$; (ii) The free energy of ES is increased at *constant* k_{cat}/K_m to a value where $K_m \simeq (S_0)$, the ambient substrate concentration *in vivo*.

bond from the proton of the backbone nitrogen of gly-193 to the carbonyl oxygen in chymotrypsin catalysis.

Having achieved optimal transition state binding it is useful to consider the converse of proposition (1) above. Let us increase K_m, thus weakening enzyme–substrate interaction. The result is that k_{cat} increases *and* (E_f) increases. Since the effects upon k_{cat} and K_m compensate, it can be seen that (E_f) increases at constant k_{cat}/K_m and thus the rate must also increase. This effect is considerable if $K_m \ll (S)$ but the "evolutionary pressure" for an increase in K_m becomes weak when $(S) \simeq K_m$ (see figure 9.6(c)). This is easily demonstrated by substitution of suitable numerical values in the equation above, maintaining constant k_{cat}/K_m.

Thus an enzyme will evolve its catalytic power by enhancing transition state complementarity and adjusting its K_m value to match that of the ambient substrate concentration. There is ample evidence that many metabolic enzymes are characterized by K_m values that approximate the substrate concentration in the milieu in which they function. The ultimate goal for enzymes not involved in metabolic control processes is for the overall rate to be limited only by diffusional encounter. The Briggs–Haldane formulation of simple enzyme kinetics (see chapter 4) yields the expression

$$k_{cat}/K_m = \frac{k_2 k_1}{k_{-1} + k_2}$$

If k_2 can evolve to be much greater than k_{-1} then

$$k_{cat}/K_m = k_1$$

which is the rate constant for diffusional encounter.

The concept of the evolution of enzyme catalytic power has been extended (by Albery and Knowles) to cover reversible reactions. The evolutionary process is considered to progress towards a diffusional limitation which may be described as "external state" limitation.

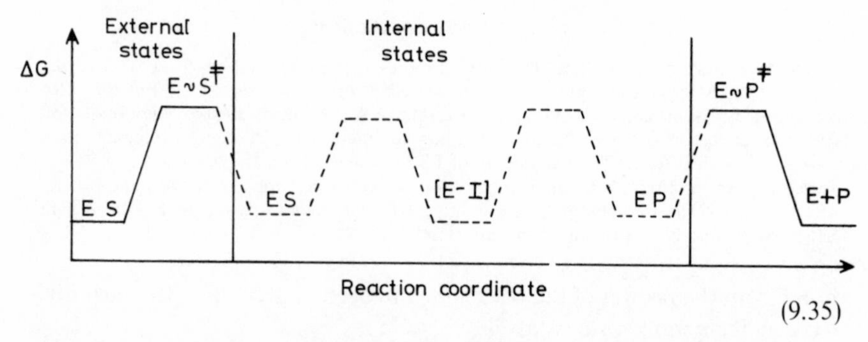

(9.35)

The rate of formation of ES and EP from free enzyme and substrate or product respectively is diffusion-limited and thus cannot be increased by any simple evolutionary process. In order that diffusional limitation should occur the maximal barrier height among the internal states must be similar to that for the diffusional processes. Little further rate enhancement can be achieved by lowering the barrier height of the internal states so that they are significantly lower than the external barriers. The free energy of the intermediates among the internal states must be optimized in order that none should accumulate to an extent that results in thermodynamic trapping of the enzyme in a bound form.

The evolutionary optimization of the internal states is seen as a result of three conceptually distinguishable processes. Adjustment of *uniform* binding involves a change in the free energy of all the internal states equally in such a way as to increase the catalytic activity. The mathematical treatment of the optimization of uniform binding is straightforward but rather cumbersome and will not be included here. A change in uniform binding may result from a change in enzyme structure that affects the binding site at a point not directly involved in chemical catalysis. For example, the binding affinity of an amino acid side chain could be altered in a protease-catalysed reaction. Thus changes in enzyme structure that may affect uniform binding are regarded as much more likely than those changes which may result in alteration of *differential* binding of the internal states. Optimization of differential binding represents another stage in the evolutionary process and presumably results from more subtle changes in structure at the catalytic site of the enzyme that directly affect the chemical catalysis. The object of the optimization of differential binding is to reduce the *kinetic significance* of intermediates and transition states. If, as a result of optimization of uniform binding, a particular intermediate or transition state has a notably higher level of kinetic significance than the others then optimization of differential binding will result in a change in the free energy of this state so that it becomes more nearly equivalent to that of the other internal states. The optimization of differential binding will be easier for dissimilar than for closely similar states, i.e. the ease of optimization will depend inversely as the distance of the species apart on the reaction coordinate. The end result of this is that the kinetically significant intermediate (if any) will be immediately adjacent to the kinetically significant transition state and together these represent the largest free energy barrier among the internal states.

The final form of optimization that may be achieved occurs as a result of *catalysis of elementary steps*. This process involves transition state stabilization by solvation or more effective catalysis. This might occur as a result of a change from mono- to bifunctional catalysis or a change in active site microenvironment. These changes will be dependent upon exploitation of the chemical properties of the transition state relative to the intermediates and the mode of such exploitation will presumably vary from one enzyme to another.

These three concepts have been elegantly applied to the evolution of the catalytic power of triosephosphate isomerase. An efficiency function has been derived that expresses the maximum possible catalytic efficiency of the enzyme at physiological substrate concentration. The evolutionary process is assessed using a comparison of the enzyme-catalysed reaction with the acetate ion-catalysed reaction (see figure 1.4, chapter 1). Table 9.8 shows the effect of each of the three conceptual stages upon the value of the efficiency function.

Table 9.8 Stages in the evolution of the catalytic power of triosephosphate isomerase[a]

	Value of the efficiency function
Acetate ion[b]	2.5×10^{-11}
Uniform binding	3×10^{-6}
Differential binding	1.5×10^{-4}
Catalysis of elementary steps (i.e. the enzyme-catalysed value)	0.6[c]

[a] The format of the table does not necessarily imply that the time course of the real evolutionary process proceeded through these stages in an ordered fashion.

[b] The concepts have been applied to, and optimized with respect to, the acetate ion-catalysed process.

[c] The maximal value of the efficiency function, "perfection", is unity, which represents diffusional limitation only.

We see from table 9.8 that triosephosphate isomerase is essentially fully evolved and that the largest improvement in the catalytic efficiency is provided as a result of optimization of uniform binding. The relatively small contribution from differential binding is a consequence of the rather even free energy distribution of the transition states and intermediates in the free energy profile of the acetate ion-catalysed reaction (figure 1.4). If the logarithm of the efficiency function is plotted against the log of the ambient substrate concentration as in figure 9.7 it is seen that there is a "plateau of perfection". The enzyme, which catalyses the reaction 10^9-fold faster than does acetate ion, is perched on the edge of this plateau since the evolutionary pressure for further improvement has virtually vanished at this stage. Although we have a clear view of the goal of enzymic evolution

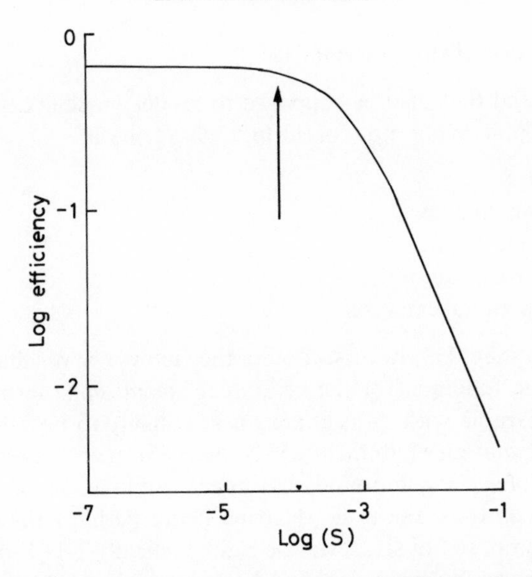

Figure 9.7 The relationship between the logarithm of the efficiency function E_f and the logarithm of the ambient substrate concentration for triosephosphate isomerase catalysis. The arrow indicates the point at which the enzyme operates *in vivo*. Redrawn from Albery and Knowles (1976).

and the means by which this goal may be achieved, it is important to remember that enzymes which play an important role in the control of metabolism do *not* conform to these criteria since the regulatory role can be predominant.

When we consider enzyme catalytic power we must bear in mind the chemical criteria outlined in chapter 2, namely approximation and strain taken together with acid-base catalysis. Approximation is undoubtedly of great importance, probably of greater importance than strain since it has been proposed that proteins are too flexible to be able to impart a great deal of strain or stress to a substrate. The components of enzyme catalytic power have not as yet been properly resolved for any particular enzyme reaction—the menu is available but the diner has not yet been able to decide upon a specific selection of dishes. Thus one of the most important challenges that face the enzymologist is the quantitative resolution of the components of catalytic power. The attainment of this goal ought to be assisted by studies of synthetic catalysts that have some of the binding and/or catalytic properties of enzymes.

Synthetic models of enzymic catalysis

Any compound that may be supposed to model enzymic catalysis will be expected to fulfil one or more of the following criteria:

(1) specificity;
(2) saturation kinetics;
(3) rate enhancement;
(4) competitive inhibition;
(5) multi-functional catalysis.

Such models should allow dissection of the factors responsible for effective catalysis. The fulfilment of the criteria in *quantitative* terms (i.e. to an extent comparable with natural enzymes) is likely to be very difficult to achieve but *qualitative* fulfilment will be useful for mechanistic analysis.

A variety of cyclic compounds have been examined as possible models of enzymic catalysis, the cyclo-dextrins being perhaps the best known. These are composed of six, seven or eight cyclically α-1–4 linked glucose residues. They are doughnut-shaped, having a cavity in the centre. Certain aromatic hydrophobic molecules are bound in this central cavity, which is relatively apolar compared with the bulk solution. The primary (C-6) hydroxyl groups project on one side (face) of the torus while the C-2, 3 hydroxyl groups are found at the other face. Cyclohexaamylose, the six-membered form, is acylated at a C-(2, 3) hydroxyl group some 230 times faster by *m*-tert-butylphenyl acetate than by the *para*-isomer. Apparently the *m*-isomer enters the cavity in such a way that its susceptible carbonyl group is oriented more favourably with respect to the C-2, 3 hydroxyl groups than in the case of the *p*-isomer. The acylation reaction shows saturation kinetics in both catalyst and substrate, the kinetic pK of the hydroxyl group responsible for the acylation reaction being 12.1.

Deacylation of cyclohexaamylose-acetate itself is uncatalysed, deacylation of benzoate esters of cyclohexaamylose is in contrast a catalysed step. The mechanism of the cyclodextrin-catalysed hydrolysis of esters has some similarity to that of chymotrypsin hydrolyses. It has been proposed that if the kinetic pK of the cyclodextrin-catalysed reactions could be reduced to 7 then a convincing model of chymotrypsin catalysis would be achieved. To this end several attempts have been made to alter the structure of cyclodextrins so as to include potentially reactive groups such as imidazole in a reactive conformation. For example histamine has been attached at the C-3 position, which greatly enhances the cyclodextrin reactivity at neutral pH. However, the rate enhancement seen in the hydrolysis of *p*-nitrophenyl acetate is 750-fold less than that achieved by chymotrypsin, for which *p*-nitrophenyl acetate is a non-specific substrate. Thus although *qualitatively* chymotrypsin-like, the *quantitative* discrepancy between the model catalysis and the catalysis of specific substrates by chymotrypsin is likely to be very large. In particular the relative effectiveness is likely to differ most in the case of non-reactive substrates such as amides since cyclodextrins will not be expected to catalyse the hydrolysis of these compounds to any measurable extent.

Cyclodextrins have been modified by "capping" of the torus in order to provide a better-defined, more hydrophobic binding site. Since favourable enzyme–transition state interaction is clearly of paramount importance it is not surprising that attempts have been made to stabilize the transition state in cyclodextrin-mediated reactions. For example, positively-charged amino groups have been symmetrically incorporated in order to stabilize the transition state in benzyl phosphate hydrolysis as well as providing productive orientation of the substrate. The complex synthetic procedures involved highlight the difficulty of working with cyclodextrins—their structure is not readily amenable to synthetic manipulation in order to optimize the catalytic properties.

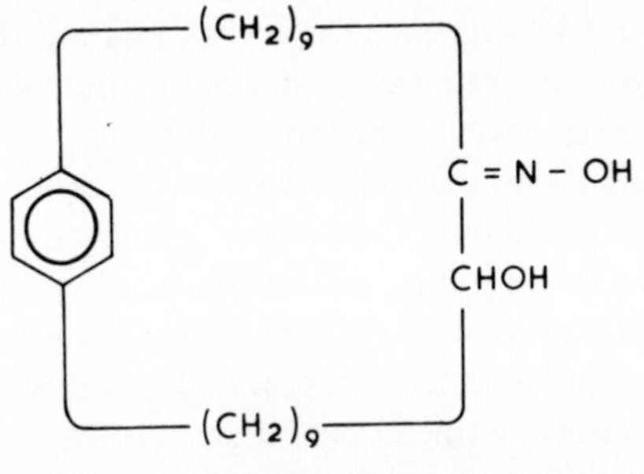

$(CH_2)_9$

C = N − OH

CHOH

$(CH_2)_9$

(9.36)

Other types of cyclic molecules have been studied as hosts for small molecules and as catalysts for the transformation of such substrates, and some examples of these are shown in (9.36) and (9.37). The factors responsible for the binding of small molecules in defined modes to these macrocycles have been intensively investigated. It seems likely that molecules of this type have the potential to represent excellent partial models of enzymic catalysis.

Polymers (both homo- and hetero-) have been very widely investigated as models of enzymic catalysis. Derivatives of polyethyleneimine (PEI) have proved to be particularly effective catalysts. Klotz has shown that a

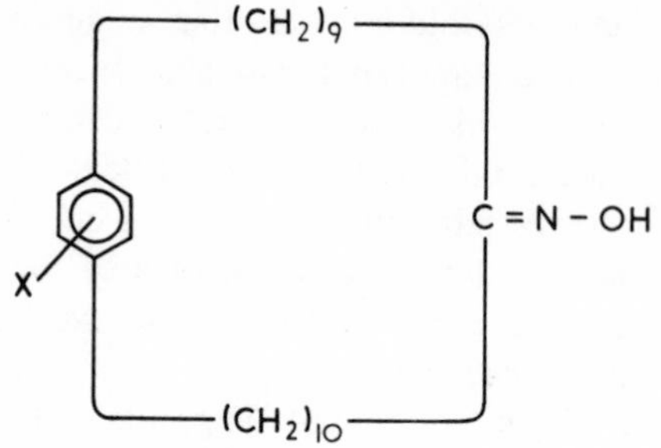

$$\text{macrocycle: benzene ring (substituent X)} -(CH_2)_9- C{=}N{-}OH -(CH_2)_{10}- \tag{9.37}$$

(i) $X = CH_2\overset{+}{N}(CH_3)_3\ Cl^-$

(ii) $= H$

(iii) $= CO_2H$

modified PEI derivative is capable of accelerating the hydrolysis of 2-hydroxy-5-nitrophenyl sulphate by the remarkable factor of 10^{12} compared with unbound imidazole. The PEI was modified by dodecylation at 10% of its nitrogen atoms and attachment of imidazole at 15% of the polymer nitrogen atoms. The rate enhancement is larger, by a factor of 10^2, than that achieved by the enzyme aryl sulphatase II. The reaction was found to be subject to both burst and saturation kinetics and clearly represents a promising model of enzymic catalysis.

Several factors must be considered when assessment is made of the effectiveness of the system described above. Firstly, an enzyme has only a single active site per molecule (or subunit) compared with the large number that are distributed structurally throughout this polymer in a statistical fashion. The detailed structure of the polymer sites will presumably vary to a considerable extent, but in contrast enzyme active sites are extremely accurately defined by elaborate and unique chain folding. Secondly, the substrate can hardly be regarded as being the natural one for aryl sulphatase II. The k_{cat} value is abnormally low ($1.7 \times 10^{-4}\ sec^{-1}$). Thirdly, we must consider the specificity of the "synzyme". Although the specificity has not been examined in detail, it is virtually certain that the synzyme is very much less specific than the natural enzyme. It seems most unlikely that it will prove possible to synthesize polymers that have statistically-defined active sites approaching the specificity of natural enzymes.

Techniques which employ substrate analogues as removable "formers" or templates are promising in that they allow a degree of design to be incorporated into the system. The aim is to achieve the synthesis of a template that is as nearly isoteric with the proposed substrate as possible and which may be incorporated covalently into a polymer matrix. After polymeric incorporation the active site is created by removal of part or all of the template using mild reaction conditions so that the polymer is undamaged. The catalytic groups may be supplied by the polymer matrix or (more usefully) may be generated upon cleavage of the bonds between the polymer and the template molecule. An example of such a system is shown in figure 9.8. The polymer matrix is cross-linked in order to minimize relaxation of the active sites when the spacer molecule *p*-hydroxybenzoate is removed. The thiol group released upon spacer group removal may be titrated using chromophoric aromatic disulphides.

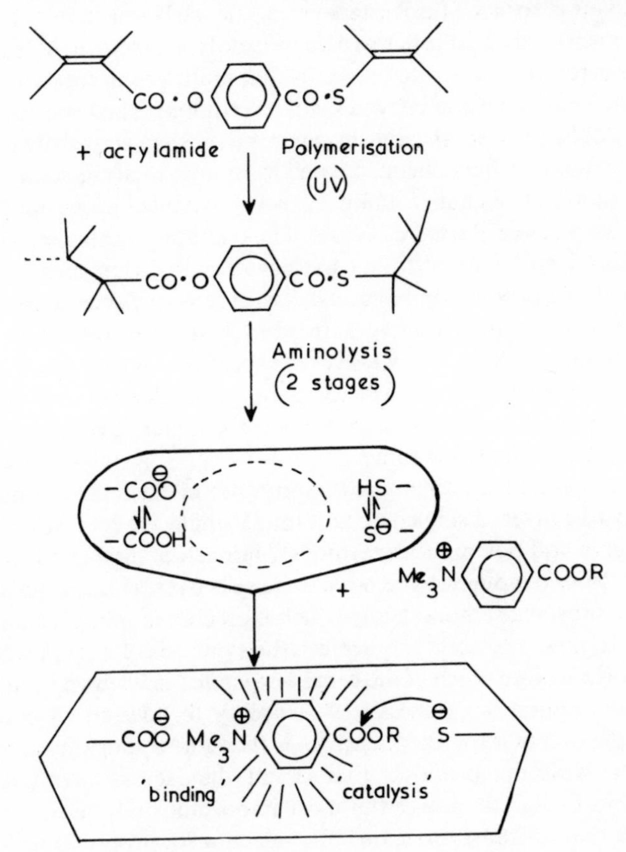

Figure 9.8 Synthesis of copolymer with active centres containing carboxyl and thiol groups.

Effective binding, but only modest catalysis, has so far been detected with such systems. A polymer having stereospecific binding sites has been synthesized using the template method and a column packed with the polymer proved capable of achieving the partial optical resolution of D- and L-glyceric acids.

The catalysts of this type so far synthesized are single group systems and are as such not capable of catalytic turnover. In the system shown in figure 9.8 acylation depends on the presence of the thiol group. This will have a high pK ($\simeq 10.5$) and is thus nucleophilic only at high pH in contrast to the sulphydryl of Cys-25 of papain which is reactive at neutral pH. The deacylation will formally be uncatalysed specific base hydrolysis, but can be enhanced artificially by incorporation of an amine in the reaction mixture. This is so since thiol esters are particularly sensitive to attack by nitrogen nucleophiles. In order to achieve catalytic turnover, at least in the case of non-reactive substrates, it seems that multi-group catalytic systems such as are seen in natural enzymes will be required. These are likely to be extremely difficult to synthesize since natural enzyme active-site geometry is very accurately defined, being created from non-adjacent residues in the protein as a result of chain folding. Closely controlled group motions are likely to be required during catalysis. Thus, although catalysts having at least qualitatively many of the properties of natural enzymes have been synthesized, the production of catalysts with sensitive specificity *and* high reactivity towards non-reactive substrates will be a very difficult and time-consuming task. We may however expect that the study of synthetic models of enzymic catalysis will assist in our understanding of natural enzymic catalysis (and vice versa) since the criteria which define the effectiveness of both are the same.

Bearing in mind the pre-eminent importance in chemical engineering of the ability to convert a single reactant into a single product we can see the wide-ranging and potentially enormously important applicability of these materials. Indeed it ought to be possible to selectively catalyse the reaction of a single substance among many which is, of course, what is achieved by natural enzymes. The ability to achieve the synthesis of a *single* product is very important since much of the energy expended in chemical engineering is used for separation processes. The ability to "design" a catalyst to enhance the rate of a specified reaction is therefore a most attractive goal.

We end with the proposal that of all the factors involved in the generation of catalytic power the most important and guiding principle should be that of transition state complementarity proposed so long ago by Linus Pauling.

CHAPTER TEN

PRACTICAL ENZYME KINETICS

Graphical display of kinetic data

As discussed in chapter 4, the Michaelis–Menten equation (4.10) predicts that a plot of v_0 against S_0 is a rectangular hyperbola, i.e. one with orthogonal asymptotes, passing through the origin (figure 10.1). One limb of the hyperbola lies completely in the second quadrant in the physically meaningless region of negative substrate concentration. The arc of the curve as usually drawn is in the first quadrant but this is only a portion of the limb which extends to infinite $-v_0$ in the third quadrant. The asymptotes are given by $S_0 = -K_m$ and $v_0 = V$. Although the S_0, v_0 relationship is simple, its hyperbolic nature gives rise to problems in presenting kinetic data and in calculating K_m and V. This is because it is difficult to extrapolate the hyperbola to estimate V, the asymptotic value of v_0. In order for v_0 to approach even to within 5% of V, S_0 must be at least 19 times K_m; apart from possible limitations in availability or solubility of the substrate, or interference with the assay method, high substrate concentrations may result in appreciable substrate inhibition (see chapter 4). As K_m is defined in terms of V an accurate estimate of V is necessary for determining K_m.

For this reason and also because scientists have an innate preference for linear relationships, the Michaelis–Menten equation is often rearranged into one of the following *linear forms* for graphical analysis:

$$\frac{1}{v_0} = \frac{K_m}{V} \cdot \frac{1}{S_0} + \frac{1}{V} \tag{10.1}$$

$$\frac{S_0}{v_0} = \frac{S_0}{V} + \frac{K_m}{V} \tag{10.2}$$

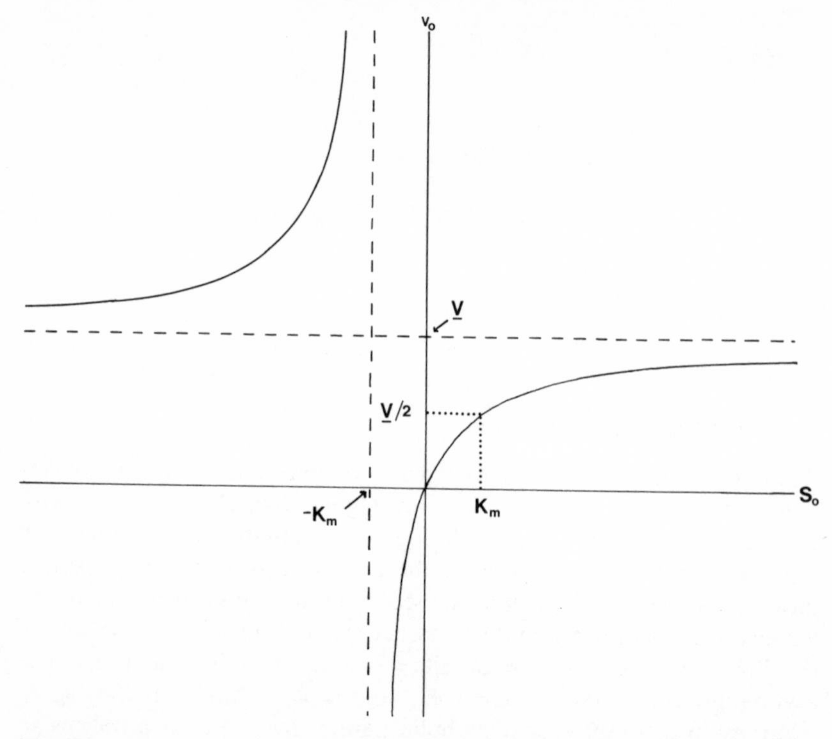

Figure 10.1 Plot of v_0 against S_0 according to the Michaelis–Menten equation showing the relationship of the rectangular hyperbola (solid curves) to its asymptotes (dashed lines).

$$v_0 = \frac{-K_m v_0}{S_0} + V \tag{10.3}$$

All three relationships are of the form $y = mx + c$ and allow estimation of K_m and V from slopes and intercepts. Equation (10.1) describes a plot of $1/v_0$ against $1/S_0$, and is known as the *double-reciprocal plot* usually attributed to Lineweaver and Burk (figure 10.2). Figure 10.3 illustrates the plot of S_0/v_0 against S_0 (equation 10.2) and is sometimes referred to as the *half-reciprocal plot*. The plot of v_0 against v_0/S_0 (equation 10.3) is shown in figure 10.4 and is known as the *Eadie* or *Hofstee plot*.

Comparison of figures 10.2, 10.3, and 10.4 shows that the double-reciprocal and half-reciprocal plots tend to "bunch up" the data, the double-reciprocal plot compressing the points at the high velocities and the half-reciprocal plot compressing the low end of the range. Only the

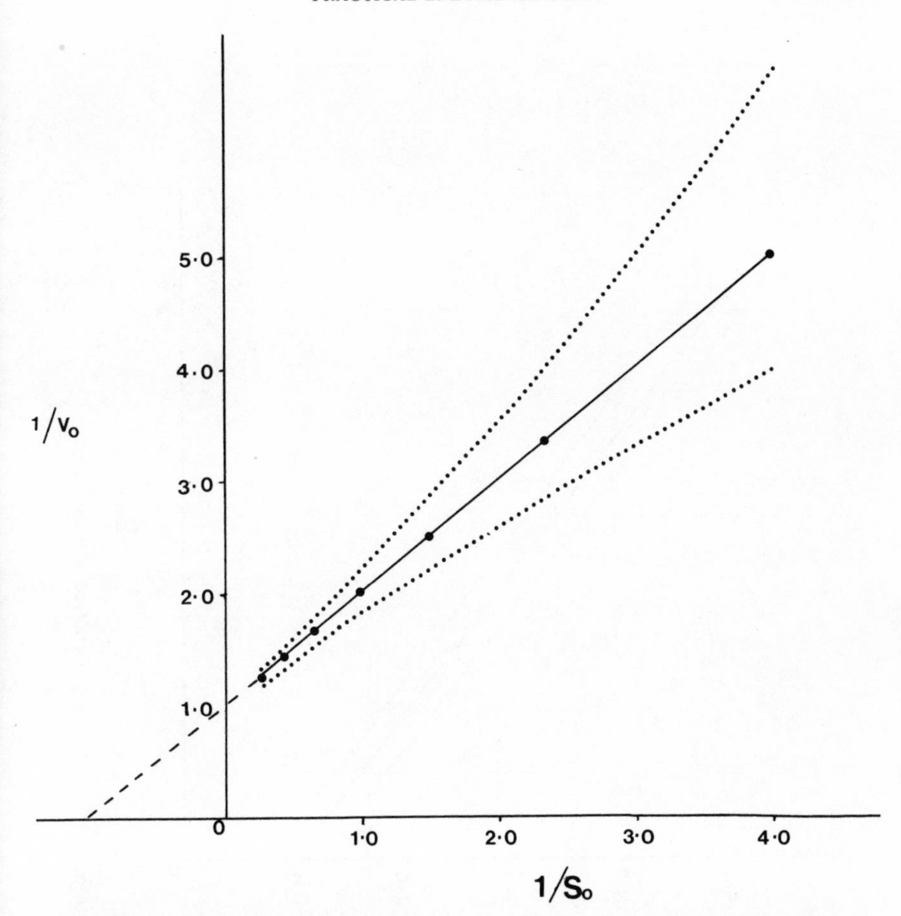

Figure 10.2 Double-reciprocal (Lineweaver–Burk) plot of $1/v_0$ against $1/S_0$. The points are calculated from the Michaelis–Menten equation assuming $K_m = V = 1$ (arbitrary units) at seven equally spaced values of v_0 from $0.2V$ to $0.8V$. The dotted lines show the effect of a constant error of ± 0.05 on v_0.

Eadie–Hofstee plot results in equal spacing of the points representing equally-spaced velocities.

The three linear forms are mathematically equivalent and in the absence of experimental error, it would make no difference which were used. In practice, however, experimental error is always present and under these circumstances the linear transformations are *not* equivalent. This is shown by the dotted lines in figures 10.2, 10.3 and 10.4 which represent the "error

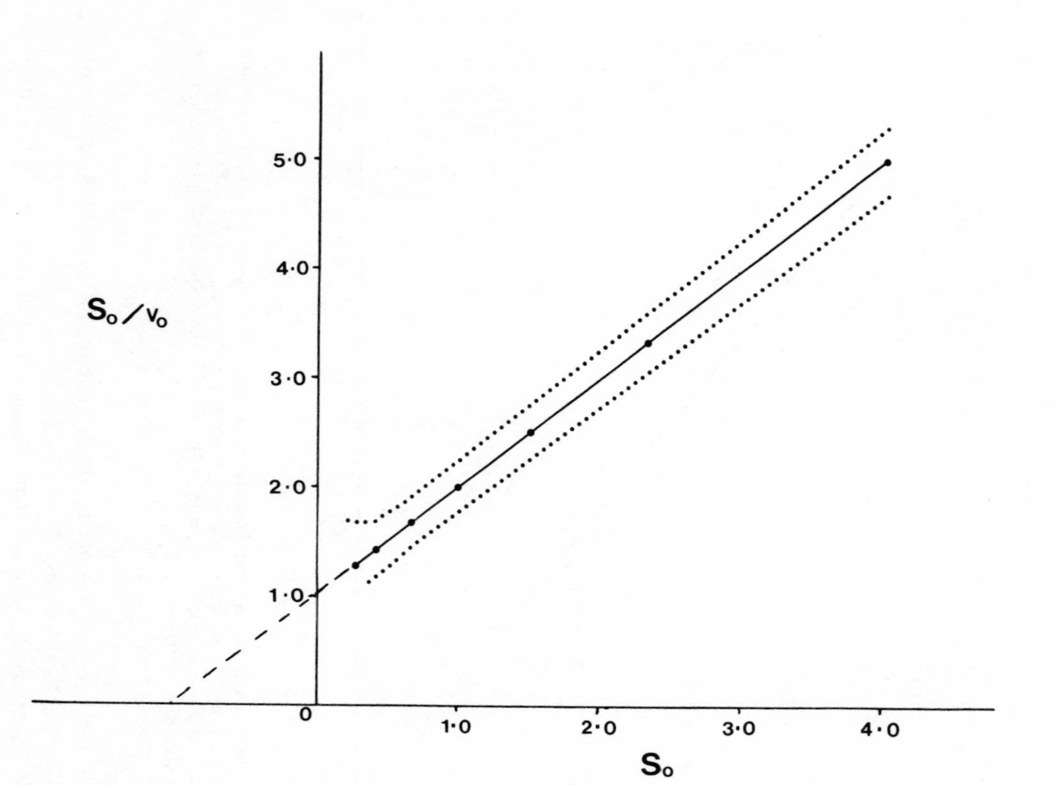

Figure 10.3 Half-reciprocal plot of S_0/v_0 against S_0. Details as in figure 10.2.

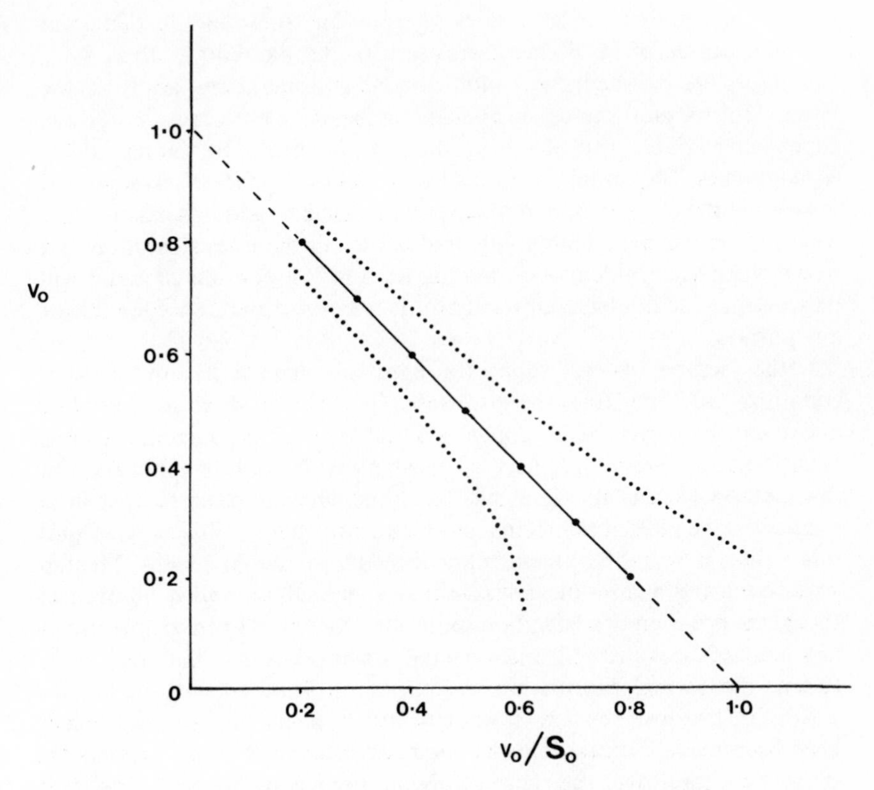

Figure 10.4 Eadie or Hofstee plot of v_0 against v_0/S_0. Details as in figure 10.2.

envelopes" resulting from the imposition of a *constant* error of $\pm 0.05V$ on v_0. It is obvious that the double-reciprocal plot most grossly distorts the error; in the velocity range $0.2V$ to $0.8V$ the constant error in v_0 gives rise to an error in $1/v_0$ which varies 18-fold from one end of the range to the other. The Eadie–Hofstee plot also distorts the error structure though not as badly. Only in the half-reciprocal plot does the error envelope reasonably reflect the constancy of the error; this is why we recommend this plotting form for the *display* of kinetic data.

Calculation of K_m and V

In analysing kinetic data v_0 is usually treated as the dependent variable subject to error, and S_0 as the independent or controlled variable. The variable S_0 appears on both sides of equation (10.2) and so there will be some degree of inevitable correlation of the plotted variables S_0/v_0 and S_0 in the half-reciprocal plot. Similar considerations apply to the Eadie–Hofstee plot where the variable v_0 appears on both axes. As v_0 is the dependent variable, this plot is unsuited to line-fitting by the method of least squares. The double-reciprocal plot has the apparent advantage of separating the dependent and independent variables and this may explain why it is the one most commonly used. However these considerations are trivial when the problem of estimating K_m and V is considered and it will be seen that the double-reciprocal plot is the worst one to use for this or any purpose.

If the method of least squares is used to calculate K_m and V from equations (10.2) or (10.3) the distortions in the error structure described above can be corrected by applying suitable weighting factors, but this negates the apparent simplicity of linear plots. Moreover application of this method often results in a calculated line which appears to have little relation to the plotted points in a double-reciprocal plot. One way around this problem is to fit kinetic data directly to the Michaelis–Menten equation using a non-linear least-squares procedure, and a number of computer programmes have been published for this. However, the use of any least-squares method relies on the assumption that the error in v_0 follows the normal distribution and requires a knowledge of the correct weighting factors. The few studies of error in enzyme kinetics which have been made indicate that the common practice of assuming that the error is constant over the range of observations is by no means justified. In most investigations it is rarely practicable to collect sufficient data to verify the normality of the error and obtain estimates of the correct weights.

We recommend using the *direct linear plot* described in chapter 4 for obtaining K_m and V from kinetic data, because it makes no assumption about the nature of experimental error except that it is as likely to be positive as to be negative. The graphical procedure is a direct replica of the computational one and the median estimates are relatively insensitive to the presence of outliers ("rogue points") which occur frequently in kinetic measurements. It is simple to apply, requiring no calculation, and as S_0 and v_0 are plotted directly there is no distortion due to transformation of data. By using a computer program, confidence limits for the kinetic parameters can be obtained. Finally, outliers are easily detected on a direct linear plot whereas the double-reciprocal plot frequently appears to provide a good fit even when the data are poor.

Systematic deviation from Michaelis–Menten kinetics

The other major use of graphical analysis of kinetic data is for detection of systematic deviation from the Michaelis–Menten equation. Many enzymes, e.g., those showing allosteric behaviour or substrate inhibition, do not follow Michaelis–Menten kinetics. As the human eye is not accustomed to dealing with hyperbolic curves, linear plots are often employed to detect non-hyperbolic behaviour. In figure 10.5a is plotted the v_0, S_0 profile for an enzyme displaying substrate inhibition. The range of substrate concentrations shown is below that in which v_0 actually decreases with increasing S_0. Examination of figure 10.5a shows that it is difficult to distinguish the shape of the v_0, S_0 profile from that of a rectangular hyperbola. The same data are plotted according to the three linear transformations in figures 10.5b, c and d.

It can be seen that the systematic deviation is most easily detectable in the Eadie–Hofstee plot of figure 10.5b. In general, plots of v_0 against v_0/S_0 exaggerate deviation of points from the "true" line because *both* plotted variables are shifted in the same direction by an error in v_0. Reference to figure 10.5d shows that again, the double-reciprocal plot performs poorly. So for detection of non-adherence of kinetic data to Michaelis–Menten behaviour, the Eadie–Hofstee plot is recommended.

Range and spacing of S_0

At high substrate concentrations (relative to K_m) the velocities will be insensitive to changes in S_0. A reasonable estimate of V can be obtained from such measurements but the estimate of K_m will be imprecise. If the substrate concentrations are much less than K_m the rate will vary almost linearly with S_0, and v_0/S_0 will be a good estimate of the ratio V/K_m, but the individual parameters K_m and V will be poorly defined. It follows that it is good practice to vary S_0 over a wide range. This will also provide a check on systematic deviation from the Michaelis–Menten equation. Factors which dictate the limitations at the high and low end of the range are solubility and availability of substrate, substrate inhibition, sensitivity of the assay, and ease of obtaining linear progress curves. A useful range of S_0 is from $0.25K_m$ to $4K_m$. This gives a 4-fold variation of v_0 from the lowest to the highest S_0 values.

Once the range has been selected the spacing of S_0 should be considered. Equally-spaced substrate concentrations will not give equally-spaced velocities. As v_0 is the variable most subject to error it is desirable that the

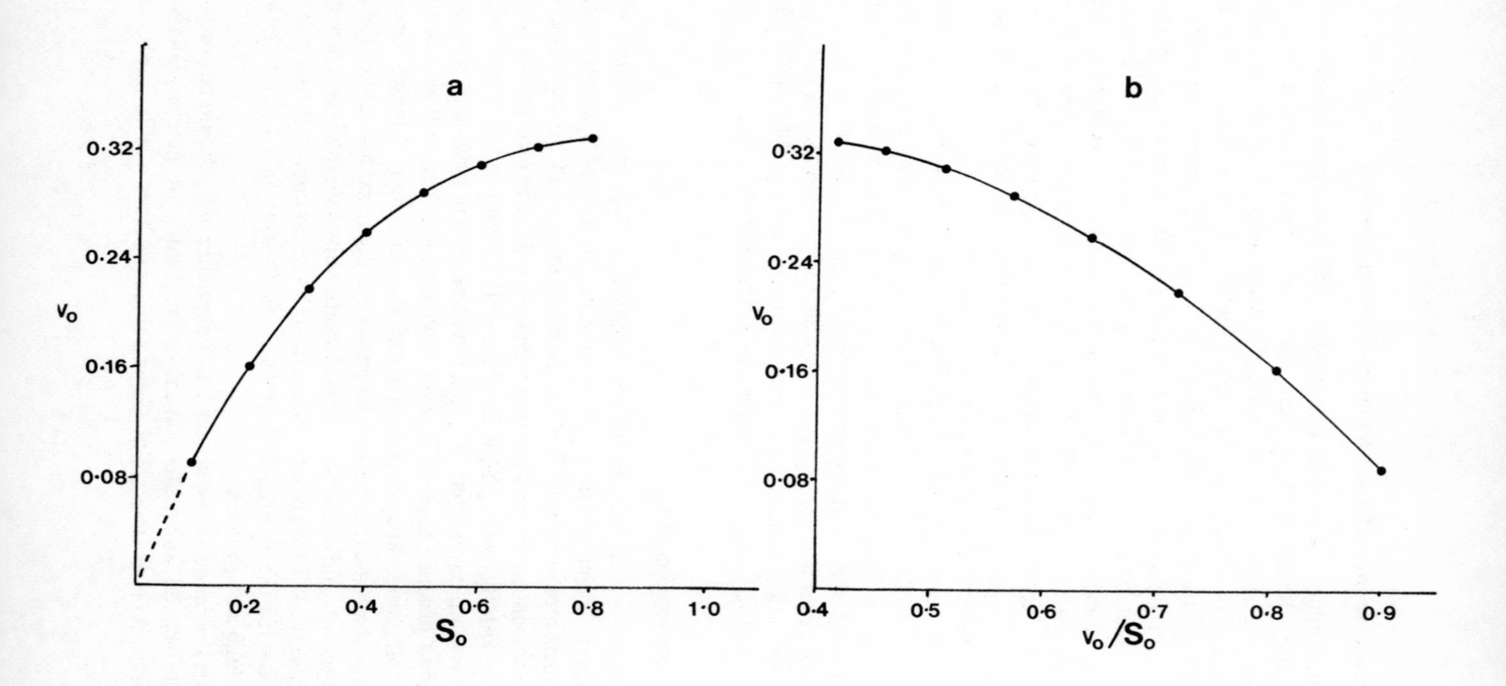
a
0·32
0·24
0·16
0·08
v₀
0·2
0·4
0·6
0·8
1·0
S₀
b
0·32
0·24
0·16
0·08
v₀
0·4
0·5
0·6
0·7
0·8
0·9
v₀/S₀

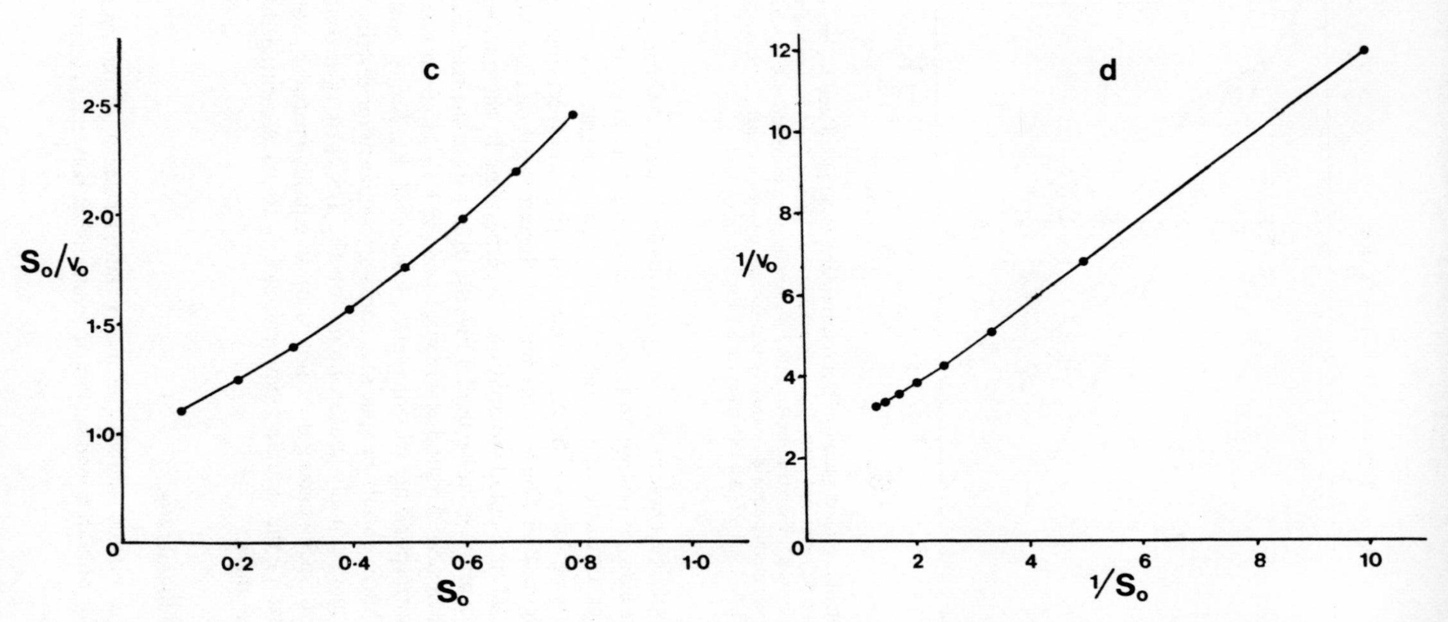

Figure 10.5 Detection of systematic deviation from the Michaelis–Menten equation. The points were calculated assuming substrate inhibition; $v_0 = V/(1+K_m/S_0+S_0/K')$ with $K_m = K' = V = 1$ (arbitrary units) at equally spaced S_0 from 0.1 to 0.8. (a) Plot of v_0 against S_0, (b) Eadie or Hofstee plot, (c) Half-reciprocal plot, (d) Double-reciprocal plot.

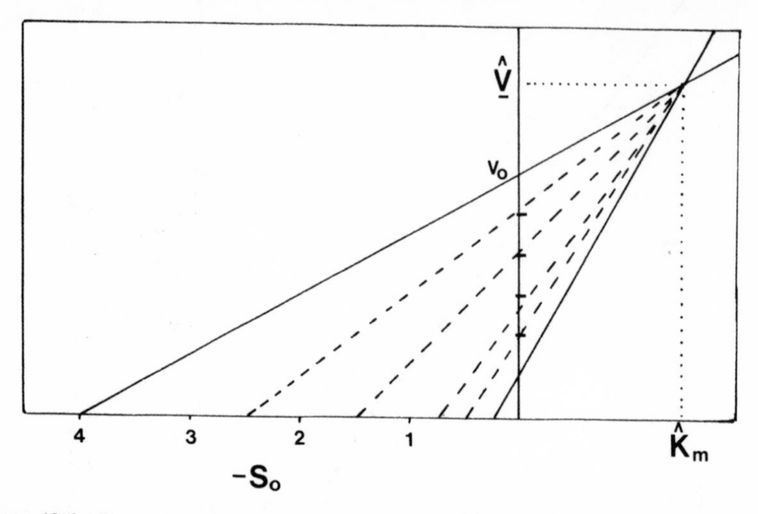

Figure 10.6 Determination of spacing of S_0 using the direct linear plot. $\hat{V}$ and $\hat{K}_m$ are rough estimates of V and K_m obtained from S_0, v_0 observations at high and low end of the range (solid lines). The dashed lines extending from the intersection point give the absolute values of S_0 which will yield roughly equally-spaced velocities throughout the range. In this example these are 0.25, 0.5, 0.75, 1.5, 2.5, 4.0 (arbitrary units). For details see text.

S_0 values should be chosen so that v_0 is equally spaced. Assuming that simple Michaelian behaviour obtains, an easy way of doing this is to obtain a rough estimate of K_m and V from measurements of v_0 at the highest and lowest values of S_0 in the range by using the direct linear plot as shown in figure 10.6. The distance between the intercepts on the V axis is then marked off into equal subdivisions corresponding to the number of substrate concentrations to be used. Lines are drawn from the intersection point through or near the subdivision marks on the V axis to points on the $-K_m$ axis corresponding to convenient values of S_0. Some authors recommend that S_0 should be spaced according to a harmonic distribution because this gives equally-spaced points on the $1/S_0$ axis of a double-reciprocal plot. Such spacing does not result in equally-spaced v_0 values, and as the use of the double-reciprocal plot is to be discouraged this practice has no merit.

Importance of initial rates

In steady-state kinetic studies the use of initial rates is of paramount importance, for only then will the linear relationship between velocity and total enzyme concentration be valid. Initial rate measurements are most often used to obtain a measure of the *amount* of active enzyme present, for example in comparisons of normal and treated tissues or to follow the activity of an enzyme during a purification procedure. As pointed out in chapter 4, proportionality of rate to enzyme concentration obtains at *any* fixed substrate concentration (provided that $S_0 \gg E_0$). In practice, saturating substrate concentrations are usually employed as the rate will then approach its maximal value, thus giving greater sensitivity. Furthermore, the effect of errors in substrate concentration due to volumetric or gravimetric errors, or to chemical instability of the substrate, becomes smaller as S_0 increases. This is illustrated in figure 10.7 which shows that a fixed error of $-0.1K_m$ in S_0 results in a velocity error of approx. $0.05V$ when $S_0 = 0.4K_m$ but only $0.004V$ when $S_0 = 4K_m$. The importance of

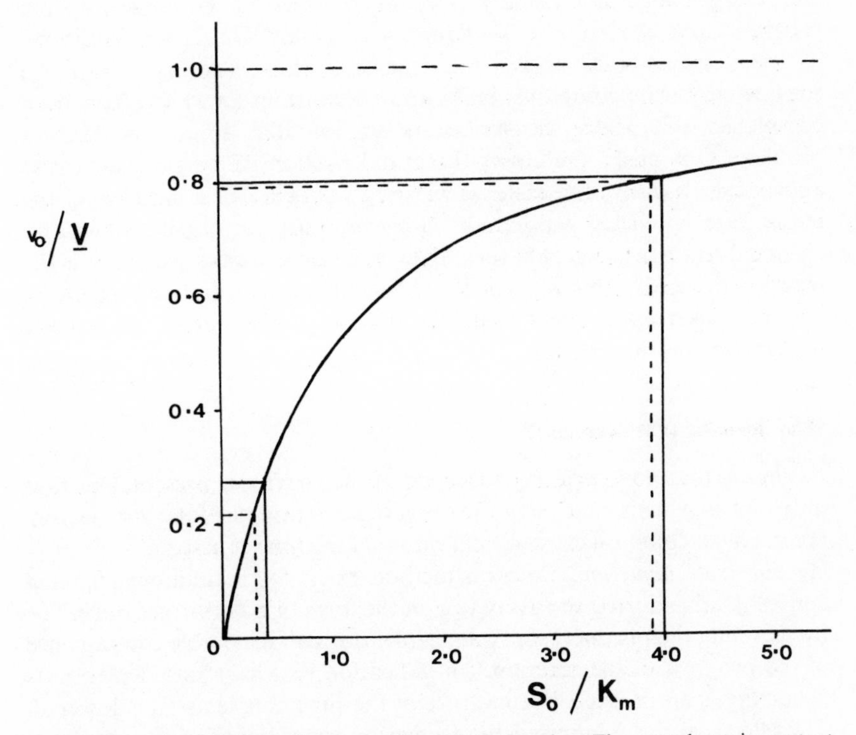

Figure 10.7 Effect of error in S_0 on v_0 at high and low S_0 values. The error shown is constant at $-0.1\ K_m$ on S_0 values of $4.0\ K_m$ and $0.4\ K_m$.

initial rate measurements is equally significant in kinetic studies where the object is the variation of rate with S_0, pH etc., because it is always inherently assumed that the steady-state velocity is proportional to the concentration of the productive enzyme–substrate complex. Finally, at zero time the concentration of substrate is most precisely known. This is especially important if the substrate is unstable under the assay conditions.

The experimental condition $S_0 \gg E_0$ used in steady-state kinetics ensures the validity of the steady-state assumption and enables the rate of substrate disappearance $-d(S)/dt$ to be equated to the rate of product appearance $d(P)/dt$ so that either substrate loss or product accumulation can be followed. Ordinarily initial rate measurements are carried out, insofar as is possible, under conditions of constant temperature, pH and ionic strength, as all these can affect the velocity.

The rate of any reaction is the slope of the tangent to the progress curve. The simplest method of determining the initial rate is to use a pencil and straight edge and visually estimate the slope of the tangent to the progress curve at zero time (see figures 4.1 and 4.2). Under the conditions of most steady-state assays the transient phase (during which the enzyme–substrate complexes build up to a constant level) will have been completed well before measurements are initiated, so this is rarely a problem. Obviously, the closer the initial portion of the progress curve approaches linearity the easier it will be to obtain a true measure of the initial rate by visual estimation. However the rate of non-zero-order chemical reactions (except autocatalytic reactions) always decreases as the reaction proceeds. This is particularly true of enzyme-catalysed reactions, and for this reason direct visual estimation of the initial rate usually underestimates the true rate.

Why does the rate decrease?

Assuming that no interfering substrate of the enzyme is present, and that only one enzyme is catalysing the observed reaction, the following factors are the most common causes of curvature in the time course.

(1) *Substrate depletion.* Unless a method exists for maintaining S_0 at a constant value during the assay (e.g. in the form of a "substrate buffer" or by recycling the product) substrate depletion is an inevitable consequence of the progress of the reaction. An indication that substrate depletion is occurring is an increase in curvature of the time course as S_0 is lowered. The effect can be minimized by decreasing the fraction of S_0 converted during the period of observation. This can be done by increasing the

sensitivity of the assay, e.g. in a spectrophotometric assay by expanding the absorbance scale or using a cuvette of longer path length. More simply, the rate can be decreased by reducing the amount of enzyme or lowering the temperature.

A more serious case of substrate depletion may result if the substrate is chemically unstable under the reaction conditions. It may be possible to correct for this by studying the kinetics of the decomposition of substrate in the absence of enzyme. But if the products of the chemical decomposition themselves effect the enzyme, the situation becomes very complex, and conditions should be sought under which the substrate is stable. For example NAD(P)H rapidly decomposes in acidic solution and assays involving $NAD(P)^+$-dependent dehydrogenases are rarely carried out at pH values much below 7.

(2) *Approach to equilibrium.* As the reaction proceeds the contribution of the back reaction to the overall rate increases. The effect of this can be minimized by employing conditions that shift the equilibrium of the reaction in the direction of product formation. Thus reactions producing protons are often run at pH $\geqslant 9.0$ provided that the enzyme is stable under these conditions. Sometimes it is possible to use trapping reagents to remove the product. The oxidation of ethanol by NAD^+ catalysed by alcohol dehydrogenase is often run in the presence of semicarbazide; this reacts with the acetaldehyde produced to form the semicarbazone and thus removes the product from the reaction mixture.

(3) *Inhibition by product.* Curvature of the time course resulting from either product inhibition or approach to equilibrium can be minimized by reducing the fraction of S_0 converted during the assay, as described above. However, product inhibition can cause curvature even when the equilibrium lies in favour of product, as discussed in chapter 4. In some cases this can be so severe that it may be necessary to sacrifice the convenience of a continuous (e.g. spectrophotometric) assay for a more sensitive (e.g. radiometric) discontinuous one.

The effect of product inhibition can be virtually eliminated if the product can be converted to a non-inhibitor. For instance, catechol-*O*-methyl transferase, which catalyses the methylation of catechols by S-adenosylmethionine, is powerfully inhibited by the product S-adenosylhomocysteine. Adenosine deaminase converts S-adenosylmethionine to S-inosylhomocysteine, which does not inhibit catechol-*O*-methyl transferase. Inclusion of adenosine deaminase in the assay thus results in a marked improvement in the linearity of the progress curve.

(4) *Enzyme inactivation.* If the cause of curvature is inactivation of the

enzyme during the assay the recommendations given above will not improve matters. Indeed some, e.g. dilution of the enzyme, may well exacerbate the situation. This cause of non-linearity can be easily detected by the method of Selwyn as follows:

The rate of an enzyme-catalysed reaction is linearly related to the concentration of active enzyme and will be a function of the concentrations of substrates (S), activators (A), inhibitors (I) and products (P) present at time (t):

$$\frac{dP}{dt} = (E) \cdot f(S)(I)(A)(P) \tag{10.4}$$

If in a series of assays all parameters (including pH, temperature, ionic strength) except (E) are held constant, the integrated form of equation (10.4) can be expressed as

$$(E) \cdot t = f(P) \tag{10.5}$$

The consequence of equation (10.5) is that no matter what the form of $f(P)$, a plot of (P) *versus* $(E) \cdot t$ obtained from a series of progress curves using different amounts of enzyme will give a set of points all of which fall on one smooth curve. This is illustrated in figure 10.8a. However, if the enzyme undergoes inactivation during the assay, equation (10.5) will no longer hold because (E) is itself time-dependent, and a plot of (P) against $(E) \cdot t$ will fall on different curves for each value of (E), as shown in figure 10.8b.

Enzyme denaturation during the assay may be a consequence of the low (usually $\ll$ nM) enzyme concentrations used in steady-state kinetic measurements. At such concentrations significant loss of activity may occur through surface denaturation or adsorption on glass. The first can sometimes be minimized by avoiding undue turbulence when mixing substrates and enzyme to start the assay. Adsorption can often be significantly reduced by using plastics or by coating glass vessels with a siliconizing solution, provided this does not interfere with the assay. Sometimes changing the buffering ions or counter ions or altering the ionic strength may improve enzyme stability.

Enzymes are often stabilized in the presence of reducing agents such as ascorbate, or thiols such as cysteine, mercaptoethanol or dithiothreitol, usually at 0.1–1.0 mM. These reagents are especially useful if the enzyme contains oxidizable thiol groups essential to its activity. However it may be necessary to deoxygenate all solutions which come into contact with the enzyme. Heavy metals can also cause enzyme inactivation. These are

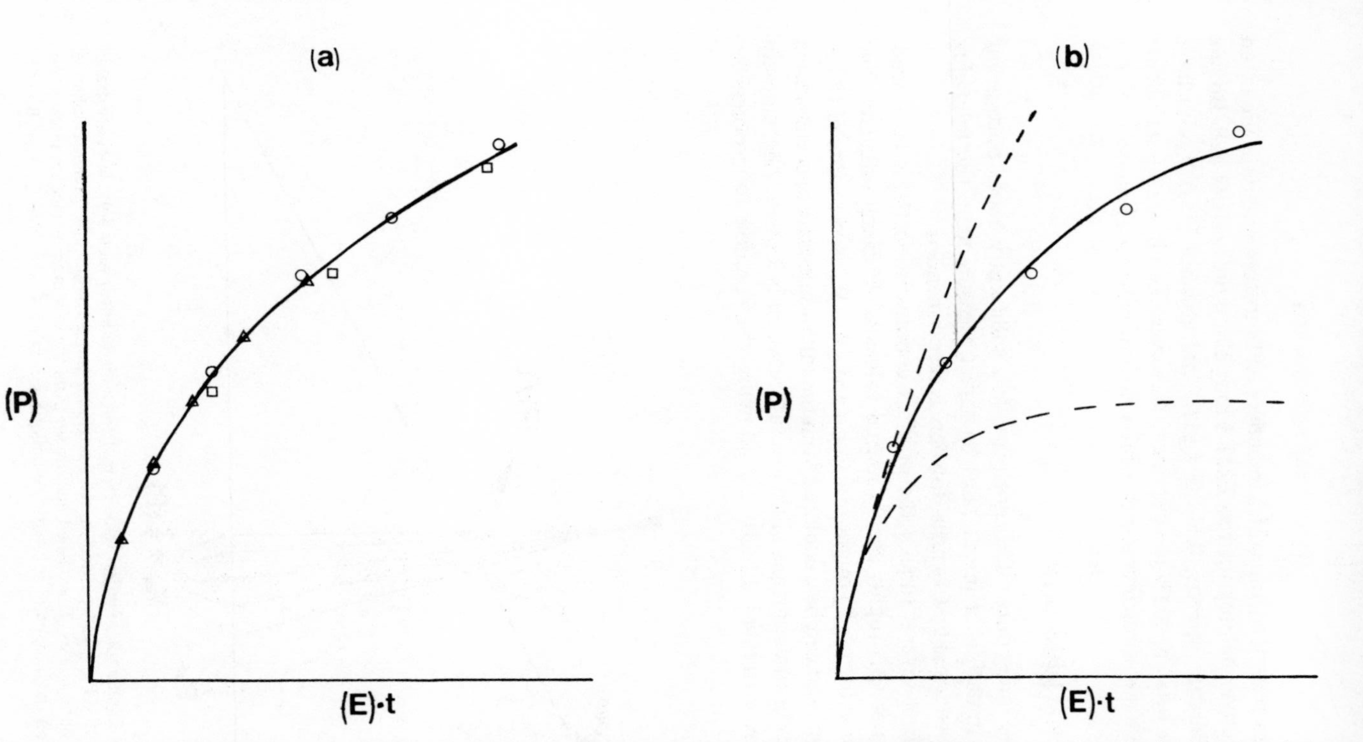

Figure 10.8 Detection of enzyme inactivation during assay. The concentration of product for an ATPase-catalysed reaction is plotted against time multiplied by amount of enzyme. (a) Curve plotted for three different amounts of enzyme (△, ○, □) where the enzyme is stable in the assay. (b) Curves calculated for three different amounts of enzyme in ratio 1 : 2 : 4 assuming that unimolecular decomposition of enzyme is responsible for decrease in rate. Taken from Selwyn (1965).

sometimes present as impurities in buffers and substrates, and the inclusion of an effective chelator such as EDTA may eliminate activity loss. Bovine serum albumin (approx. 0.1–1.0 mg/ml) and glycerol (5–20 % v/v) have also been used to stabilize enzymes. However the choice of stabilizing agent is often empirical and only trial will show which is effective.

Estimation of initial rates

Where the time course does not provide a sufficiently linear portion for reliable estimation of initial rates by direct measurement, the following method, proposed by Cornish-Bowden, is recommended.

At each of several times t the product concentration (P) is measured, about five or six roughly evenly spaced values of (P) being sufficient. The maximum value of (P) ought not to exceed $\frac{1}{2}P_{\infty}$. P_{∞} is the value of (P) at equilibrium and can be calculated from the initial substrate concentrations if the equilibrium constant of the overall reaction is known. Alternatively P_{∞} can be estimated directly by allowing the reaction to proceed for

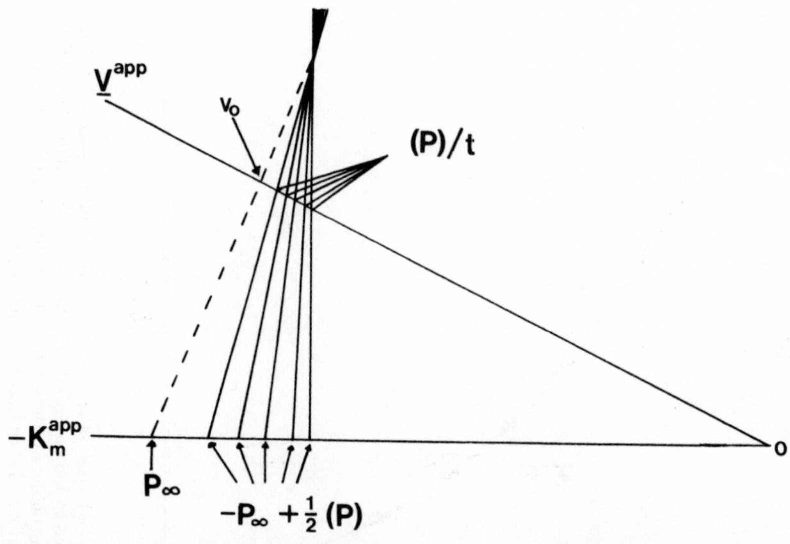

Figure 10.9 Estimation of initial velocity, v_0, from a direct linear plot. Solid lines are drawn from values of $-P_{\infty}+\frac{1}{2}(P)$ plotted on the horizontal axis through corresponding values of $(P)/t$ on the oblique axis. The dashed line drawn from the common or median intersection point to P_{∞} on the horizontal axis cuts the oblique axis at v_0. Taken from Cornish-Bowden (1975).

sufficient time to approach equilibrium. Ordinate and abscissa axes are constructed at about 20–30° to each other as in figure 10.9 and marked V^{app} and $-K_m^{app}$ respectively. For each $(P)/t$ observation, $-P_\infty+\frac{1}{2}(P)$ is plotted on the K_m^{app} axis and the corresponding values of $(P)/t$ are marked off on V^{app} axis. Straight lines are then drawn through each pair of points as in a direct linear plot. These will form a cluster of intersections usually in the first quadrant but occasionally in the third quadrant. Finally a straight line is drawn from the middle of the intersection cluster through P_∞ on the K_m^{app} axis. This line cuts the V^{app} axis at v_0, the initial velocity (see figure 10.9).

The method is based on equation (10.6).

$$V^{app}t = K_m^{app} \cdot \ln\frac{P_\infty}{P_\infty-(P)} + (P) \tag{10.6}$$

which is similar to equation (4.48), the integrated form of the simple Michaelis–Menten equation. In fact, in the single substrate irreversible case with no product inhibition $P_\infty = S_0$, and K_m^{app} and V^{app} are identical to K_m and V respectively. However, equation (10.6) is applicable to a number of more realistic circumstances including competitive product inhibition and reversibility. In such cases K_m^{app} and V^{app} are complex constants whose meaning will depend on the mechanism and may even take on negative values. Differentiation of equation (10.6) with respect to t gives

$$v_0 = \frac{V^{app}P_\infty}{P_\infty + K_m^{app}} \tag{10.7}$$

and so V^{app} and K_m^{app}, whatever their meanings, can be used to estimate the initial velocity v_0. By rearranging equation (10.6) to

$$\frac{V^{app}}{(P)/t} - \frac{K_m^{app}}{(P)/\ln[P_\infty/(P_\infty-(P))]} = 1$$

it can be seen that the median of the intersections of a direct linear plot of $(P)/t$ on the V^{app} axis against $-(P)/\ln[(P_\infty/(P_\infty-(P))]$ on the K_m^{app} axis gives the point (V^{app}, K_m^{app}). A straight line drawn from this point to $-P_\infty$ on the K_m^{app} axis crosses the V^{app} axis at $V^{app}P_\infty/(P_\infty+K_m^{app})$. From equation (10.7) this is seen to be v_0.

The graphical application of the method as described above uses $-P_\infty+\frac{1}{2}(P)$ as an approximation to $-(P)/\ln[P_\infty/(P_\infty-(P))]$ which is easier to calculate and gives negligible errors in v_0 provided that $(P)/P_\infty < 0.5$. The method is surprisingly insensitive to small inaccuracies in P_∞ and can also be applied with good results to situations which are not described by equation (10.6), such as mixed product inhibition and inactivation of enzyme during assay. Cornish-Bowden's method is as easily applied to continuous as to discontinuous assays. However where a large number of discontinuous assays is to be performed under varying conditions, e.g. of substrate or inhibitor concentrations, determination of a progress curve of several points for each assay may involve an impracticably large number of experiments. In such cases it is usually worthwhile to try to linearize the initial portion of the progress curve by altering the fixed parameters of the assay system. Such experiments should be carried out under conditions of the variable giving the highest and lowest enzyme activities in the experimental series. If the progress curves are linear for these rates it is usually safe to assume that they will be linear over that time period for all assays measuring intermediate activities. Single-point assays during the linear portion should then suffice. As a further check, proportionality of rate to enzyme concentration should also be determined for the highest and lowest enzyme activities.

Techniques of assay

(a) *Direct assays*

The methods used for measuring the rate of an enzyme-catalysed reaction will depend on the nature of the chemical change and the ingenuity of the investigator. If product and substrate differ in an easily-measured physical property it may be possible to follow the reaction continuously. *Continuous assays* always are to be preferred as they give a better picture of any curvature or irregularity in the time course. Where the reaction cannot be followed continuously, a *discontinuous* or *stopped assay* must be used in which the reaction is terminated at a given time and the reaction mixture analysed. For estimation of initial rates, discontinuous assays are inherently less accurate than continuous assays and should only be used as a last resort. It is especially important that the control experiments described below and earlier in this chapter be carried out to confirm that a true estimate of the initial rate is obtained. Usually enzyme-catalysed reactions are stopped by lowering or raising the pH to a value where the enzyme is known to be inactive or by precipitating the protein. However it is always best to check (by running a "zero-time" assay) that the reaction is truly stopped by the quenching method. When using an assay method for the first time it is also essential to check for the presence of an uncatalysed rate. This is best done by performing a "boiled enzyme" assay which is run by substituting boiled enzyme (inactivated by heating at 100° for 15 min) for the enzyme preparation. Finally the stability of the product (or substrate) whose change is being followed must be established under the assay conditions. Instability may be due to chemical side-reactions or to a reaction catalysed by another enzyme present in an impure preparation. For example the assay of NAD^+-dependent dehydrogenases in a crude cell extract may be complicated by the presence of an NADH oxidase. If the reaction is followed in the direction NADH → NAD^+ this will lead to high blank rates. If the change being monitored is in the NAD^+ → NADH direction, the presence of an NADH oxidase will result at best in artefactual curvature of the time course and consequent underestimation of the true rate. At worst the relative activity of the contaminating enzyme may be high enough to completely mask the presence of the dehydrogenase. In such cases it may be necessary to incorporate an inhibitor of the NADH oxidase (such as cyanide) or to purify the enzyme.

(b) *Coupled assays*

It frequently happens that the substrates and products do not differ sufficiently in an easily measured property to make direct continuous assay possible. In such cases a coupled assay is often used. This consists in linking the reaction under study to one or more subsequent reactions which can be followed continuously.

For example, glycerol kinase catalyses the phosphorylation of glycerol by ATP; in this reaction there is no appreciable difference in the UV/visible spectra of substrates and products. However in the presence of glycerol phosphate dehydrogenase and NAD, the product L-α-glycerol phosphate is oxidized to dihydroxyacetone phosphate, and NADH is formed; the NAD → NADH conversion can be followed spectrophotometrically at 340 nm (figure 10.10a). The assay is usually run at pH > 9.0 to drive the coupling reaction to the right.

Coupled systems need not be restricted to one auxiliary enzyme. Many ATP-dependent kinases can be assayed spectrophotometrically in the presence of phosphoenolpyruvate, pyruvate kinase, NADH and lactate dehydrogenase. The pyruvate kinase catalyses the reaction of ADP, formed in the primary reaction, with phosphoenolpyruvate giving pyruvate which is then reduced to lactate by NADH in the presence of lactate dehydrogenase, the NADH → NAD^+ conversion being followed as a decrease in absorbance at 340 nm. The system is shown for glycerol kinase in figure 10.10b. Because of the favourable equilibrium of the second

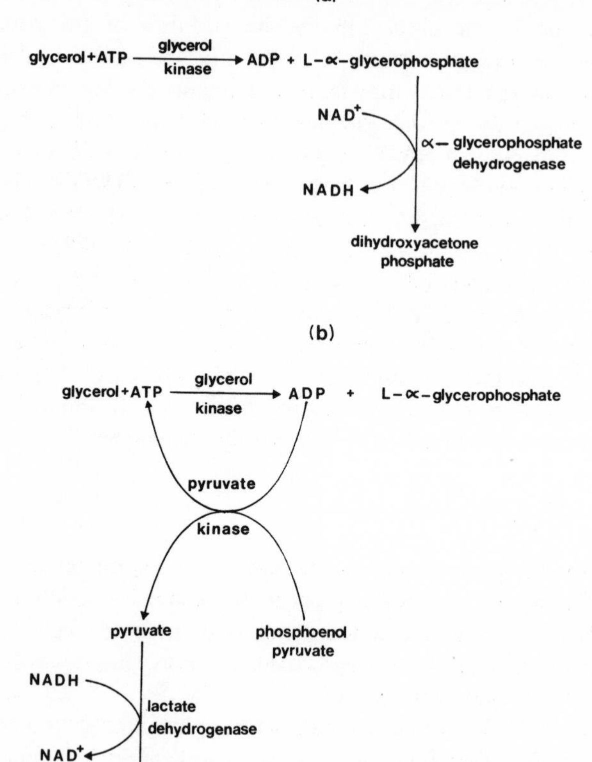

Figure 10.10 Coupled assays for glycerol kinase.

coupling reaction, this assay is not restricted to high pH values. Also, the product ADP is recycled to ATP so that problems that may arise from depletion of ATP are circumvented.

Non-enzymic reactions are also used in coupled assays. The condensation of oxaloacetate with acetyl coenzyme A catalysed by citrate synthase yields citrate and coenzyme A. In the presence of 5,5′-dithio*bis*(2-nitrobenzoate)(DTNB) the thiol group of coenzyme A reacts rapidly to form a disulphide and 5-thio-2-nitrobenzoate anion which absorbs strongly at 412 nm so the reaction can be followed spectrophotometrically (figure 10.11).

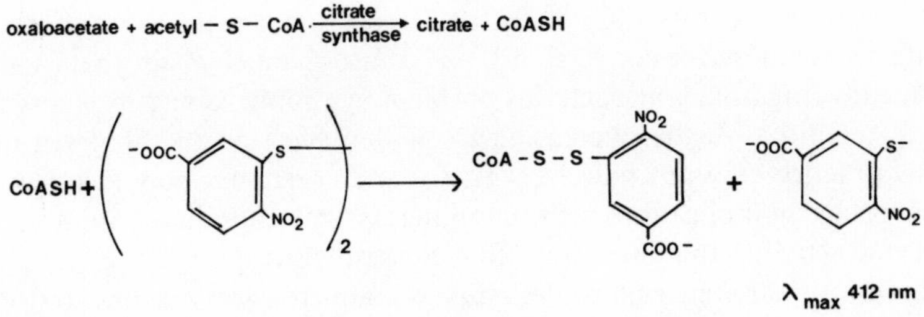

$$\text{oxaloacetate} + \text{acetyl}-\text{S}-\text{CoA} \xrightarrow[\text{synthase}]{\text{citrate}} \text{citrate} + \text{CoASH}$$

$$\text{CoASH} + \left({}^{-}\text{OOC-C}_6\text{H}_3(\text{NO}_2)\text{-S}- \right)_2 \longrightarrow \text{CoA}-\text{S}-\text{S}-\text{C}_6\text{H}_3(\text{NO}_2)\text{-COO}^{-} + {}^{-}\text{OOC-C}_6\text{H}_3(\text{NO}_2)\text{-S}^{-}$$

λ_{max} 412 nm

Figure 10.11 Coupled assay for citrate synthase.

A simple coupled system can be described by

$$A \xrightarrow{E_1} B \xrightarrow{E_2} C$$

where (A) is the concentration of the substrate of the *primary enzyme* E_1. (B) is the concentration of the product of the primary reaction which is converted to C by the action of the *coupling enzyme* E_2, (C) being the concentration whose change is monitored in the assay. Initially (at time zero) (B)=0 and as (B) builds up to its steady-state value, $(B)_{SS}$, $d(C)/dt$ will increase and reach a constant value when $(B) = (B)_{SS}$. Thus the appearance of C will typically show a lag as seen in figure 10.12. In order

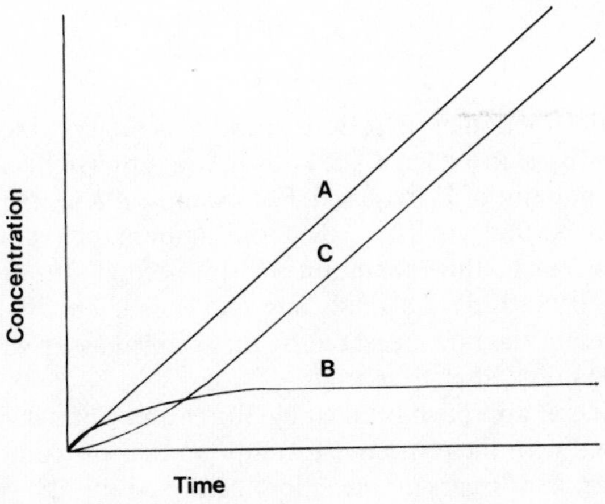

Figure 10.12 Progress curves for reactants in a coupled assay $A \xrightarrow{E_1} B \xrightarrow{E_2} C$. Adapted from McClure (1969).

for the observed rate $d(C)/dt$ to truly reflect $-d(A)/dt$, the rate of the primary reaction, the conversion B → C must be much faster than A → B. Some investigators approach this problem by simply using a "vast excess" of E_2 such that further increasing E_2 has no effect on the observed rate. This practice may not only be wasteful and costly but may also lead to unexpected complications if the coupling enzyme contains contaminating enzyme activities that interfere with the overall assay.

In a rational approach to designing a coupled assay one has to decide what difference between $d(C)/dt$, the measured rate, and $-d(A)/dt$, the true rate, one is willing to tolerate, and also the length of the time lag required for (B) to reach that fraction of $(B)_{SS}$ that will give the desired observed rate. McClure has proposed a method for doing this which relies on the assumption that the primary reaction is zero order and irreversible. Zero order kinetics can be reasonably assumed to hold if only a small fraction of A is converted to B during the period of observation; irreversibility is assured by the continuous removal of B by the coupling enzyme. The method also assumes that the coupling reaction is irreversible and is first order with respect to B. Irreversibility is approximated if less than 5% of the total extent of the second reaction is followed. If the coupling enzyme follows the Michaelis–Menten equation $-d(B)/dt = V_B(B)/(K_B + (B))$, where V_B is the maximum velocity and K_B the Michaelis constant of E_2 for B, the first order condition will be reasonably met if $(B) < 0.05K_B$, for under such conditions the Michaelis–Menten equation for B reduces to $-d(B)/dt = V_B(B)/K_B$. McClure's equation is

$$V_B = -\frac{K_B \cdot \ln(1 - F_B)}{t^*} \tag{10.8}$$

where F_B is the fraction of $(B)_{SS}$ desired and t^* is the time required for (B) to attain $F_B \cdot (B)_{SS}$. Provided K_B is known, equation (10.8) can be used to calculate the amount of E_2 required. For example if $K_B = 0.5$ mM and we desire (B) to be 98% of $(B)_{SS}$ then the amount of coupling enzyme necessary to reach this condition in 15 seconds ($t^* = 0.25$ min) is $-(0.5)\ln(1 - 0.98)/0.25 = 7.8$ μmols/min/ml. Thus if 7.8 U/ml of coupling enzyme is present the rate measured by the coupled assay will be in error by no more than 2% after 15 seconds.

A more general approach is taken by Storer and Cornish-Bowden who have dispensed with the restrictive assumption that the coupling reaction is first order and treat the reaction catalysed by E_2 as following Michaelis–Menten kinetics. They show that the time required for v_2, the velocity of the second reaction, to reach a given value at fixed values of v_1,

V_B and K_B is given by

$$t = \frac{V_B K_B}{(V_B - v_1)^2} \cdot \ln \frac{v_1(V_B - v_2)}{V_B(v_1 - v_2)} - \frac{K_B v_2}{(V_B - v_2)(V_B - v_1)} \quad (10.9)$$

where v_1 is the velocity of the primary reaction. Equation (10.9) can be written in the form

$$t = \frac{\phi K_B}{v_1} \quad (10.10)$$

ϕ is a dimensionless quantity which is a function only of v_2/v_1 (this corresponds to F_B in McClure's treatment) and v_1/V_B; table 10.1 gives some values of ϕ at various values of v_2/v_1 and v_1/V_B. To use this table ϕ is calculated from equation (10.10) with t as the desired lag time and v_1 as the maximum expected rate of the primary reaction. The ϕ value is then located in the table under the appropriate column of v_2/v_1, interpolating if necessary, and the corresponding value of v_1/V_B is read off the left-hand column. From V_B the required amount of coupling enzyme can be obtained.

Coupled assays are necessarily more complex than direct assays and consequently are more liable to interference and misinterpretation. The added substrates and cofactors for the coupling systems should not interfere with the enzyme under study. Possible inhibition of coupling enzymes by substrates of the primary enzyme can be easily tested. Finally, linearity of the progress curve in a coupled assay should not be taken as evidence that sufficient coupling enzyme is present. If the build-up of (B) to its steady-state concentration is slow the progress curve may appear

Table 10.1 Values of ϕ for calculation of lag times. ϕ values are given for $v_2/v_1 = 0.95$ and 0.99. For details see text. Adapted from Storer and Cornish-Bowden (1974).

v_1/V_B	$\phi(v_2/v_1 = 0.95)$	$\phi(v_2/v_1 = 0.99)$
0.0	0.00	0.00
0.1	0.35	0.54
0.2	0.81	1.31
0.3	1.46	2.42
0.4	2.39	4.12
0.5	3.80	6.86
0.6	6.08	11.7
0.7	10.2	21.4
0.8	18.7	45.5
0.9	42.8	141

substantially linear but will not necessarily reflect the true rate of the primary reaction. Reliability of a coupled assay system should always be confirmed by ensuring that the observed rate is directly proportional to the concentration of the primary enzyme.

(c) *Methods*

Spectrophotometry. At a single wavelength, the absorbance or optical density, A, of a sample can be defined as

$$\mathrm{A} = \log\frac{I_0}{I} = \varepsilon lc \tag{10.11}$$

where I_0 is the intensity of the incident light or the light through a reference, I is the light intensity through the sample, l is the length of light path through the sample and c the sample concentration. The proportionality constant ε is known as the absorption or extinction coefficient and is dependent on the nature of the absorbing species and the wavelength.

If the extinction coefficients of product and substrate differ sufficiently at a given wavelength, it will be possible to follow the reaction spectrophotometrically. For example, fumarase activity may be assayed at 300 nm due to the presence of the conjugated double bond system of fumarate which is absent in malate.

$$\mathrm{^-O{-}C({=}O){-}CHOH{-}CH_2{-}C({=}O){-}O^-} \underset{}{\overset{\text{fumarase}}{\rightleftharpoons}} \mathrm{^-O{-}C({=}O){-}CH{=}CH{-}C({=}O){-}O^-}$$

Because it is convenient and non-destructive, spectrophotometry is the most commonly used method of continuous assay.

Use of spectrophotometry depends on observance of *Beer's Law* (equation 10.11) which states that the absorbance of a sample is directly proportional to the concentration of the chromophore(s), the path length being held constant. It is therefore essential to confirm that Beer's Law is followed by the system under study. Deviations occasionally arise at high concentrations of absorbing substance due to chemical association, hydrogen bonding, ion pair formation or solvation effects. Instrumental factors that can contribute to deviation are stray light, which is light from wavelengths outside those of spectral interest, resolution and wavelength and absorbance calibration. Turbid samples or solutions which contain dust particles will scatter the incident light. Another cause is air bubbles resulting from dissolved air in the sample.

If Beer's Law is obeyed, the sensitivity of a spectrophotometric assay will depend only on the extinction coefficient of the chromophore, the path length and the sensitivity of the instrument. The most frequently used chromophore in spectrophotometric assays is NADH, which has an extinction coefficient of 6200 $M^{-1} cm^{-1}$ at 340 nm. Thus, using an instrument able to detect an absorbance change of 0.005 min^{-1} and a 10 mm cuvette containing 1.0 ml solution, as little as 0.8 mUnits of enzyme activity can be assayed. This lower limit can be extended by using cuvettes with a longer light path.

Spectrophotometric assays can sometimes be applied to the assay of enzymes whose natural substrates and products have no appreciable differences in extinction coefficients. If the specificity of the enzyme is sufficiently broad, it may accept a synthetic *chromogenic* substrate. For example the absorbance of *p*-nitrophenolate ion at pH 7.2 is appreciable at 400 nm, whereas the extinction coefficient of *p*-nitrophenyl esters is low at that wavelength. Many proteases also act as esterases, and so chromogenic substrates of this type are often used for enzyme assay as well as for kinetic studies.

Fluorimetry. Fluorescence is the characteristic re-emission of absorbed light by a sample. The optimum wavelength for the incident (or *exciting*) light usually corresponds to the absorbance maximum of the compound; fluorescence *emission* always occurs at a higher wavelength. Because fluorescence emission from small molecules is radiated equally in all directions, fluorescence measurements are usually made at right angles to the incident light beam to reduce interference from reflected light and from light transmitted through the sample.

The relationship between the intensity of the incident light, I_0, and the emitted fluorescence, F, is given by

$$F = I_0(1 - 10^{-\varepsilon lc})Q \qquad (10.12)$$

where Q is known as the *quantum efficiency* and has a value between 0 and 1. The quantum efficiency is the fraction of absorbed energy which can be re-emitted as fluorescence and depends upon the electronic structure of the molecule and its environment. For most absorbing molecules Q is very low and relatively few of the many compounds which absorb light will re-emit it as fluorescence. While limiting the general applicability of fluorimetry this also confers upon it useful selectivity. Equation (10.12) shows that the intensity of the emitted light is proportional to that of the incident light; this accounts for the use of highly intense light sources such as xenon arcs in spectrofluorimeters. However, photodecomposition of the sample and light scattering impose upper limits on the intensity of the excitation light beam.

Measurements of absorbance involve comparison of the intensities of two light beams I_0 and I (see equation 10.11), and the lower limit of sensitivity depends upon the accuracy with which two similar intensities can be compared. Estimation of fluorescence emission (equation 10.12) involves a single measurement against a low background intensity. This accounts for the more than 100-fold greater sensitivity of fluorescence over absorbance measurements. If the optical density (εlc) is small (< 0.02–0.05) equation (10.12) reduces to

$$F = I_0(2.3\ \varepsilon lc)Q$$

This predicts a linear relationship between fluorescence intensity and concentration but also places an upper limit on the useful concentration range. For NADH an optical density of 0.05 in a 1 cm cuvette corresponds to about 8 μM. As this is also the lower limit for convenient spectrophotometric measurements, the two techniques are seen to be complementary.

NADH, which absorbs at 340 nm and emits at 460 nm, is the most commonly used fluorophore in enzyme studies. Coupled fluorimetric assay systems can be set up in the same manner as in spectrophotometry. If neither substrate nor product is fluorescent it may be possible to use a *fluorogenic* substrate, analogous to the chromogenic substrates described in the section on spectrophotometry. Fluorogenic substrates frequently used for the assay of hydrolytic enzymes are based on esters of phenolic compounds such as 1- and 2-naphthol and of the highly fluorescent umbelliferone (7-hydroxy-coumarin) and its derivatives. The high sensitivity of fluorescence measurements has been found useful for detecting enzyme activities in small amounts of tissue but it is absolutely required if one wishes to determine small ($<10^{-6}$ M) K_m values. As discussed elsewhere in this chapter accurate determination of K_m requires substrate concentrations in the region of the K_m itself. For a K_m of 10^{-6} M one would need to measure concentration changes as small as 5×10^{-8} M which is below the range of spectrophotometry for most chromophores.

The great sensitivity of fluorimetry relies on low blank readings, and greater care must be taken to exclude fluorescent impurities from reagents and buffers than in spectrophotometry. High blanks can also result from light scattering due to dust particles and scratched or dirty cuvettes. The water should be glass distilled and it may be necessary to filter all solutions through membrane filters. A further interference results from *quenching*. Iodide, for example, is a notorious quencher and its inclusion in a fluorimetric assay must be avoided. However any substance which absorbs the incident light will reduce its intensity along the light path and hence also the intensity of the fluorescence. This includes the fluorophore itself and results in the so-called "inner-filter" effect or concentration quenching; this places a further restriction on the upper concentration limit analytically amenable to fluorescence measurement. Quenching of the emitted light is seldom a problem because fluorescence occurs in the visible region and interfering substances can be detected visually.

As mentioned above both the intensity and the wavelength of fluorescence emission can depend on the environment of the fluorophore. Thus it is often found that the fluorescence of ligands is *enhanced* (and shifted) when binding to proteins occurs. Tryptophan residues are also fluorescent and enhancement or quenching can occur as a result of conformational alterations in the protein. While not strictly relevant to rate measurements, such studies have provided evidence on the environment and strength of binding and the nature and extent of conformational changes in proteins.

pH-Potentiometry. Because the activity of enzymes is sensitive to changes in hydrogen ion concentration, it is desirable to maintain constant pH in enzyme assays. This is usually done by using buffered solutions. Some reactions, e.g. those catalysed by many kinases and hydrolases, themselves involve net production or consumption of hydrogen ions. The resulting pH change can be used to assay such reactions by running them in *unbuffered medium*. Sensitive instruments have been constructed which give full scale deflections corresponding to very small pH changes. Alternatively the progress of the reaction can be followed by maintaining constant pH by the addition of base or acid as appropriate. The *rate* of addition will then reflect the time course of the reaction. The device most commonly used in such studies is the *pH-stat*.

A pH-stat consists of a pH-electrode train in a thermostatted reaction vessel, and a pH meter which is connected to a motor-driven burette via a controller. The contents of the vessel are protected from atmospheric CO_2 by a blanket of nitrogen. The controller is adjusted to a preset pH value, i.e. the desired pH for the reaction under study. The pH of the reaction mixture is also adjusted to the desired value. The reaction is started and as the pH of the reaction mixture drifts from the preset value, the controller actuates the motor-driven burette so that base or acid is delivered to the reaction vessel until the pH is restored to its original value. In modern instruments the rate of addition is proportional to the amount by which the pH of the reaction mixture deviates from its preset value. By recording the amount of titrant added as a function of time, the pH-stat provides a direct and continuous record of the progress curve.

The sensitivity of pH-stat assays is around 50–100 times less than spectrophotometric assays using NADH. This is partially due to the difficulties of working with CO_2-free solutions of standardized base more dilute than 10 mM; also reaction volumes of much less than 1 ml are not compatible with commercially available electrodes. Nevertheless pH-stat assays are simple and convenient and have been applied to coupled systems. For example an assay for phosphohexose isomerase (PHI) uses a pH-stat to measure the rate of production of protons from the coupling reaction catalysed by phosphofructokinase (PFK):

$$\text{glucose-6-phosphate}^{2-} \xrightarrow{\text{PHI}} \text{fructose-6-phosphate}^{2-}$$

$$\text{ATP}^{4-} + \text{fructose-6-phosphate}^{2-} \xrightarrow{\text{PFK}} \text{fructose-1,6-diphosphate} + \text{ADP}^{3-} + \text{H}^{-}$$

This assay is simpler than the coupled spectrophotometric assay which requires two additional enzymes and their associated cosubstrates.

Polarography and the oxygen electrode. In polarographic measurements a potential difference is applied across a circuit consisting of a polarizable electrode and a reference electrode. The substance to be detected is electrolytically reduced or oxidized at the polarizable electrode and the resulting current flow between the two electrodes is measured. The currents actually used are of the order of μamps, or sufficiently small to ensure that the amount of material electrolysed is usually insignificant compared to its bulk concentration. Under suitable conditions the limiting or diffusion current will be directly proportional to the concentration of the electroactive species. The system can then be calibrated and used to assay enzyme activities involving that species as a substrate or product.

The most common application of polarographic assays is in the measurement of the concentration of dissolved oxygen. For this the Clark oxygen electrode is the most convenient, but other oxygen electropodes are available. The Clark electrode employs platinum as the polarizable electrode with a silver–silver chloride reference. The electrode is isolated from the bulk solution by a thin membrane of plastic film (polyethylene, teflon or sandwich wrap) through which oxygen, but not other electro-reducible substances, can diffuse. The bulk solution must be stirred to replenish oxygen at the membrane surface but the system must also be protected from diffusion of atmospheric oxygen. The response time of the Clark electrode is limited by the rate of diffusion of oxygen through the membrane; however this is rarely a problem in steady-state kinetic measurements.

Polarographic assays are not limited to oxygen measurement. Coenzyme A, but not its S-acyl derivatives, gives an anodic polarographic wave at a dropping mercury electrode in deoxygenated solutions, the magnitude of which is proportional to coenzyme A concentration. Thus the reactions catalysed by malate synthase

$$\text{acetyl CoA} + \text{glyoxylate} \rightarrow \text{malate} + \text{coenzyme A}$$

and citrate synthase

$$\text{acetyl CoA} + \text{oxaloacetate} \rightarrow \text{citrate} + \text{coenzyme A}$$

can be followed polarographically. The polarographic method has certain advantages over spectrophotometric assays. For example, the above reactions can also be followed spectrophotometrically using the chromophore produced by reaction of DTNB with the —SH group of coenzyme A (see figure 10.11). However spectrophotometric assays cannot easily be used with turbid solutions or in the presence of high background absorbance. Furthermore, malate synthase is inactivated by DTNB as a result of reaction with -SH groups on the enzyme itself; similarly the allosteric properties of citrate synthases from some sources are modified by reaction of the enzyme with DTNB. Polarographic assays do not suffer from these disadvantages and are also quite sensitive, rates as low as 10^{-9} moles min^{-1} being measurable. Other electroactive species include lipoamide, pyruvate and $NAD(P)^+$. As the method is continuous it should be possible to devise coupled polarographic assays for many enzyme-catalysed reactions which do not themselves involve electroactive substrates.

Radiochemical methods. Radiochemical assays are the most sensitive available to the enzymologist. Sensitivity is limited ultimately by the specific radioactivity of the isotope. However for a number of reasons, not the least being expense, it is not usual to use radiolabelled compounds as substrates at the high specific radioactivities as supplied and they are usually diluted with unlabelled ("cold") carrier. Normally fractions of a picomole can easily be detected and sensitivities higher than 10^{-15} mole are possible. Turbidity does not usually present a problem and crude cell extracts can be assayed without difficulty. A further advantage of radiochemical assays is that very small volumes can be used, the lower limit being determined mainly by the accuracy of pipetting. A wide variety of radioactively labelled substrates is available from commercial sources, and preparative methods have been published for most of those that are not. Thus radiochemical assays can be devised for most enzyme-catalysed reactions.

Radiochemical assays involve stopping the reaction at specified times, separating labelled product from substrate and measuring the radioactivity, usually by liquid scintillation counting. Each of these steps introduces its own problems. As for any stopped assay the efficacy of the method used to terminate the reaction must be confirmed by running a zero-time blank. The stability of the labelled materials in the terminated reaction mixture must also be determined.

The separation method will depend on the physical and/or chemical differences between substrate and product. One of the simplest is solvent extraction. The procedure can be speeded up by centrifuging or by freezing the aqueous phase. If a suitable organic phase can be used it may be possible to extract the labelled substance directly into the scintillant. A further simplification may be possible if a tritium-labelled substrate is used because the energy of β-emission from tritium is so weak that it is effectively quenched in aqueous solution. In some cases it may be possible to run the assay, extract and count directly in the scintillation vial. Such an assay has been employed for catechol-*O*-methyl transferase, using (^{3}H-methyl)-S-adenosylmethionine as the methyl donor and extracting the *O*-methylated product into isoamyl alcohol–toluene containing scintillator. Decarboxylases are frequently assayed by terminating the reaction with acid and selectively absorbing the released $^{14}CO_2$ on to a filter wick soaked in base. Chromatography and selective adsorption are also commonly used separation procedures. Whatever separation method is used, its efficiency should be checked by running standard mixtures of substrate and product through the assay.

The determination of radioactivity in the separated product or substrate is usually carried out by liquid scintillation counting. In this technique the energy of the ionizing radiation is transferred to the solvent and thence to the phosphor in the scintillant which emits a burst of light photons (scintillation). This is detected by a photomultiplier and is registered as a *count* by the associated circuitry. The *efficiency* is defined as the ratio of observed *counts per minute* to the actual *disintegrations per minute* and should be determined if the enzyme activity is to be expressed in molar terms. Various factors can affect the photon yield from a radioactive source. The most common of these is called *chemical quenching* which is due to interference with the energy transfer between solvent and phosphor by components of the sample, possibly including the substance itself. Determination of counting efficiency requires quench correction and several methods for this are available.

Colour quenching is identical to the quenching of fluorescence emission and results from absorption of the scintillation emission by the sample. The use of "secondary solutes" which shift the wavelength of fluorescence emission is sometimes advantageous in such cases. If the sample is very highly coloured it may be necessary to decolourize it with bleaching agents. In extreme cases, sample oxidation, in which the sample is burned to CO_2 and H_2O (for ^{14}C or ^{3}H) may be the only alternative.

The major disadvantage of radiochemical assays is that they are inherently discontinuous and require separation of labelled product from labelled substrate. As in any stopped assay it is necessary to establish conditions such that linear rates are being measured and the necessity for carrying out appropriate controls described elsewhere in this chapter cannot be overstressed. A further complication is the instability of radiolabelled compounds; the most obvious cause is natural isotopic decay. This is not usually a problem with long-lived isotopes such as ^{14}C (half-life 5730 years) or ^{3}H (12.3 years) but can be significant with ^{32}P (14.3 days) which decays to sulphur. Radiation energy will also cause self-radiolysis and chemical instability (resulting from generation of free radicals due to interaction of ionizing radiation with the solvent). As it is the *labelled* compound which is used as a tracer, it is essential that the absence of labelled impurities be checked.

Derivation of rate equations

To derive the overall rate equation from a mechanism containing n enzyme forms, rate equations for all individual enzyme intermediates but one are written and equated to zero (steady-state assumption). This gives a set of $n-1$ linear simultaneous equations in n unknowns. These are then solved to give expressions for each enzyme form in terms of rate constants, concentrations of substrates and modifiers and the concentration of one chosen enzyme intermediate. (In theory it does not matter which enzyme form is used but it is usually convenient to choose the form(s) which appear in the rate expression for product formation.) The expressions are then substituted into the conservation equation for enzyme which is solved to give an expression for the chosen form. Substitution of this expression into the rate equation for product formation then gives the overall rate equation. The method is illustrated below for an ordered mechanism for the irreversible reaction $A + B \rightarrow P$:

$$E \underset{k_{-1}}{\overset{k_1(A)}{\rightleftharpoons}} EA \underset{k_{-2}}{\overset{k_2(B)}{\rightleftharpoons}} EAB \xrightarrow{k_3} E + P$$

The rate of product formation is

$$\frac{d(P)}{dt} = k_3(EAB) \tag{10.13}$$

The rate equations for (EA) and (EAB) are

$$\frac{d(EA)}{dt} = 0 = k_1(E)(A) + k_{-2}(EAB) - (k_{-1} + k_2(B))(EA) \tag{10.14}$$

and

$$\frac{d(EAB)}{dt} = 0 = k_2(EA)(B) - (k_{-2} + k_3)(EAB) \tag{10.15}$$

From equation (10.15)

$$k_2(EA)(B) = (k_{-2} + k_3)(EAB)$$

and

$$(EA) = \frac{(k_{-2} + k_3)(EAB)}{k_2(B)} \tag{10.16}$$

From equations (10.14) and (10.16)

$$(E) = \left(\frac{k_{-1}(k_{-2} + k_3)}{k_2(B)} + k_3\right)\frac{(EAB)}{k_1(A)} \tag{10.17}$$

The conservation equation is

$$E_0 = (E) + (EA) + (EAB) \tag{10.18}$$

Substitution of equations (10.16) and (10.17) in (10.18) gives, after factoring out (EAB) and rearranging,

$$(EAB) = \frac{E_0}{\dfrac{k_{-1}(k_{-2} + k_3)}{k_1 k_2(A)(B)} + \dfrac{k_3}{k_1(A)} + \dfrac{k_{-2} + k_3}{k_2(B)} + 1} \tag{10.19}$$

and the final rate equation is obtained by substituting equation (10.19) into (10.13):

$$\frac{d(\mathrm{P})}{dt}=\frac{k_3\mathrm{E}_0}{\dfrac{k_{-1}(k_{-2}+k_3)}{k_1k_2(\mathrm{A})(\mathrm{B})}+\dfrac{k_3}{k_1(\mathrm{A})}+\dfrac{k_{-2}+k_3}{k_2(\mathrm{B})}+1} \tag{10.20}$$

or

$$\frac{d(\mathrm{P})}{dt}=\frac{V}{\dfrac{K_{\mathrm{AB}}}{(\mathrm{A})(\mathrm{B})}+\dfrac{K_{\mathrm{A}}}{(\mathrm{A})}+\dfrac{K_{\mathrm{B}}}{(\mathrm{B})}+1} \tag{10.21}$$

where V, K_{AB}, K_{A} and K_{B} are defined according to equation (10.20).

While this method will always give the correct rate equation, it becomes extremely tedious when applied to complex mechanisms and errors can easily occur. In order to obviate this difficulty King and Altman suggested a schematic method for solving the simultaneous equations. To use the method the mechanism is written in such a manner that each enzyme form appears but once. All possible combinations of reaction steps that lead to the formation of a particular enzyme species are written eliminating combinations which form loops or cycles. This is then repeated for each enzyme species. King and Altman's scheme is certainly easier to apply than solving the simultaneous equations by elimination but the process of obtaining all possible combinations can be very tedious and liable to errors of omission.

Orsi has suggested a method based on King and Altman's which is simple and less liable to errors than other methods. It is illustrated by applying it to the mechanism described above with the additional complication of an inhibitor, I, which forms a dead-end complex with free enzyme. The mechanism is written out in King–Altman form, each enzyme species appearing only once.

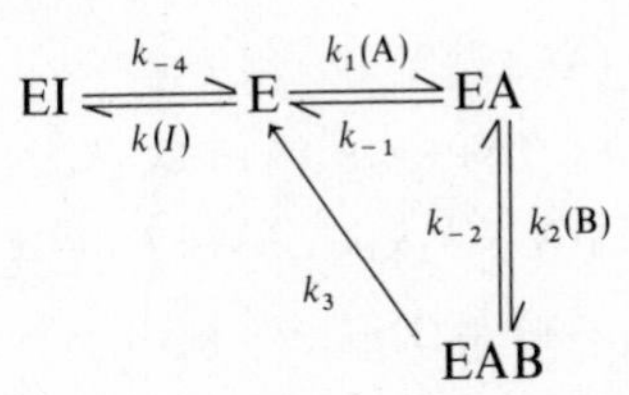

For each enzyme species write the sum of the rate constants (and associated concentration term if any) of all the steps that lead *away* from that species. These are called *Reaction Terms.*

Reaction Terms

For EI: k_{-4}
E: $k_1(\mathrm{A})+k_4(I)$
EA: $k_{-1}+k_2(\mathrm{B})$
EAB: $k_{-2}+k_3$

Using the reaction terms, the distribution equations for each enzyme species can be calculated. The distribution equations show how the total enzyme concentration E_0 is partitioned amongst the different species and can be written in the form:

$$\frac{(\mathrm{E})}{\mathrm{E}_0}=\frac{\{\mathrm{E}\}}{D};\quad \frac{(\mathrm{EA})}{\mathrm{E}_0}=\frac{\{\mathrm{EA}\}}{D};\quad \frac{(\mathrm{EAB})}{\mathrm{E}_0}=\frac{\{\mathrm{EAB}\}}{D};\quad \frac{(\mathrm{EI})}{\mathrm{E}_0}=\frac{\{\mathrm{EI}\}}{D},$$

where $D=\{\mathrm{E}\}+\{\mathrm{EA}\}+\{\mathrm{EAB}\}+\{\mathrm{EI}\}$, the terms in braces being called *Distribution Terms.* Once these have been calculated, the problem is essentially solved.

The Distribution Terms can be expressed as the product of all the Reaction Terms barring the one associated with the enzyme species being considered.

Distribution Terms

For
$$\begin{aligned}
\{\mathrm{EI}\} &= (k_1(\mathrm{A})+k_4(\mathrm{I}))(k_{-1}+k_2(\mathrm{B}))(k_{-2}+k_3)\\
&= \underline{(k_1k_{-1}(\mathrm{A})}+k_1k_2(\mathrm{A})(\mathrm{B})+k_2k_4(\mathrm{B})(\mathrm{I})+k_{-1}k_4(\mathrm{I}))(k_{-2}+k_3)\\
&= \underline{k_1k_2k_{-2}(\mathrm{A})(\mathrm{B})}+\underline{k_2k_{-2}k_4(\mathrm{B})(\mathrm{I})}+k_1k_{-2}k_4(\mathrm{I})\\
&\quad +\underline{k_1k_2k_3(\mathrm{A})(\mathrm{B})}+k_2k_3k_4(\mathrm{B})(\mathrm{I})+k_{-1}k_3k_4(\mathrm{I})\\
\{\mathrm{E}\} &= (k_{-4})(k_{-1}+k_2(\mathrm{B}))(k_{-2}+k_3)\\
&= (k_{-4})(k_{-1}k_{-2}+k_{-1}k_3+\underline{k_2k_{-2}(\mathrm{B})}+k_2k_3(\mathrm{B}))\\
&= k_{-1}k_{-2}k_{-4}+k_{-1}k_3k_{-4}+k_2k_3k_{-4}(\mathrm{B})\\
\{\mathrm{EA}\} &= (k_{-4})(k_1(\mathrm{A})+k_4(\mathrm{I}))(k_{-2}+k_3)\\
&= (k_{-4})(k_1k_{-2}(\mathrm{A})+k_1k_3(\mathrm{A})+k_{-2}k_4(\mathrm{I})+k_3k_4(\mathrm{I}))\\
&= k_1k_{-2}k_{-4}(\mathrm{A})+k_1k_3k_{-4}(\mathrm{A})+\underline{k_{-2}k_4k_{-4}(\mathrm{I})}+k_2k_4k_{-4}(\mathrm{I})\\
\{\mathrm{EAB}\} &= (k_{-4})(k_1(\mathrm{A})+k_4(\mathrm{I}))(k_{-1}+k_2(\mathrm{B}))\\
&= (k_{-4})(\underline{k_{-1}k_1(\mathrm{A})}+k_1k_2(\mathrm{A})(\mathrm{B})+k_{-1}k_4(\mathrm{I})+k_2k_4(\mathrm{I})(\mathrm{B}))\\
&= k_1k_2k_{-4}(\mathrm{A})(\mathrm{B})+\underline{k_{-1}k_4k_{-4}(\mathrm{I})}+\underline{k_2k_4k_{-4}(\mathrm{I})}
\end{aligned}$$

Any term which contains both the forward and reverse constants for the same step constitutes a one-step cycle and must be discarded. Such terms can be advantageously eliminated as the product is expanded. Similarly any multistep cycle (such as $k_1k_2k_3(\mathrm{A})(\mathrm{B})$ in the distribution term for $\{\mathrm{EI}\}$) is redundant. The redundant terms are underlined above. Each individual rate constant term will contain $n-1$ rate constants at least one of which will be associated with a step leading *to* the species being considered.

The expression for (EAB) can now be evaluated from $(\mathrm{EAB}) = \mathrm{E}_0\{\mathrm{EAB}\}/D$ where D is the sum of the Distribution Terms, and substituted into equation (10.13) to give the overall rate equation.

Collecting terms and rationalizing gives

$$v = \frac{V}{\dfrac{K_{\mathrm{AB}}}{(\mathrm{A})(\mathrm{B})}\left(1+\dfrac{(\mathrm{I})}{K_{\mathrm{i}}}\right)+\dfrac{K_{\mathrm{A}}}{(\mathrm{A})}\left(1+\dfrac{(\mathrm{I})}{K_{\mathrm{i}}}\right)+\dfrac{K_{\mathrm{B}}}{(\mathrm{B})}+1}$$

where K_{AB}, K_{A}, K_{B} and V are defined as in equations (10.20) and (10.21), and $K_{\mathrm{i}} = k_{-4}/k_4$. The method can be applied equally well to reversible systems.

BIBLIOGRAPHY

General references

Bio-organic Mechanisms, 2 vols., Bruice, T. C. and Benkovic, S. J. (1966), Benjamin, New York.

Catalysis in Chemistry and Enzymology, Jencks, W. P. (1969), McGraw-Hill, New York.

The Chemical Kinetics of Enzyme Action, Laidler, K. J. and Bunting, P. S. (1973), Clarendon Press, Oxford.

Enzymatic Reaction Mechanisms, Walsh, C. (1979), Freeman, San Francisco.

The Enzymes, 3rd ed., Ed. Boyer, P. D., Academic Press, New York.

Enzymes, Dixon, M., Webb, E. C., Thorne, C. J. R. and Tipton, K. F. (1979), Longman, London.

Enzyme Structure and Function, Fersht, A. R. (1977), Freeman, San Francisco.

Fundamentals of Enzyme Kinetics, Cornish-Bowden, A. (1979), Butterworth, London.

Chapter 1—The Nature of Catalysis

Albery, W. J. and Knowles, J. R. (1976) Evolution of enzyme function and the development of catalytic efficiency, *Biochemistry*, **15**, 5631–5640.

Glasstone, S., Laidler, K. J. and Eyring, H. (1941) *The Theory of Rate Processes*, McGraw-Hill, New York and London.

Gutfreund, H. (1972) *Enzymes: Physical Principles*, Wiley, London.

Hammes, G. G. (1978) *Principles of Chemical Kinetics*, Academic Press, New York.

Hinshelwood, C. N. (1955) *The Kinetics of Chemical Change*, Oxford University Press, London.

Laidler, K. J. (1969) *Theories of Chemical Reaction Rates*, McGraw-Hill, New York.

Chapter 2—Chemical Catalysis

Aldersley, M. F., Kirby, A. J., Lancaster, P. W., McDonald, R. S. and Smith, C. R. (1974) Intramolecular catalysis of amide hydrolysis by the carboxyl group. Rate determining proton transfer from external general acids in the hydrolysis of substituted maleamic acids, *J. Chem. Soc. Perkin* II, 1487–1495.

Bell, R. P. and Higginson, W. C. E. (1949) The catalysed dehydration of acetaldehyde hydrate and the effect of structure on the velocity of protolytic reactions, *Proc. Roy. Soc.* **A**197, 141–159.

Bender, M. L. and Turnquest, B. W. (1957) The imidazole-catalysed hydrolysis of *p*-nitrophenyl acetate, *J. Amer. Chem. Soc.*, **79**, 1652–1655.

Cleland, W. W., O'Leary, M. H. and Northrop, D. B. (Eds.) (1977) *Isotope Effects on Enzyme-Catalysed Reactions*, University Park Press, Baltimore.

Danforth, C., Nicholson, A. W., James, J. C. and Loudon, G. M. (1976) Steric acceleration of lactonization reactions: an analysis of "stereo-population control", *J. Amer. Chem. Soc.*, **98**, 4275–4281.

Fedor, L. R. and Bruice, T. C. (1965) Nucleophilic displacement reactions at the thiolester bond. General base catalyzed hydrolysis of ethyl trifluorothiolacetate, *J. Amer. Chem. Soc.*, **87**, 4138–4147.

Fersht, A. R. and Kirby, A. J. (1980) Intramolecular catalysis and the mechanism of enzyme action, *Chemistry in Britain*, **16**, 136–142.

Hammond, G. S. (1955) A correlation of reaction rates, *J. Amer. Chem. Soc.*, **77**, 334–338.

Hine, J. (1962) *Physical Organic Chemistry*, McGraw-Hill, New York.

Kirby, A. J. and Lancaster, P. W. (1972) Structure and efficiency in intramolecular catalysis. Catalysis of amide hydrolysis by the carboxy group of substituted maleamic acids, *J. Chem. Soc. Perkin* II, 1206.

Kirby, A. J. and Lloyd, G. J. (1974) Intramolecular general base catalysis of intramolecular nucleophilic catalysis of ester hydrolysis, *J. Chem. Soc. Perkin* II, 637.

Kirby, A. J. and Lloyd, G. J. (1976) Structure and efficiency in intramolecular and enzymic catalysis: intramolecular general base catalysis. Hydrolysis of monoaryl malonates, *J. Chem. Soc. Perkin* II, 1753–1761.

Kirby, A. J., McDonald, R. S. and Smith, C. R. (1974) Intramolecular catalysis of amide hydrolysis by two carboxy groups, *J. Chem. Soc. Perkin* II, 1495–1504.

Milstein, S. and Cohen, L. A. (1970) Concurrent general acid and general base catalysis of esterification, *J. Amer. Chem. Soc.*, **92**, 4377–4382.

Milstein, S. and Cohen, L. A. (1970) Rate acceleration by stereopopulation control: models for enzyme action, *Proc. Natl. Acad. Sci. USA*, **67**, 1143–1147.

Milstein, S. and Cohen, L. A. (1972) Stereopopulation control. I. Rate enhancement in the lactonisations of *o*-hydroxyhydrocinnamic acids, *J. Amer. Chem. Soc.*, **94**, 9158–9165.

Page, M. I. (1973) The energetics of neighbouring group participation, *Chem. Soc. Rev.*, **2**, 295–323.

Chapter 3—Protein Structure

Adams, M. J., McPherson, A., Rossmann, M. G., Schevitz, R. W. and Wonacot, A. J. (1970) The structure of the nicotinamide-adenine dinucleotide coenzyme when bound to lactate dehydrogenase, *J. Mol. Biol.*, **51**, 31–38.

Anfinsen, C. B. and Scheraga, H. A. (1975) Experimental and theoretical aspects of protein folding, *Adv. Protein Chem.*, **29**, 205–300.

Blake, C. C. F. (1979) X-ray crystallography of biological macromolecules, in *Companion to Biochemistry*, vol. **2**, ch. 10, Bull, A. T., Lagnado, J. R., Thomas, J. O. and Tipton, K. F. (Eds.), Longmans, London.

Blum, A. D., Smallcombe, S. H. and Baldwin, R. L. (1978) Nuclear magnetic resonance evidence for a structural intermediate at an early stage in the refolding of ribonuclease A, *J. Mol. Biol.*, **118**, 305–316.

Creighton, T. E. (1979) Intermediates in the refolding of reduced ribonuclease A, *J. Mol. Biol.*, **129**, 411–431.

Dickerson, R. E. and Geis, I. (1969) *The Structure and Action of Proteins*, Harper and Row, New York.

Garel, J. R. (1978) Early steps in the refolding reaction of reduced ribonuclease A, *J. Mol. Biol.*, **118**, 331–345.

Gurney, R. W. (1962) *Ionic Processes in Solution*, Dover, New York.

Holbrook, J. J., Liljas, A., Steindel, J. and Rossmann, M. G. (1975) Lactate dehydrogenase, in *The Enzymes*, 3rd ed., vol. **11**, ch. 4, Boyer, P. D. (Ed.), Academic Press, New York.

McCammon, J. A. and Karplus, M. (1980) Simulation of protein dynamics, *Ann. Rev. Phys. Chem.*, vol. **31**.

Metcalfe, J. C. (1974) Nuclear magnetic resonance spectroscopy of proteins, in *Companion to*

Biochemistry, vol. **1**, ch. 3, Bull, A., Lagnado, J. R., Thomas, J. O. and Tipton, K. F. (Eds.), Longmans, London.

Munro, I., Pecht, I. and Stryer, L. (1979) Subnanosecond motions of tryptophan residues in proteins, *Proc. Natl. Acad. Sci. USA*, **76**, 56–60.

Niekamp, C. W., Hinz, H. J., Jaenicke, R., Woenckhaus, C. and Jeck, R. (1980) Correlations between tertiary structure and energetics of coenzyme binding in pig heart lactate dehydrogenase, *Biochemistry*, **19**, 3144–3152.

Phillips, D. C. and Richards, F. M. (Eds.) (1973) *Atlas of Molecular Structures in Biology*, vol. **1**. Ribonuclease-S, Oxford University Press, London.

Ribiero, A. A., King, R., Restivo, C. and Jardetzky, O. (1980) An approach to the mapping of internal motions in proteins. Analysis of ^{13}C NMR relaxation in the bovine pancreatic trypsin inhibitor, *J. Amer. Chem. Soc.*, **102**, 4040–4051.

Richards, F. M. and Wyckoff, H. W. (1971) Bovine pancreatic ribonuclease, in *The Enzymes*, 3rd ed., vol. **4**, ch. 24, Boyer, P. D. (Ed.), Academic Press, New York.

Sdhmid, F. X. and Baldwin, R. L. (1979) Detection of an early intermediate in the folding of ribonuclease by protection of amide protons against exchange, *J. Mol. Biol.*, **135**, 199–215.

Sternberg, M. J. E., Grace, D. E. P. and Phillips, D. C. (1979) Dynamic information from protein crystallography, *J. Mol. Biol.*, **130**, 231–253.

Chapter 4—Simple Enzyme Kinetics

Cornish-Bowden, A. and Eisenthal, R. (1978) Estimation of Michaelis constant and maximum velocity from the direct linear plot, *Biochim. Biophys. Acta*, **523**, 268–272.

Eisenthal, R. and Cornish-Bowden, A. (1974) The direct linear plot, *Biochem. J.*, **139**, 715–720.

Griffiths, J. R. (1978) A more general definition of K_m. *Biochem. Soc. Trans.*, **6**, 258–260.

Haldane, J. B. S. (1930) *Enzymes*, Longmans Green, London.

Kuhn, R. (1923) Saccharase- und Raffinase-wirking des Invertins, *Zeitschr. Physiol. Chem.*, **125**, 28–92.

Reiner, J. M. (1969) *Behaviour of Enzyme Systems*, 2nd ed., Van Nostrand Reinhold, New York.

Segal, H. L. (1959) The development of enzyme kinetics, in *The Enzymes*, 2nd ed., vol. **1**, ch. 1, Boyer, P. D., Lardy, H. and Myrbäck, K. (Eds.), Academic Press, New York.

Tipton, K. F. (1980) Kinetic mechanism and enzyme function, *Biochem. Soc. Trans.*, **8**, 242–245.

Westley, J. (1969) *Enzyme Catalysis*, Harper and Row, New York.

Chapter 5—Coenzyme Mechanisms

Benkovic, S. J. and Schray, K. J. (1973) Chemical basis of biological phosphoryl transfer, in *The Enzymes*, 3rd ed., vol. **8**, ch. 6, Boyer, P. D. (Ed.), Academic Press, New York.

Benkovic, S. J. (1980) On the mechanism of action of folate- and biopterin-requiring enzymes, *Ann. Rev. Biochem.*, **49**, 227–251.

Bruice, T. C. (1976) Some pertinent aspects of mechanism as determined with small molecules, *Ann. Rev. Biochem.*, **45**, 331–373.

Davis, L. and Metzler, D. E. (1976) Pyridoxal-linked elimination and replacement reactions, in *The Enzymes*, 3rd ed., vol. **7**, ch. 2, Boyer, P. D. (Ed.), Academic Press, New York.

Knowles, J. R. (1980) Enzyme-catalysed phosphoryl transfer reactions, *Ann. Rev. Biochem.*, **49**, 877–919.

Metzler, D. E. (1979) Tautomerism in pyridoxal phosphate and in enzymatic catalysis, *Advan. Enzymol.*, **50**, 1–40.

Mildvan, A. S. (1979) The role of metals in the enzyme-catalysed nucleophilic substitutions at the phosphorous atoms of ATP, *Advan. Enzymol.*, **49**.

Parker, D. M. and Holbrook, J. J. (1977) *Pyridine Nucleotide-Dependent Dehydrogenases*, Proc. 2nd Int. Symp. Konstanz, FEBS Symp., No. **49**.

Popjak, G. (1970) Stereospecificity of enzymic reactions, in *The Enzymes*, 3rd ed., vol. **2**, ch. 3, Boyer, P. D. (Ed.), Academic Press, New York.

Scheffers-Sap, M. M. E. and Buck, H. M. (1979) Quantum-chemical considerations on the acidity of thiamine pyrophosphate and related systems, *J. Amer. Chem. Soc.*, **101**, 4807–4811.

Williams, C. H. (1976) Flavin-containing dehydrogenases, in *The Enzymes*, 3rd ed., vol. **13**, ch. 2, Boyer, P. D. (Ed.), Academic Press, New York.

Wood, H. G. and Barden, R. E. (1977) Biotin enzymes, *Ann. Rev. Biochem.*, **46**, 385–413.

Chapter 6—Effects of Inhibitors and pH

Inhibitors

Cornish-Bowden, A. (1979) *Fundamentals of Enzyme Kinetics*, Butterworth, London.

Jencks, W. P. (1966) Strain and conformation change in enzymatic catalysis, in *Current Aspects of Biochemical Energetics*, pp. 273–298, Kaplan, N. O. and Kennedy, E. P. (Eds.), Academic Press, New York.

Leinhard, G. E. (1973) Enzymatic catalysis and transition-state theory, *Science*, **180**, 149–154.

Price, N. C. (1979) What is meant by "Competitive Inhibition"? *Trends Biochem. Sci.* **4**, N272–N273.

Rudnick, G. and Abeles, R. H. (1975) Reaction mechanism and structure of the active site of proline racemase, *Biochemistry*, **14**, 4515–4522.

Webb, J. L. (1963) *Enzyme and Metabolic Inhibitors*, vol. **1**, Academic Press, New York.

Wolfenden, R. (1969) Transition state analogues for enzyme catalysis, *Nature*, **223**, 704–705.

Zeller, E. A., Ramachander, G., Fleisher, G. A., Ishimaru, T. and Zeller, B. (1965) Ophidian L-amino acid oxidase, *Biochem. J.*, **95**, 262–269.

Effects of pH

Alberty, R. A. and Massey, V. (1954) On the interpretation of the pH variation of the maximum initial velocity of an enzyme-catalysed reaction, *Biochim. Biophys. Acta*, **13**, 347–353.

Cleland, W. W. (1977) Determining the chemical mechanisms of enzyme-catalysed reactions by kinetic studies, *Advan. Enzymol.*, **45**, 273–387.

Dixon, H. B. F. (1973) Shapes of curves of pH-dependence of reactions, *Biochem. J.*, **131**, 149–154.

Dixon, H. B. F. (1976) The unreliability of estimates of group dissociation constants, *Biochem. J.*, **153**, 627–629.

Dixon, H. B. F. (1979) Derivation of molecular p*K* values from pH-dependences, *Biochem. J.*, **177**, 249–250.

Engel, P. C. (1977) *Enzyme Kinetics*, Chapman and Hall, London.

Friedenwald, J. S. and Maengwyn-Davies, G. D. (1954) Elementary kinetic theory of enzymatic activity: influence of pH, in *The Mechanism of Enzyme Action*, pp. 191–208, McElroy, W. D. and Glass, B. (Eds.), Johns Hopkins Press, Baltimore.

Hare, M. L. C. (1928) Tyramine oxidase, *Biochem. J.*, **22**, 968–979.

Herries, D. G., Mathias, A. P. and Rabin, B. R. (1962) The active site and mechanism of action of bovine pancreatic ribonuclease, *Biochem. J.*, **85**, 127–134.

Tipton, K. F. and Dixon, H. B. F. (1979) Effects of pH on enzymes, *Methods in Enzymol.*, **63**, 183–234.

Waley, S. G. (1953) Some aspects of the kinetics of enzymic reactions, *Biochim. Biophys. Acta* **10**, 27–34.

Chapter 7—Complex Kinetics and Cooperativity

Bisubstrate Kinetics

Anderson, S. R., Florini, J. R. and Vestling, C. S. (1964) Rat liver lactate dehydrogenase. Kinetics and specificity, *J. Biol. Chem.*, **239**, 2991–2997.

Cleland, W. W. (1963) The kinetics of enzyme-catalysed reactions with two or more substrates or products, *Biochim. Biophys. Acta*, **67**, 104–137.

Cleland, W. W. (1970) Steady-state kinetics, in *The Enzymes*, 3rd ed., vol. **2**, pp. 1–65, Boyer, P. D. (Ed.), Academic Press, New York.

Dixon, M., Webb, E. C., Thorne, C. J. R. and Tipton, K. F. (1979) *Enzymes*, 3rd ed., Longman, London.

Morrison, J. F. (1965) Kinetic methods for the determination of enzyme reaction mechanisms, *Austral. J. Sci.*, **27**, 317–327.

Morrison, J. F. and James, E. (1965) The mechanism of the reaction catalysed by adenosine triphosphate-creatine phospho-transferase, *Biochem. J.*, **97**, 37–52.

Segel, I. H. (1975) *Enzyme Kinetics*, Wiley, New York.

Silverstein, E. and Boyer, P. D. (1964) Equilibrium reaction rates and the mechanisms of bovine heart and rabbit muscle lactate dehydrogenases, *J. Biol. Chem.*, **239**, 3901–3907.

Tipton, K. F. (1974) Enzyme kinetics, in *Companion to Biochemistry*, ch. 6, Bull, A. T. *et al.* (Eds.), Longman, London.

Velick, S. F. and Vavra, J. (1962) A kinetic and equilibrium analysis of the glutamic oxalacetate transaminase mechanism, *J. Biol. Chem.*, **237**, 2109–2122.

Cooperativity and Allostery

Dalziel, K. (1973) Kinetics of control enzymes, *Symp. Soc. Exptl. Biol.*, **27**, 21–48.

Frieden, C. (1979) Slow transitions and hysteretic behaviour in enzymes, *Ann. Rev. Biochem.*, **48**, 471–489.

Tipton, K. F. (1979) Kinetic properties of allosteric and cooperative enzymes, in *Companion to Biochemistry*, vol. **2**, ch. 11, Bull, A. T., Lagnado, J. R., Thomas, J. O. and Tipton, K. F. (Eds.), Longmans, London.

Chapter 8—Rapid Reaction Techniques

Douzou, P. (1977) Enzymology at sub-zero temperatures, *Advan. Enzymol.*, **45**.

Halford, S. E. (1974) Rapid reaction techniques, in *Companion to Biochemistry*, vol. **1**, ch. 5, Bull, A. T., Lagnado, J. R., Thomas, J. O. and Tipton, K. F. (Eds.), Longmans, London.

Hammes, G. G. and Schimmel, P. R. (1970) Rapid reactions and transient states, in *The Enzymes*, 3rd ed., vol. **2**, ch. 2, Boyer, P. D. (Ed.), Academic Press, New York.

Hollaway, M. R. and White, H. A. (1975) A double-beam rapid-scanning stopped-flow spectrophotometer, *Biochem. J.*, **149**, 221–231.

Kungi, S., Hirohara, H. and Ise, N. (1979) Kinetic and thermodynamic study of specificity in the elementary steps in α-chymotrypsin-catalysed hydrolyses, *J. Amer. Chem. Soc.*, **101**, 3640–3646.

Wolfman, N. M., Storer, A. C. and Hammes, G. G. (1979) Temperature-jump study of the interaction of rabbit muscle phosphofructokinase with adenylyl imidodiphosphate and adenosine 5′-triphosphate, *Biochemistry*, **18**, 2451–2456.

Chapter 9—Enzyme Mechanisms

Albery, W. J. and Knowles, J. R. (1976) Evolution of enzyme function and the development of catalytic efficiency, *Biochemistry*, **15**, 5631–5640.

Angelides, K. J. and Fink, A. T. (1978) Cryoenzymology of papain: reaction mechanisms with an ester substrate, *Biochemistry*, **17**, 2659–2668.

Angelides, K. J. and Fink, A. L. (1979) Mechanism of action of papain with a specific anilide substrate, *Biochemistry*, **18**, 2355–2363.

Bachovchin, W. W. and Roberts, J. D. (1978) Nitrogen-15 nuclear magnetic resonance spectroscopy. The state of histidine in the catalytic triad of α-lytic protease, *J. Amer. Chem. Soc.*, **100**, 8041–8047.

Baker, E. N. (1980) Structure of actinidin, after refinement at 1.7 Å resolution, *J. Mol. Biol.*, **141**, 441–484.

Berger, A. and Schecter, I. (1970) The subsite specificity of papain, *Phil. Trans. R. Soc.*, **B 257**, 249.

Birktoft, J. J., Kraut, J. and Freer, S. T. (1976) A detailed structural comparison between the charge relay system in chymotrypsinogen and in α-chymotrypsin, *Biochemistry*, **15**, 4481.

Blow, D. M. (1971) The structure of chymotrypsin, in *The Enzymes*, 3rd ed., vol. **3**, ch. 6, Boyer, P. D. (Ed.), Academic Press, New York.

Branden, C-I., Jornvall, H., Eklund, H. and Furugren, N. (1975) Alcohol dehydrogenases, in *The Enzymes*, 3rd ed., Boyer, P. D. (Ed.), Academic Press, New York.

Chao, Y., Weisman, G. R., Sogah, G. D. Y. and Cram, D. J. (1979) Host–guest complexation. 21. Catalysis and chiral recognition through designed complexation of transition states in transacylations of amino ester salts, *J. Amer. Chem. Soc.*, **101**, 4948–4958.

Drenth, J., Jansonius, J. N., Koekoek, R. and Wolthers, B. G. (1971) The structure of papain, *Adv. Protein. Chem.*, **25**, 79–115.

Drenth, J., Kalk, K. H. and Swen, H. M. (1976) Binding of chloromethyl ketone substrate analogues to crystalline papain, *Biochemistry*, **15**, 3731–3738.

Fastrez, J. and Fersht, A. R. (1973) Mechanism of chymotrypsin. Structure, reactivity and nonproductive binding relationships, *Biochemistry*, **12**, 1067–1074.

Fersht, A. R. (1974) Catalysis, binding and enzyme–substrate complementarity, *Proc. Roy. Soc. Lond.* **B**, **187**, 397–407.

Fink, A. L. (1973) The α-chymotrypsin-catalysed hydrolysis of *N*-acetyl-L-tryptophan *p*-nitrophenyl ester in dimethyl sulphoxide at subzero temperatures, *Biochemistry*, **12**, 1736–1742.

Gandour, R. D. and Schowen, R. L. (Eds.) (1978) *Transition States of Biological Processes*, Plenum Press, New York.

Gertler, A. G., Walsh, K. A. and Neurath, H. (1974) Catalysis by chymotrypsinogen—demonstration of an acyl-zymogen intermediate, *Biochemistry*, **13**, 1302–1310.

Glazer, A. N. and Smith, E. L. (1972) Papain and other plant sulphydryl proteolytic enzymes, in *The Enzymes*, 3rd ed., vol. **3**, ch. 14, Boyer, P. D. (Ed.), Academic Press, New York.

Harris, J. I. and Waters, M. (1976) Glyceraldehyde-3-phosphate dehydrogenase, in *The Enzymes*, 3rd ed., vol. **13**, ch. 1, Boyer, P. D. (Ed.), Academic Press, New York.

Hess, G. P. (1972) Chymotrypsin, chemical properties and catalysis, in *The Enzymes*, 3rd ed., vol. **3**, ch. 7, Boyer, P. D. (Ed.), Academic Press, New York.

Hollaway, M. R., Antonini, E. and Brunori, M. (1969) The ficin-catalysed hydrolysis of *p*-nitrophenyl hippurate, *Eur. J. Biochem.*, **32**, 537–546.

Hunkapiller, M. W., Smallcombe, S. H., Whitaker, D. R. and Richards, J. H (1973) Carbon nuclear magnetic resonance studies of the histidine residue in α-lytic protease, *Biochemistry*, **12**, 4732–4743.

Hunkapiller, M. W., Forgac, M. D. and Richards, J. H. (1976) Mechanism of action of serine proteases: tetrahedral intermediate and concerted proton transfer, *Biochemistry*, **15**, 5581–5588.

Ingles, D. W. and Knowles, J. R. (1967) Specificity and stereospecificity of α-chymotrypsin, *Biochem. J.*, **104**, 369–377.

Jencks, W. P. (1975) Binding energy, specificity and enzyme catalysis: the circe effect, *Adv. Enzymol.*, **43**, 219–410.

Kezdy, F. J. and Bender, M. L. (1962) The kinetics of the chymotrypsin-catalysed hydrolysis of *p*-nitrophenyl acetate, *Biochemistry*, **1**, 1097–1106.

Koehler, K. A. and Lienhard, G. E. (1971) 2-Phenylethaneboronic acid, a possible transition-state analog for chymotrypsin, *Biochemistry*, **10**, 2477–2483.

Kraut, J. (1977) Serine proteases: structure and mechanism of catalysis, *Ann. Rev. Biochem.*, **46**, 331–358.

Lewis, S. D., Johnson, F. A. and Shafer, J. A. (1976) Potentiometric determination of ionisations at the active site of papain, *Biochemistry*, **15**, 5009–5017.

Lowe, G. (1976) The cysteine proteinases, *Tetrahedron*, **32**, 291–302.

Markley, J. L. and Ibanez, I. B. (1978) Zymogen activation in serine proteinases. Proton magnetic resonance pH titration studies of the two histidines of bovine chymotrypsinogen A and chymotrypsin A_α, *Biochemistry*, **17**, 4627–4640.

Matthews, D. A., Alden, R. A., Birktoft, J. J., Freer, S. T. and Kraut, J. (1977) Re-examination of the charge relay system in subtilisin and comparison with other serine proteases, *J. Biol. Chem.*, **252**, 8875–8883.

O'Leary, M. H. and Kluetz, M. D. (1972) Nitrogen isotope effects on the chymotrypsin-catalysed hydrolysis of *N*-acetyl-L-tryptophanamide, *J. Amer. Chem. Soc.*, **94**, 3585–3589.

O'Leary, M. H., Urberg, M. and Young, A. P. (1974) Nitrogen isotope effects on the papain-catalysed hydrolysis of *N*-benzoyl-L-argininamide, *Biochemistry*, **13**, 2077–2081.

Oppenheimer, H. L., Labouesse, B. and Hess, G. P. (1966) Implication of an ionising group in the control of conformation and activity of chymotrypsin, *J. Biol. Chem.*, **241**, 2720–2730.

Pocker, Y. and Sarkanen, S. (1978) Carbonic anhydrase. Structure, catalytic versatility and inhibition, *Advan. Enzymol.*, **47**, 149–274.

Polgar, L. (1974) Mercaptide-imidazolium ion-pair: the reactive nucleophile in papain catalysis, *FEBS Lett.*, **47**, 15–18.

Richards, F. M. and Wyckoff, H. W. (1971) Bovine pancreatic ribonuclease, in *The Enzymes*, 3rd ed., vol. **4**, ch. 24, Boyer, P. D. (Ed.), Academic Press, New York.

Shea, K. J., Thompson, E. A., Pandey, S. D. and Beauchamp, P. S. (1980) Template synthesis of macromolecules. Synthesis and chemistry of functionalised macroporous polydivinylbenzene, *J. Amer. Chem. Soc.*, **102**, 3149–3155.

Shipton, M. and Brocklehurst, K. (1978) Characterisation of the papain active centre by using two-protonic-state electrophiles as reactivity probes, *Biochem. J.*, **171**, 385–401.

Sluyterman, L. A. AE. and De Graff, M. J. M. (1969) The activity of papain in the crystalline state, *Biochim. Biophys. Acta*, **171**, 277–287.

Thompson, R. C. (1974) Binding of peptides to elastase: implications for the mechanism of substrate hydrolysis, *Biochemistry*, **13**, 5495.

Wharton, C. W. (1979) Synthetic polymers as models of enzyme catalysis—a review, *Int. J. Biolog. Macromol.*, **1**, 3–15.

Chapter 10—Practical Enzyme Kinetics

Ainsworth, S. (1977) *Steady-State Enzyme Kinetics*, Macmillan, London.

Allison, R. D. and Purich, D. L. (1979) Practical considerations in the design of initial velocity enzyme rate assay, *Methods in Enzymol.*, **63**, 3–22.

Cornish-Bowden, A. (1975) The use of the direct linear plot for determining initial velocities, *Biochem. J.*, **149**, 305–312.

Cornish-Bowden, A. and Eisenthal, R. (1974) Statistical considerations in the estimation of enzyme kinetic parameters by the direct linear plot and other methods, *Biochem. J.*, **139**, 721–730.

Cornish-Bowden, A., Porter, W. R. and Trager, W. F. (1978) Evaluation of distribution-free confidence limits for enzyme kinetic parameters, *J. Theor. Biol.*, **74**, 163–175.

Dowd, J. E. and Riggs, D. S. (1965) A comparison of estimates of Michaelis–Menten kinetic constants from various linear transformations, *J. Biol. Chem.*, **240**, 863–869.

Gulliver, P. A. and Tipton, K. F. (1978) Direct extraction radioassay for catechol-*O*-methyltransferase activity, *Biochem. Pharmacol.*, **27**, 773–775.

Henderson, P. J. F. (1978) Statistical analysis of enzyme kinetic data, in *Techniques in Protein and Enzyme Biochemistry*, vol. **B113**, pp. 1–43, Kornberg, H. L. *et al.* (Eds.), Elsevier/North Holland, Amsterdam.

Jacobsen, C. F., Leonis, J., Linderstrøm-Lang, K. and Ottesen, M. (1957) The pH-stat and its use in biochemistry, in *Methods of Biochemical Analysis*, vol. **4**, pp. 171–210, Glick, D. (Ed.), Wiley, New York.

McClure, W. R. (1969) A kinetic analysis of coupled enzyme assays, *Biochemistry*, **8**, 2782–2786.

Oldham, K. G. (1973) Radiometric methods of enzyme assay, in *Methods of Biochemical Analysis*, vol. **21**, pp. 191–286, Glick, D. (Ed.), Wiley, New York.

Orsi, B. A. (1972) A simple method for the derivation of the steady-state rate equation for an enzyme mechanism, *Biochim. Biophys. Acta*, **258**, 4–8.

Rudolph, F. B., Baugher, B. W. and Beissner, R. S. (1979) Techniques in coupled enzyme assays, *Methods in Enzymol.*, **63**, 22–41.

Selwyn, M. J. (1965) A simple test for inactivation of an enzyme during assay, *Biochim. Biophys. Acta*, **105**, 193–195.

Storer, A. C. and Cornish-Bowden, A. (1974) The kinetics of coupled enzyme reactions, *Biochem. J.*, **141**, 205–209.

Tipton, K. F. (1978) Enzyme assay and kinetic studies, in *Techniques in Protein and Enzyme Biochemistry*, vol. **B112**, pp. 1–56, Kornberg, H. L. *et al.* (Eds.), Elsevier/North Holland, Amsterdam.

Udenfriend, S. (1962) *Fluorescence Assay in Biology and Medicine*, Academic Press, New York.

Weitzman, P. D. J. (1976) Assay of enzymes by polarography, *Biochem. Soc. Trans.*, **4**, 724–726.

Index